AF689702

Robotics and Autonomous Systems: Perception to Planning

by

ant

Front Matter

The field of robotics and autonomous systems is advancing at an unprecedented pace, weaving itself into the fabric of our daily lives and reshaping industries at a fundamental level. From the intricate dance of robotic arms on assembly lines to the sophisticated navigation of autonomous vehicles, these intelligent machines are no longer confined to science fiction but are becoming integral components of our modern society. This textbook is born from a desire to provide a comprehensive and rigorous foundation for students, researchers, and practitioners eager to understand, design, and deploy these complex systems. We aim to bridge the gap between theoretical principles and practical implementation, offering a clear, accessible, and engaging exploration of the core concepts that drive modern robotics.

Our journey begins with the foundational elements of robotics, dissecting the essential components that enable intelligent machines to perceive, reason, and act. We will then delve into the intricate world of sensing modalities, exploring how robots gather information about their surroundings, and the crucial processes of calibration and data preprocessing that ensure the accuracy and reliability of this information. The subsequent chapters will unravel the power of perception, particularly through the lens of machine learning and deep learning, and explore advanced techniques for understanding complex scenes and objects in 3D.

A significant portion of this book is dedicated to state estimation and localization, two paramount challenges in robotics that allow systems to know where they are and build representations of their environment. We will meticulously examine Simultaneous Localization and Mapping (SLAM), a cornerstone problem that enables robots to map unknown spaces while simultaneously navigating within them. The critical aspect of motion planning will be explored, from fundamental pathfinding algorithms to sophisticated techniques for navigating dynamic and kinematically constrained environments. We will also venture into the realm of robot manipulation and grasping, understanding how robots interact physically with objects, and then expand to the complex domain of multi-agent systems and coordination, where multiple robots collaborate to achieve shared objectives. Finally, we will address the increasingly vital areas of human-robot interaction, and the profound ethical and societal implications that accompany the rise of autonomous technologies. Throughout this text, we have striven to maintain a balance between theoretical rigor and practical applicability, incorporating real-world examples and case studies to illustrate abstract concepts and inspire further exploration.

The landscape of robotics and autonomous systems is undergoing a profound transformation, moving beyond the realm of industrial automation into nearly every facet of human endeavor. This evolution is driven by a confluence of advancements in sensing, computation, artificial intelligence, and control theory, enabling robots to perform increasingly complex tasks with greater autonomy, intelligence, and adaptability. This textbook is designed to equip advanced undergraduate and graduate students, as well as practicing researchers and engineers, with the essential knowledge and skills to navigate this dynamic field. Our objective is to provide a robust understanding of the theoretical underpinnings, coupled with practical insights into the implementation of robotic systems.

We begin by laying the groundwork, defining what constitutes an autonomous system and exploring the core components that form the backbone of any robotic endeavor: sensors for perception, processors for decision-making, and effectors for interaction with the physical world. The subsequent chapters systematically delve into the critical aspects of robotic operation. We will investigate the diverse array of sensing modalities, from cameras and LiDAR to IMUs, and the indispensable processes of sensor calibration and data preprocessing that ensure the fidelity of the information gathered. A significant focus will be placed on perception, leveraging the power of machine learning and deep learning to enable robots to interpret their environment, recognize objects, and understand complex scenes.

The challenge of knowing one's location and building a map of the surroundings is addressed through in-depth discussions on state estimation, localization, and the seminal problem of Simultaneous Localization and Mapping (SLAM). Furthermore, we will explore the art and science of motion planning, detailing algorithms that enable robots to navigate from point A to point B while avoiding obstacles and respecting their own kinematic and dynamic constraints. The book also covers the critical domain of robot manipulation and grasping, essential for robots that interact physically with objects, and extends to the sophisticated topic of multi-agent systems and coordination, where robots collaborate to achieve common goals. Finally, we will examine the vital area of human-robot interaction, exploring how robots can communicate and collaborate effectively with humans, and address the significant ethical and societal implications that arise from the widespread deployment of autonomous technologies. Throughout this text, the emphasis is on providing a coherent narrative that connects these diverse topics, illustrated with concrete examples from self-driving cars, industrial robots, and aerial drones, to foster a deep and applicable understanding of the field.

Table of Contents

Chapter 1: Introduction to Robotics and Autonomous Systems

The field of robotics and autonomous systems stands as a testament to humanity's enduring quest to extend its capabilities and understand the principles of intelligence and agency. What began as rudimentary mechanical automatons designed for repetitive tasks has blossomed into a sophisticated, interdisciplinary domain encompassing artificial intelligence, advanced sensing, intricate control systems, and seamless human-robot interaction. This evolution is not merely an academic pursuit; it is a tangible force reshaping industries, enhancing human lives, and opening new frontiers for exploration. As we embark on this exploration, it is crucial to appreciate the breadth and dynamism of this field, understanding its historical roots to better comprehend its trajectory and profound implications for the future.

The genesis of robotics can be traced back to early concepts of artificial beings and automated machines. Early mechanical devices, often powered by intricate clockwork or steam, demonstrated the desire to replicate or automate human actions. The Industrial Revolution, with its emphasis on mass production and efficiency, provided a fertile ground for the development of more practical automated systems. Pneumatic and hydraulic actuators, coupled with programmable logic controllers (PLCs), ushered in an era of industrial automation. These early robots were largely confined to repetitive, high-precision tasks on assembly lines, performing actions like welding, painting, and material handling with unwavering consistency. Their "intelligence" was rudimentary, dictated by pre-programmed sequences, and their interaction with the environment was minimal and highly constrained. They were extensions of human labor, designed to increase throughput and improve safety by removing humans from hazardous or monotonous tasks. The emphasis was squarely on mechanical execution and predictable operation within highly structured environments.

A significant paradigm shift occurred with the advent of computing power and the nascent field of artificial intelligence. The integration of microprocessors and advanced algorithms allowed robots to move beyond simple pre-programmed routines. This period saw the emergence of robots capable of basic sensing and feedback control, enabling them to adapt, to a limited extent, to variations in their environment. The development of early vision systems, for instance, allowed robots to perform rudimentary object recognition and sorting. Simultaneously, advancements in control theory provided mechanisms for robots to maintain balance, follow trajectories, and execute more complex movements. This era marked the transition from pure mechanical automation to the early stages of "intelligent" automation, where robots began to exhibit a degree of adaptability and decision-making capability.

The latter half of the 20th century and the dawn of the 21st century witnessed an explosion of progress, largely fueled by breakthroughs in computer science, sensing technologies, and machine learning. The miniaturization of sensors, coupled with dramatic increases in computational power and algorithmic sophistication, has led to the development of robots that are increasingly sophisticated, versatile, and autonomous. We have moved from fixed-platform industrial arms to highly mobile robots capable of navigating complex, dynamic environments, to dexterous manipulators that can perform delicate surgical procedures, and to multi-robot systems that can collaborate to achieve large-scale tasks. The concept of the "intelligent agent" has become central to modern robotics, where robots are designed not just to perform actions but to perceive, reason, plan, and act in an intelligent manner, often with minimal human intervention.

The integration of artificial intelligence, particularly machine learning and deep learning, has been a transformative force. Robots are no longer solely reliant on explicitly programmed rules. Instead, they can learn from data, adapt their behavior, and generalize their knowledge to novel situations. This has enabled significant advancements in areas such as perception (allowing robots to understand their environment through vision, LiDAR, and other sensors with unprecedented accuracy), decision-making (enabling robots to make more nuanced and context-aware choices), and natural human-robot interaction (allowing for more intuitive and fluid communication). The symbiotic relationship between robotics and AI is thus a defining characteristic of the current landscape, pushing the boundaries of what autonomous systems can achieve.

The increasing sophistication of robotics has led to their proliferation across an astonishingly diverse range of societal applications. In manufacturing, robots continue to enhance efficiency and safety, performing tasks ranging from heavy lifting and assembly to intricate quality inspection. Beyond traditional manufacturing, robotics is revolutionizing healthcare. Surgical robots offer enhanced precision and minimally invasive procedures, improving patient outcomes. Assistive robots are being developed to support the elderly and individuals with disabilities, promoting independence and improving quality of life. In logistics and warehousing, autonomous mobile robots (AMRs) are optimizing inventory management, order fulfillment, and delivery processes.

The impact extends further into domains that were once the sole purview of human exploration and operation. Drones, a ubiquitous form of aerial robotics, are employed for aerial surveying, agricultural monitoring, infrastructure inspection, and disaster response, providing critical data and enabling operations in hazardous or inaccessible areas. Even our homes are becoming increasingly populated by robots, from vacuum cleaners and lawnmowers to more sophisticated personal assistants that can manage schedules, provide information, and interact with smart home devices. This integration into domestic life raises new considerations about human-robot relationships, privacy, and the definition of companionship.

The interdisciplinary nature of robotics is a critical factor in its rapid advancement. It is a field that inherently draws upon and integrates knowledge from multiple established disciplines, creating a synergistic effect. Computer science provides the foundational algorithms for perception, planning, and control, along with the software architectures that govern robot behavior. Electrical engineering contributes expertise in sensor design, actuator development, power systems, and embedded systems. Mechanical engineering is crucial for the physical design of robots, including their structure, kinematics, dynamics, and the manipulation of materials. Artificial intelligence, particularly machine learning, imbues robots with the ability to learn, adapt, and reason, enabling them to tackle complex and uncertain environments. Furthermore, fields such as cognitive science, psychology, and even sociology contribute valuable insights into human behavior, perception, and interaction, which are essential for designing robots that can effectively and ethically operate alongside humans. This cross-pollination of ideas and expertise is what allows the field to tackle increasingly complex challenges and to innovate at such a rapid pace.

Looking ahead, several key trends are shaping the future trajectory of robotics. The drive towards greater autonomy, where robots can operate with minimal or no human supervision, remains a central theme. This necessitates advancements in areas such as robust perception in unstructured environments, sophisticated decision-making under uncertainty, and reliable self-localization and mapping. The development of more dexterous and adaptable manipulators, capable of handling a wider variety of objects and performing more complex manipulation tasks, is also a significant area of research. Soft robotics, inspired by biological systems, promises robots that are more compliant, adaptable, and safe for interaction with humans and delicate environments.

Human-robot interaction (HRI) is another area poised for significant growth. As robots become more integrated into our daily lives, the ability for them to understand and respond to human intentions, emotions, and social cues will become paramount. This involves developing more natural modes of communication, building trust between humans and robots, and designing collaborative systems where humans and robots can work together seamlessly. The ethical and societal implications of increasingly capable robots are also a growing area of concern and research. Issues of job displacement, privacy, algorithmic bias, and the responsible development of autonomous systems require careful consideration and proactive policy-making.

The journey from simple mechanical automatons to sophisticated intelligent agents has been remarkable, and the pace of innovation shows no signs of slowing. The integration of robots into every facet of society, from industrial floors and operating rooms to our homes and workplaces, is a testament to their increasing utility and capability. Understanding the historical evolution, the interdisciplinary foundations, and the burgeoning trends is essential for anyone seeking to engage with this transformative field. The subsequent chapters will delve deeper into the core components, principles, and applications that define modern robotics and autonomous systems, building upon this foundational understanding of the evolving landscape.

The evolution of robots from mere mechanical extensions of human labor to sophisticated entities capable of independent action and decision-making hinges on the fundamental concept of autonomy. To truly grasp the capabilities and potential of modern autonomous systems, it is imperative to establish a precise understanding of what autonomy entails in this context. Autonomy, in the realm of robotics, refers to the capacity of a system to operate and make decisions without direct human intervention, adapting to its environment and achieving its goals. This is not a binary state but rather a spectrum, allowing for a nuanced classification of robotic systems based on their level of independent operation.

At the lowest end of this spectrum lies **automation**. Automated systems are designed to perform a specific set of tasks in a predetermined sequence. They operate based on pre-programmed instructions and are typically confined to highly structured and predictable environments. For instance, a traditional industrial robot arm on an assembly line, programmed to perform welding at specific points, exemplifies automation. While it executes tasks flawlessly and efficiently, it lacks the ability to deviate from its programmed path or react to unforeseen circumstances. If an object is out of place or a component is misaligned, the automated system will likely stop, signal an error, or continue its operation incorrectly, unable to self-correct. Its "intelligence" is purely in its programming; it does not perceive, reason, or adapt beyond its explicit instructions. The core characteristic is the absence of variability in its response to environmental conditions that fall within its narrow operational parameters. These systems excel at repetitive, high-volume tasks where consistency is paramount and environmental variations are minimal or non-existent. Think of conveyor belt systems in factories, traffic lights that operate on fixed timers, or even simple household appliances like a toaster. They follow a script, and any deviation from that script requires external human intervention to reset or recalibrate. The definition of "operation" here is strictly adherence to a pre-defined operational envelope.

Moving along the spectrum, we encounter **semi-autonomous** systems. These robots possess a degree of independence, capable of making certain decisions and performing tasks with limited human oversight. They often incorporate sensors that allow them to perceive aspects of their environment and react to predefined conditions. For example, a semi-autonomous vacuum cleaner might use infrared sensors to detect obstacles and change its direction, or a semi-autonomous vehicle might be equipped with adaptive cruise control and lane-keeping assist features. These systems can handle some level of environmental variability, but their decision-making capabilities are typically constrained to a set of rules or algorithms designed by humans. They might be able to decide *how* to perform a task or *when* to adjust a parameter, but the ultimate goals and the overarching strategy are usually determined by a human operator. The human still plays a crucial role, either in defining the operational context, setting high-level objectives, or intervening when the system encounters situations beyond its programmed capabilities. This level of autonomy is often characterized by a "human-in-the-loop" or "human-on-the-loop" paradigm, where human supervision or intervention is an integral part of the operational cycle. Consider a remotely operated vehicle (ROV) in underwater exploration. While the ROV can navigate and perform certain maneuvers independently based on sensor readings, a human operator on the surface makes critical decisions about its path, scientific objectives, and response to unexpected discoveries or hazards. The semi-autonomous system acts as an intelligent tool, extending the operator's reach and capabilities while still requiring direct guidance for complex or critical actions. Another illustrative example is an advanced autopilot system in an aircraft. It can manage navigation, altitude, and speed with great precision, but the pilots remain in command, making strategic decisions, monitoring system performance, and taking over in unforeseen emergencies. The intelligence here is reactive rather than truly proactive or strategic.

The pinnacle of this classification is **full autonomy**. A fully autonomous system is capable of operating independently, perceiving its environment, making complex decisions, planning its actions, and executing them to achieve its objectives with minimal or no human intervention. These systems are not merely following pre-programmed instructions; they possess a degree of "intelligence" that allows them to learn, adapt, and reason about novel situations. A fully autonomous agent can perceive its surroundings through a suite of sensors, process this information to form an understanding of its state and the state of its environment (perception), devise a plan to reach a desired state (planning), and then execute the necessary actions (actuation) while continuously monitoring and adapting to changes. This level of autonomy is what drives research in areas like self-driving cars, advanced drones for complex missions, and sophisticated robotic assistants capable of performing tasks in unpredictable human environments. The key differentiator here is the ability to handle uncertainty, ambiguity, and novelty. A fully autonomous system is not just reacting to predefined stimuli; it is proactively seeking to achieve goals in a dynamic and often unpredictable world. It embodies a level of self-direction and problem-solving that transcends mere programmed responses. For example, a fully autonomous robot designed for disaster relief would need to navigate through rubble, identify survivors, assess structural integrity, and coordinate its actions with other robots or human rescue teams, all without explicit step-by-step guidance for each scenario. This requires sophisticated perception, robust decision-making under extreme uncertainty, and the ability to learn from past experiences to improve future performance.

The term "intelligent agent" is intrinsically linked to the concept of autonomy, particularly at higher levels of the autonomy spectrum. An intelligent agent can be formally defined as any entity that perceives its environment through sensors and acts upon that environment through actuators. The "intelligence" of the agent lies in its ability to make choices and take actions that are conducive to achieving its goals. This requires a sophisticated interplay of several key characteristics:

Sensing and Perception: This is the agent's ability to gather information about its environment. Sensors are the "eyes," "ears," and "touch" of the robot, translating physical phenomena into data that the system can process. This includes cameras for visual input, LiDAR and radar for depth and distance measurement, microphones for auditory information, tactile sensors for physical contact, and inertial measurement units (IMUs) for tracking motion and orientation. However, simply gathering raw data is insufficient. **Perception** is the process of interpreting this sensory data to create a meaningful representation of the environment. This involves tasks such as object recognition, scene understanding, localization (determining the agent's position and orientation), and mapping (creating a model of the environment). For a robot navigating a busy street, perception would involve identifying pedestrians, other vehicles, traffic signals, road boundaries, and understanding their intentions and potential movements. Machine learning, particularly deep learning, has revolutionized perception, enabling robots to process complex visual and auditory information with accuracy previously thought impossible for machines.

Decision-Making and Planning: Once the agent has a perception of its environment and its own state, it needs to decide what to do next. This involves evaluating potential actions based on its goals, current situation, and learned knowledge. **Decision-making** can range from simple rule-based logic to complex probabilistic reasoning and optimization algorithms. For more complex tasks, the agent needs to engage in **planning**. Planning involves devising a sequence of actions or a strategy that will lead from the current state to a desired goal state. This might involve path planning (finding the best route), task planning (determining the order of operations), or even contingency planning (preparing for potential failures or unexpected events). A fully autonomous vehicle, for instance, must continuously plan its trajectory, considering factors like speed limits, traffic flow, pedestrian crossings, and the actions of other road users, all while ensuring passenger safety and comfort. The intelligence here is in the ability to foresee consequences and choose the most advantageous course of action.

Actuation and Execution: After a decision has been made and a plan formulated, the agent must translate these internal commands into physical actions. **Actuation** refers to the use of actuators – motors, servos, pneumatic cylinders, etc. – to move the robot's limbs, wheels, or other effectors. This allows the robot to interact with and manipulate its environment. For example, a robotic arm executing a pick-and-place operation uses its actuators to grip an object, move it to a new location, and release it. The effectiveness of actuation is directly tied to the precision and robustness of the control systems that govern it. Sophisticated control algorithms ensure that actions are executed smoothly, accurately, and safely, even in the face of disturbances or uncertainties.

The role of
intelligence, particularly in the context of artificial intelligence (AI) and machine learning (ML), is fundamental to achieving advanced autonomy. While early robots were largely governed by explicit programming, modern autonomous systems leverage AI to imbue them with capabilities that go far beyond pre-defined scripts. Machine learning, a subset of AI, allows systems to learn from data without being explicitly programmed for every possible scenario. This learning can take many forms:

Supervised Learning: The agent learns from labeled data, essentially being trained on examples of correct input-output pairs. For instance, a robot learning to identify different types of objects might be shown thousands of images labeled with the correct object names.
Unsupervised Learning: The agent learns to find patterns and structures in unlabeled data. This can be useful for tasks like clustering similar data points or identifying anomalies.
Reinforcement Learning: This is perhaps the most relevant paradigm for advanced autonomy. In reinforcement learning, an agent learns to make decisions by trial and error, receiving rewards or penalties based on its actions. The goal is to learn a policy that maximizes cumulative reward over time. This is crucial for tasks where the optimal strategy is not known beforehand and must be discovered through interaction with the environment. For example, a robot learning to walk or a self-driving car learning to navigate complex traffic situations might use reinforcement learning.

Deep learning, a subfield of ML that utilizes artificial neural networks with multiple layers, has enabled significant breakthroughs in perception and decision-making for autonomous systems. These deep neural networks can automatically learn hierarchical features from raw data, allowing them to achieve remarkable performance in tasks like image recognition, natural language processing, and speech synthesis.

Therefore, intelligence, as embodied by AI and ML, is not merely an add-on for autonomous systems; it is an essential enabler. It provides the framework for perception systems to interpret complex sensory inputs, for decision-making modules to reason under uncertainty, and for learning algorithms to adapt and improve performance over time. Without these intelligent capabilities, robots would remain confined to predictable environments and simple, repetitive tasks, unable to truly navigate and operate in the complex, dynamic world that surrounds us. This section has aimed to establish a shared vocabulary and conceptual framework, clarifying the distinctions between automation, semi-autonomy, and full autonomy, and outlining the core characteristics of intelligent agents. This foundational understanding is crucial for delving deeper into the specific technologies, algorithms, and applications that constitute the exciting field of robotics and autonomous systems.

A robotic system, at its core, is an intricate orchestration of hardware and software designed to perceive, reason, and act within its environment. While the complexity and sophistication of these systems can vary dramatically, from the simple automated arms on an assembly line to the advanced autonomous vehicles navigating our cities, they all share a common architectural blueprint comprising several indispensable components. Understanding these fundamental building blocks is paramount to appreciating how robots function and evolve. These core constituents can be broadly categorized into sensing and perception, computation and decision-making, and actuation and control, all woven together by an underlying software architecture.

The first critical element of any robotic system is its ability to understand its surroundings – a capability enabled by its
sensors. Sensors are the robot's conduits to the physical world, translating external stimuli into electrical signals that the robot's internal processing units can interpret. The diversity of sensors employed is vast, reflecting the myriad environments and tasks robots are designed to handle. For instance, **vision sensors**, such as cameras (both monochromatic and color, and increasingly, depth-sensing cameras like stereo cameras or structured light sensors), provide the robot with visual information. This allows for tasks like object recognition, navigation, and inspection. The data from cameras can be processed to detect features, identify objects, and even reconstruct three-dimensional scenes, a process that bridges the gap between raw sensor data and meaningful environmental understanding. Beyond vision, **LiDAR (Light Detection and Ranging)** and **radar** sensors are crucial for robust environmental perception, particularly in applications requiring accurate distance and spatial mapping. LiDAR emits laser pulses and measures the time it takes for them to return after reflecting off objects, creating a precise 3D point cloud of the surroundings. Radar, using radio waves, is effective at detecting objects over longer distances and in adverse weather conditions where optical sensors might struggle.

Auditory sensors, or
microphones, allow robots to perceive sound, which can be used for communication, detecting unusual noises (e.g., an alarm or a malfunction), or even for sound-based localization. For robots that physically interact with their environment, **tactile sensors** and **force/torque sensors** are essential. Tactile sensors provide information about contact, pressure, and texture, enabling robots to "feel" objects they are handling, which is vital for delicate manipulation tasks. Force/torque sensors, often integrated into robotic joints or end-effectors, measure the forces and torques applied during interaction, allowing for precise control during assembly, grinding, or any task involving physical contact.

Beyond these,
inertial Measurement Units (IMUs), which typically combine accelerometers and gyroscopes, are fundamental for estimating a robot's orientation, acceleration, and angular velocity. This information is critical for maintaining balance (in humanoid or legged robots), tracking motion, and fusing with other sensor data for more accurate state estimation. **GPS (Global Positioning System)** receivers are indispensable for outdoor mobile robots, providing absolute position information, though their accuracy can be limited in urban canyons or indoor environments. Other specialized sensors might include temperature sensors, chemical sensors, or proximity sensors, each tailored to specific operational requirements. The raw data from these sensors, however, is just the beginning. It must be

processed and interpreted through the **perception system**. Perception is the cognitive process by which the robot makes sense of the sensory input, transforming it into a representation of its environment and its own state within that environment. This involves tasks like sensor fusion (combining data from multiple sensors to gain a more comprehensive understanding), object detection and tracking, scene segmentation, and mapping. For example, a self-driving car might fuse data from its cameras, LiDAR, and radar to build a detailed, real-time model of the road, including other vehicles, pedestrians, road signs, and lane markings. This perceived world model then serves as the foundation for subsequent decision-making.

The second pillar of a robotic system is **computation and decision-making**. This is the "brain" of the robot, responsible for processing the perceived information, reasoning about the current situation, and determining the appropriate course of action. At the heart of this component are the **processors** – microprocessors, microcontrollers, or even powerful GPUs (Graphics Processing Units) for computationally intensive tasks like deep learning inference. These processors execute the software that governs the robot's behavior. The software architecture can range from simple, monolithic programs for basic automation to highly complex, modular, and distributed systems for advanced autonomy.

The
decision-making module takes the processed sensor data and the robot's internal state (e.g., its current

position, velocity, battery level) and, based on its programming and objectives, decides what to do next. This can involve simple rule-based logic, such as "if obstacle detected within 1 meter, stop," or much more sophisticated algorithms. For tasks requiring a sequence of actions to achieve a goal, a **planning module** comes into play. Path planning algorithms, for instance, calculate an optimal or feasible trajectory for a mobile robot to navigate from point A to point B while avoiding obstacles. Task planning, on the other hand, determines the sequence of operations needed to complete a more complex objective, such as assembling a product.

In advanced autonomous systems, these decision-making and planning processes are heavily influenced by artificial intelligence (AI) and machine learning (ML). As discussed previously, ML algorithms allow robots to learn from data, adapt their behavior, and make predictions about future events. For example, a robot might use reinforcement learning to learn how to walk efficiently on uneven terrain or how to optimally grasp a variety of objects. The computational demands of these AI/ML algorithms, especially deep neural networks, often necessitate specialized hardware like GPUs or TPUs (Tensor Processing Units) to achieve real-time performance. The software architecture is crucial here, often employing frameworks like ROS (Robot Operating System) or specialized middleware to manage the flow of data and commands between different software modules, enabling modularity, reusability, and scalability.

Finally, the third fundamental component is **actuation and control**. Once a decision has been made and a plan formulated, the robot must be able to execute the intended actions in the physical world. This is the role of **effectors**, which are the mechanisms that allow the robot to interact with its environment. The most common type of effector is an **actuator**. Actuators are the "muscles" of the robot, converting electrical energy into mechanical motion. Electric motors (DC motors, servo motors, stepper motors) are widely used due to their precision and controllability. Hydraulic and pneumatic actuators are also employed, especially in industrial settings where high power or specific motion profiles are required.

The actuators are connected to the robot's physical structure, such as its limbs, wheels, or end-effectors (like grippers or tools). For a mobile robot, wheels driven by motors are the primary effectors for locomotion. For a robotic manipulator arm, a series of joint actuators (often servo motors) allow each link to move, enabling the arm to reach different positions and orientations. End-effectors, attached to the end of the arm, perform the actual task: a gripper might pick and place objects, a welding torch might perform welding, or a camera might capture images.

Crucially, the actuators do not operate in isolation. They are governed by **control systems**. The control system is a sophisticated loop that takes the desired action commands from the decision-making module and translates them into precise signals for the actuators. It also uses feedback from sensors (often encoders on motors, or force sensors) to monitor the actual motion and adjust the actuator commands to ensure the action is performed accurately and safely. For example, if a robotic arm is commanded to move to a specific position, the control system will send power to the joint motors. As the arm moves, encoders report the joint angles. If the arm encounters resistance or deviates from the planned path, the control system will adjust the motor power to correct the error and achieve the target position. PID (Proportional-Integral-Derivative) controllers are a classic and widely used type of control algorithm, but more advanced techniques, including model-predictive control and adaptive control, are employed for complex dynamics and uncertain environments.

The seamless integration of these three core components – sensing and perception, computation and decision-making, and actuation and control – is orchestrated by the robot's
software architecture. This architecture defines how data flows between modules, how tasks are scheduled, and how different functionalities are integrated. A well-designed software architecture is crucial for developing, testing, and maintaining complex robotic systems. It promotes modularity, allowing individual components to be developed and updated independently. It also facilitates the integration of new sensors, algorithms, or actuators. Frameworks like ROS provide a standardized set of tools and libraries that simplify the development of distributed robotic software systems, enabling

communication between different processes running on one or more computers.

To illustrate these concepts, let's consider a few common robotic platforms. A simple **mobile robot** designed for indoor navigation might utilize wheel encoders and an IMU for odometry (estimating its movement based on wheel rotation and inertia), combined with ultrasonic or infrared sensors for short-range obstacle detection. Its processor would run software for basic path planning (e.g., moving in a straight line until an obstacle is detected, then turning) and motor control. A more advanced mobile robot, such as a warehouse logistics robot, would incorporate LiDAR for detailed mapping and localization, cameras for reading barcodes or identifying specific items, and more sophisticated planning algorithms to navigate complex environments efficiently and cooperatively with other robots. Its actuators would be powerful drive motors and potentially a lifting mechanism for carrying goods.

A **robotic manipulator arm**, commonly found in manufacturing, relies heavily on joint encoders and force/torque sensors for precise positioning and interaction. Vision systems (cameras mounted on the arm or in the environment) are often used for object recognition, pose estimation, and quality inspection. The computation module would run algorithms for inverse kinematics (calculating the required joint angles to reach a desired end-effector pose), trajectory generation, and task-specific control logic. Its actuators would be high-precision servo motors for each joint.

Humanoid robots represent a pinnacle of complexity, integrating locomotion (legged robots), manipulation (arms), and often, sophisticated human-robot interaction capabilities. They require a vast array of sensors: IMUs for balance, vision systems for perception, tactile sensors on their hands and feet for safe interaction and stability, and microphones for speech. Their computation and decision-making systems must handle real-time control of dozens of degrees of freedom, complex path planning in dynamic environments, and potentially, natural language understanding. Their actuators are typically electric motors, often designed to mimic the power and efficiency of biological muscles.

In summary, every functional robotic system, regardless of its specific application, is a cohesive entity built upon the foundational pillars of sensing for environmental awareness, computation for intelligent processing and decision-making, and actuation for physical interaction. The seamless interplay between these hardware components, guided by a robust software architecture, dictates the robot's ability to perceive its world, reason about it, and act within it to achieve its designated goals. The ongoing advancements in each of these areas – more sophisticated sensors, more powerful and efficient processors, more dexterous actuators, and more intelligent software – are continuously pushing the boundaries of what is possible in robotics and autonomous systems.

The pervasive integration of robotics and autonomous systems into the fabric of modern society is no longer a futuristic concept; it is a rapidly unfolding reality. These sophisticated machines, once confined to the realm of science fiction and specialized industrial settings, are now demonstrably transforming nearly every sector of human endeavor. Their impact is multifaceted, driving unprecedented gains in efficiency, safety, precision, and convenience, while simultaneously reshaping economic landscapes and prompting profound societal dialogues. This section delves into the diverse applications that showcase the current capabilities and burgeoning potential of robotics and autonomous systems, illustrating their tangible influence across a spectrum of fields.

One of the most mature and profoundly impacted domains is
manufacturing and industrial automation. For decades, robotic arms have been the workhorses of assembly lines, performing repetitive, strenuous, or hazardous tasks with unwavering consistency. Today's industrial robots are far more advanced than their predecessors. They are increasingly collaborative, designed to work alongside human counterparts in a shared workspace, a concept known as cobots. These cobots are equipped with advanced safety features, such as force and torque sensors that allow them to detect unexpected contact and immediately halt or adjust their movement, preventing injury to human

workers. In automotive manufacturing, for example, robots are indispensable for tasks like welding, painting, and heavy lifting. They can perform these operations with micron-level precision, significantly enhancing product quality and reducing defects. Beyond the assembly line, robots are being deployed for intricate tasks like the inspection of components, ensuring adherence to stringent quality control standards. Automated guided vehicles (AGVs) and autonomous mobile robots (AMRs) are revolutionizing factory floor logistics, transporting materials, tools, and finished products between different workstations, thereby streamlining supply chains and minimizing downtime. The adoption of robotics in manufacturing has not only boosted productivity and reduced operational costs but has also led to the creation of safer working environments by removing humans from the most dangerous and ergonomically challenging roles. This has necessitated a shift in the human workforce, with a growing demand for skilled technicians and engineers who can program, maintain, and oversee these automated systems, rather than performing the manual labor themselves.

The
healthcare sector is another area witnessing a dramatic surge in robotic applications, promising to revolutionize patient care, surgical procedures, and medical diagnostics. **Surgical robots** have emerged as powerful allies for surgeons, enabling minimally invasive procedures with enhanced precision and dexterity. Systems like the da Vinci Surgical System allow surgeons to operate through small incisions,

manipulating robotic arms that provide magnified 3D vision and instruments with a greater range of motion than human hands. This leads to reduced patient trauma, shorter recovery times, and fewer complications. Beyond the operating room, robots are assisting in rehabilitation, providing patients with personalized physical therapy exercises and monitoring their progress. Exoskeletons, robotic suits worn by individuals with mobility impairments, are helping them to walk again or regain strength. In diagnostic imaging, robots can precisely position imaging equipment or even perform biopsies under remote supervision, increasing accuracy and safety. Furthermore, robotic assistants are being developed to aid nurses and caregivers with tasks such as lifting and repositioning patients, delivering medications, and even providing companionship. The integration of AI and robotics in healthcare is also paving the way for automated drug discovery, personalized treatment plans, and advanced diagnostics through the analysis of vast datasets. While the initial cost of these advanced systems can be a barrier, their long-term benefits in terms of patient outcomes, reduced hospital stays, and improved efficiency are substantial.

The world of
logistics and warehousing is undergoing a profound transformation driven by robotics. The explosion of e-commerce has created an unprecedented demand for efficient and rapid fulfillment of orders, a challenge that traditional manual warehousing struggled to meet. Autonomous mobile robots (AMRs) are now a common sight in large distribution centers. These robots can

navigate complex warehouse layouts, pick and sort items, and transport them to packing stations with remarkable speed and accuracy. Systems like Amazon's Kiva robots have demonstrated the scalability of robotic solutions in managing massive inventories and processing millions of orders daily. These robots work in concert with human operators, creating a hybrid workforce where robots handle the heavy lifting, repetitive picking, and swift transport, while humans focus on more complex tasks like quality control, order verification, and customer interaction. Beyond indoor warehousing, the last-mile delivery of goods is also being explored through robotic solutions. Autonomous delivery vehicles and drones are being tested and deployed in various regions, promising to reduce delivery times and costs, especially in dense urban areas and remote locations. The efficiency gains from robotic logistics are not only benefiting businesses through reduced operational expenses and faster delivery cycles but are also leading to improved customer satisfaction.

The domestic sphere, once considered the exclusive domain of human interaction, is also increasingly being touched by robotic innovation.
Domestic robots, ranging from simple automated vacuum cleaners to more sophisticated personal assistants, are becoming commonplace in households. Robot vacuums, equipped with sensors to navigate around furniture and map rooms, have freed up considerable time for busy individuals and families. As technology advances, we are seeing the emergence of robots capable of performing a wider array of chores,

such as lawn mowing, window cleaning, and even cooking. Beyond task-oriented robots, there is a growing interest in **social robots** designed for companionship and assistance, particularly for the elderly and individuals with disabilities. These robots can provide reminders for medication, facilitate communication with loved ones, engage in conversation, and offer a sense of presence, combating loneliness and promoting independence. The development of intuitive human-robot interfaces, including natural language processing and gesture recognition, is crucial for making these domestic robots accessible and user-friendly for a broad population. The ethical implications of integrating robots into our homes, particularly concerning privacy, security, and the nature of human-robot relationships, are subjects of ongoing discussion and research.

The applications of robotics and autonomous systems extend far beyond these core sectors. In **agriculture**, robots are being developed for precision farming tasks such as automated harvesting, targeted spraying of pesticides and fertilizers, and even soil analysis. These "agribots" can significantly reduce the need for manual labor, optimize resource usage, and improve crop yields. In **exploration and disaster response**, ruggedized robots are deployed in environments too dangerous or inaccessible for humans. This includes exploring the depths of the ocean, investigating nuclear accident sites, and searching for survivors in collapsed buildings after earthquakes. Their ability to operate in hazardous conditions, equipped with advanced sensors, makes them invaluable for gathering critical information and performing rescue operations. **Transportation** is perhaps one of the most anticipated frontiers for autonomous systems. Self-driving cars, trucks, and buses promise to revolutionize personal mobility and freight transport, potentially reducing accidents caused by human error, easing traffic congestion, and increasing accessibility for those unable to drive. While significant regulatory and technological hurdles remain, the development of autonomous vehicles is progressing rapidly. Even in areas like **entertainment and education**, robotics is making its mark, with robotic toys that teach programming concepts to children and robotic performers in theme parks and shows.

The pervasive adoption of these technologies has profound
societal and economic implications. One of the most widely discussed impacts is on the **job market**. While fears of widespread job displacement due to automation are valid, the reality is more nuanced. Historically, technological advancements have tended to transform rather than eliminate jobs. Robotics is no exception. Many existing roles are being augmented or redefined, requiring new skills. There is a growing demand for professionals in areas such as robotics engineering, AI development, data science, and the maintenance and supervision of automated systems. Simultaneously, new industries and services are emerging, creating novel employment opportunities. The economic benefits are also substantial, with increased productivity, reduced costs, and the potential for new markets and innovative products and services.

However, ensuring that the benefits of automation are shared broadly and that displaced workers have opportunities for retraining and re-skilling is a critical societal challenge.

Furthermore, the increasing sophistication of robots and AI raises significant
ethical considerations. Questions surrounding accountability for the actions of autonomous systems, particularly in critical applications like self-driving cars or medical robots, are paramount. The potential for bias in AI algorithms, which can perpetuate and even amplify existing societal inequalities, requires careful attention and mitigation strategies. The development of ethical frameworks and regulatory guidelines is crucial to ensure that these powerful technologies are developed and deployed in a responsible and beneficial manner. Issues of privacy, as robots equipped with sensors collect vast amounts of data, and security, to prevent malicious actors from compromising autonomous systems, are also key concerns. The very nature of human-robot interaction and the potential impact on social connections and human identity are also subjects of deep philosophical and ethical inquiry.

In conclusion, the applications of modern robotics and autonomous systems are incredibly diverse and their impact is far-reaching. From enhancing the efficiency and safety of industrial manufacturing and revolutionizing healthcare with precision surgery, to streamlining logistics and offering assistance in our homes, these technologies are reshaping our world. As development continues at an exponential pace, driven by advancements in artificial intelligence, sensor technology, and materials science, we can anticipate even more transformative applications and deeper integration into every facet of our lives. Navigating this evolving landscape requires a comprehensive understanding of not only the technological capabilities but also the profound societal, economic, and ethical implications, ensuring that the development and deployment of robotics and autonomous systems ultimately serve the betterment of humanity. The journey from industrial automation to sophisticated personal assistants and intelligent explorers is a testament to human ingenuity and the relentless pursuit of augmenting our capabilities through intelligent machines. The continued exploration and thoughtful implementation of these technologies will undoubtedly define the progress and well-being of future generations.

This textbook has been meticulously structured to provide a comprehensive and progressive learning experience in the field of robotics and autonomous systems. Our approach is rooted in the belief that a deep understanding of foundational principles is essential before delving into more complex applications. Therefore, the book begins by establishing a solid theoretical bedrock, gradually building upon it with increasingly sophisticated concepts and practical implementations. This logical flow is designed to cater to a diverse audience, from students with limited prior exposure to seasoned professionals seeking to broaden their expertise.

The initial chapters are dedicated to laying the groundwork. We commence with a broad overview of what constitutes a robot and an autonomous system, exploring the historical evolution of these technologies and the key driving forces behind their current rapid advancement. This foundational section aims to demystify the subject matter, providing a common language and conceptual framework for the subsequent discussions. We then embark on a detailed exploration of the core components that comprise robotic systems. This includes delving into the mechanics of robotic manipulators, the principles of sensor technologies that enable robots to perceive their environment, the intricacies of actuation mechanisms that allow for movement and interaction, and the fundamental concepts of control systems that govern robotic behavior. Each of these areas is presented with a balance of theoretical underpinnings and illustrative examples, ensuring that abstract principles are readily connected to tangible realities.

Following the establishment of these fundamental building blocks, the narrative transitions to the computational intelligence that breathes life into robotic systems. This involves an in-depth examination of algorithms for perception, which enable robots to interpret sensory data and build an understanding of their surroundings. We explore various approaches to localization and mapping, crucial for robots to know where they are and to navigate unfamiliar environments. Path planning and motion control are then addressed, focusing on how robots can devise and execute sequences of movements to achieve specific goals while avoiding obstacles. A significant portion of the book is dedicated to artificial intelligence (AI) and machine learning (ML) techniques, as these are increasingly central to the development of truly autonomous systems. We introduce core ML concepts, such as supervised, unsupervised, and reinforcement learning, and illustrate how they are applied in areas like object recognition, decision-making, and adaptive control.

As the reader progresses through the text, the focus gradually shifts towards the integration of these components into functional robotic systems and the development of increasingly complex autonomous behaviors. This involves exploring different types of robotic platforms – from mobile robots and drones to humanoids and collaborative robots – and discussing the unique challenges and opportunities associated with each. We examine how these systems interact with their environments, both physical and digital, and how they are being deployed in real-world scenarios. The latter chapters of the book are dedicated to advanced topics and emerging trends, offering insights into areas such as multi-robot systems, human-robot interaction, robot ethics, and the future of artificial intelligence in robotics. These sections aim to inspire further exploration and critical thinking about the broader implications of this rapidly evolving field.

The pedagogical philosophy underpinning this textbook emphasizes a hands-on, learn-by-doing approach, complemented by robust theoretical grounding. We firmly believe that true comprehension of robotics and autonomous systems is achieved not solely through passive reading, but through active engagement with the material. To this end, the book integrates a rich array of pedagogical tools designed to facilitate this active learning process. Each chapter is carefully crafted to move from the general to the specific, beginning with a clear statement of learning objectives and concluding with a summary of key takeaways. Throughout the text, numerous worked examples are provided to illustrate theoretical concepts in practice. These examples range from straightforward calculations to more complex scenario analyses, demonstrating how the principles discussed can be applied to solve practical problems.

Crucially, the textbook incorporates a series of progressive exercises and project-based learning opportunities. These are designed to gradually build the reader's skills and confidence. Early chapters will feature smaller, focused exercises that reinforce specific concepts, such as writing basic control algorithms or analyzing sensor data. As the reader advances, these exercises will evolve into more substantial mini-projects. These projects are structured to mimic real-world development cycles, often requiring the integration of multiple concepts learned in previous sections. For instance, a project might involve designing a simple navigation system for a simulated mobile robot, requiring the student to apply principles of mapping, localization, and path planning.

A cornerstone of our pedagogical strategy is the extensive use of simulation. Recognizing the inherent costs and complexities associated with acquiring and maintaining physical robotic hardware, we leverage simulation environments as a powerful tool for learning and experimentation. The book will guide readers through the setup and use of popular simulation platforms, such as Gazebo, CoppeliaSim, or ROS-integrated simulators. These environments allow for the rapid prototyping, testing, and debugging of algorithms in a safe and cost-effective manner. Readers will be encouraged to implement and test the algorithms and concepts discussed in the book within these simulated worlds. This hands-on experience with simulation not only solidifies theoretical understanding but also develops crucial practical skills in software development, debugging, and system integration. The progression of simulation-based projects will mirror the book's structure, starting with simple single-robot tasks and moving towards more complex multi-robot scenarios and intricate environmental interactions.

While simulation is an invaluable tool, we also underscore the indispensable nature of hands-on hardware experience. The nuances of real-world robotic systems – the unexpected sensor noise, the limitations of actuators, the physical constraints of the environment – cannot always be perfectly replicated in simulation. Therefore, where feasible, the book will suggest opportunities for readers to engage with physical hardware. This might involve using accessible development kits, such as the Raspberry Pi or Arduino, for basic robotics projects, or working with more advanced robotic platforms if available through academic institutions or personal investment. The aim is to bridge the gap between the simulated and the real, allowing readers to appreciate the challenges and intricacies of deploying algorithms on physical robots. For those undertaking capstone projects or advanced coursework, the book provides guidance on integrating simulation-based development with hardware implementation, ensuring a well-rounded and robust learning experience.

The book is designed to be a versatile resource, adaptable to various learning contexts. For students enrolled in formal courses, it serves as a comprehensive textbook that can guide lectures, assignments, and laboratory sessions The structured progression of topics and integrated projects make it ideal for building a curriculum that spans an academic semester or year. Instructors will find that the depth and breadth of the content, combined with the pedagogical tools, provide a solid foundation for teaching both theoretical and practical aspects of robotics and autonomous systems.

For individuals pursuing self-study, the book offers a clear roadmap for acquiring knowledge and practical skills independently. The self-contained nature of the chapters, along with the readily available simulation tools and suggested project ideas, empowers learners to progress at their own pace. The emphasis on practical implementation through simulation and incremental projects ensures that self-learners can develop tangible capabilities and build a portfolio of work.

For students undertaking capstone projects or thesis work, this textbook provides a rich source of foundational knowledge, advanced concepts, and potential project directions. The structure encourages readers to identify research gaps and to leverage the concepts presented to develop novel solutions. The detailed explanations of algorithms and techniques can serve as starting points for deeper investigation, while the emphasis on both simulation and hardware provides a framework for designing and executing comprehensive research projects. The book aims to equip readers with the critical thinking skills and technical proficiency necessary to not only understand the current state of robotics and autonomous systems but also to contribute to its future advancement. The learning journey outlined within these pages is intended to be both intellectually stimulating and practically rewarding, preparing individuals to tackle the exciting challenges and opportunities that lie ahead in this dynamic field.

Chapter 2: Sensing Modalities and Data Acquisition

Robots, to be effective, must possess the ability to sense their surroundings and their own internal states. Without sensory input, a robot is effectively blind and deaf, incapable of making informed decisions or executing meaningful actions. Sensors act as the robot's interface with the world, translating physical phenomena into electrical signals that can be processed by the robot's computational systems. This perception is paramount, underpinning everything from basic navigation and object manipulation to complex human-robot interaction and autonomous decision-making. Just as humans rely on their five primary senses – sight, hearing, touch, smell, and taste – to understand and navigate their environment, robots employ a diverse array of sensing technologies to gather information. The quality, type, and interpretation of this sensory data directly dictate the robot's capabilities and its potential for success in any given task. Therefore, a foundational understanding of robotic sensors, their principles of operation, and their limitations is an indispensable prerequisite for anyone aspiring to design, build, or program intelligent robotic systems.

The vast array of sensors available for robotic applications can be broadly categorized based on the nature of the information they gather. A fundamental distinction is made between **proprioceptive sensors** and **exteroceptive sensors**. Proprioceptive sensors, often referred to as internal sensors, provide information about the robot's own state. This includes data about the configuration and dynamics of its own body. Examples include motor encoders, which measure the rotation of motor shafts to determine joint angles in robotic arms or wheel rotations in mobile robots, thus informing the robot about its kinematic state. Inertial Measurement Units (IMUs) are another critical class of proprioceptive sensors. An IMU typically comprises accelerometers and gyroscopes, which measure linear acceleration and angular velocity, respectively. By integrating these measurements over time, an IMU can provide estimates of the robot's orientation (pitch, roll, yaw), velocity, and position, though drift in integration can be a significant challenge. Other proprioceptive sensors might include force/torque sensors embedded in robotic joints or end-effectors, providing feedback on the forces and torques being exerted or experienced. Understanding the robot's own state is crucial for control, for example, to ensure that actuators are operating within their limits or to maintain balance.

In contrast,
exteroceptive sensors are designed to gather information about the robot's external environment. These sensors allow the robot to perceive the world around it, detecting objects, surfaces, and events. This category encompasses a wide spectrum of technologies, each with its unique strengths and weaknesses. Vision sensors, such as cameras, are perhaps the most ubiquitous exteroceptive sensors. They capture visual information, providing rich data that can be used for object recognition, scene understanding, and navigation. Range sensors are another vital group, providing information about distances to objects. Sonar sensors (ultrasonic sensors) emit sound waves and

measure the time it takes for the echoes to return, thereby determining the distance to nearby obstacles. While relatively inexpensive and robust to some environmental conditions, sonar sensors typically offer lower resolution and can be affected by the acoustic properties of surfaces. Lidar (Light Detection and Ranging) sensors, on the other hand, use lasers to measure distances. They typically scan a scene with a laser beam and record the time-of-flight of the reflected light. Lidar systems can provide highly accurate and dense 3D point clouds of the environment, making them invaluable for mapping, localization, and obstacle avoidance in complex settings. Infrared (IR) sensors can also be used for distance measurement, often for shorter ranges, and can detect the presence or absence of objects. Tactile sensors, embedded in grippers or other contact points, provide information about contact forces, texture, and shape when the robot interacts physically with its environment. Odor sensors (electronic noses) and microphones (for auditory sensing) are also employed in specific robotic applications, enabling robots to perceive chemical or acoustic phenomena.

The selection of appropriate sensors for a robotic system is a critical design decision that depends heavily on the specific application requirements. Factors such as the operating environment, the tasks the robot needs to perform, the desired level of accuracy and precision, cost constraints, and computational resources all play a significant role. For instance, a robot designed for indoor navigation in a well-lit office environment might rely heavily on cameras and infrared sensors for close-range obstacle detection and visual odometry. In contrast, a self-driving car operating in diverse weather conditions and at high speeds would necessitate a robust suite of sensors, likely including lidar for precise 3D mapping and obstacle detection, radar for detecting objects through fog or rain, and cameras for recognizing traffic signs and lane markings.

Each sensor type operates on distinct physical principles, leading to inherent trade-offs. For example, cameras provide rich visual data but are highly sensitive to lighting conditions and can struggle with depth perception without stereo vision or other depth-sensing mechanisms. Lidar offers excellent range and accuracy but can be expensive and may have difficulty with certain surfaces (e.g., highly reflective or absorptive materials) or in adverse weather. Sonar sensors are cost-effective and relatively robust but provide lower resolution and can suffer from specular reflections, where the sound wave bounces off a surface at an angle, leading to inaccurate readings. Inertial Measurement Units are excellent for measuring short-term changes in motion and orientation but accumulate errors over time due to integration, requiring periodic recalibration using other sensors, such as GPS or visual odometry.

The data generated by these sensors is rarely perfect. It is often subject to noise, biases, and uncertainties. Noise refers to random fluctuations in the sensor readings that are not related to the actual physical quantity being measured. This can arise from various sources, including thermal effects within the sensor, electromagnetic interference, or inherent limitations in the sensor's resolution. Biases are systematic errors that cause the sensor readings to consistently deviate from the true value in a particular direction. For example, a camera might have lens distortion that systematically warps images, or an accelerometer might have a zero-offset error.

Understanding and mitigating these sensor imperfections is a crucial aspect of robotics. This involves techniques such as filtering, calibration, and sensor fusion. Filtering, such as Kalman filtering or its variants, can be used to estimate the true state of a system by combining noisy sensor measurements with a model of the system's dynamics. Calibration is the process of determining the specific parameters of a sensor that characterize its behavior and to correct for known biases and systematic errors. For example, calibrating a camera involves determining its intrinsic parameters (focal length, principal point, lens distortion coefficients) and extrinsic parameters (position and orientation relative to the robot's base frame). Sensor fusion, as mentioned earlier, involves combining data from multiple sensors, often of different types, to obtain a more accurate, reliable, and comprehensive understanding of the environment or the robot's state than could be achieved with any single sensor alone. This is because different sensors may have complementary strengths and weaknesses. For instance, fusing lidar data (for precise geometry) with camera data (for color and texture information) can lead to richer and more robust object recognition and scene understanding.

The choice of sensor modality also dictates the type of information available for downstream processing. Proprioceptive sensors provide internal feedback vital for closed-loop control. For instance, motor encoders in a robot arm allow a controller to precisely regulate the position of each joint, enabling the robot to perform tasks requiring high kinematic accuracy. An IMU's data on angular velocity and acceleration is essential for stabilizing a drone's flight or for detecting sudden changes in a mobile robot's motion. Exteroceptive sensors, on the other hand, facilitate interaction with the external world. Cameras enable object detection and tracking, allowing a robot to identify and follow a target. Lidar's 3D point clouds can be used to construct detailed maps of an environment, which are then used for navigation. Sonar sensors can provide simple proximity alerts, preventing collisions with walls or other obstacles.

The fidelity of the sensor data directly impacts the performance of any robotic system. Inaccurate or unreliable sensor readings can lead to incorrect decisions, poor control, and ultimately, mission failure. A robot tasked with picking up a delicate object will fail if its tactile sensors provide inaccurate force readings, leading it to apply too much pressure and crush the object. Similarly, a self-driving car relying on erroneous depth information from its sensors might fail to detect an obstacle, with potentially catastrophic consequences. Therefore, careful consideration of sensor capabilities, limitations, and the necessary signal processing techniques is paramount in robotic system design. The subsequent sections will delve deeper into specific types of sensors, exploring their underlying principles, mathematical models, and common applications in robotics.

Cameras represent one of the most versatile and information-rich exteroceptive sensing modalities available to robotic systems. Their ability to capture detailed visual information about the environment has made them indispensable for a wide array of applications, from industrial automation and autonomous navigation to human-robot interaction and augmented reality. Unlike sensors that provide a single scalar value (e.g., distance from a sonar sensor) or a limited set of values (e.g., angular velocity from a gyroscope), cameras produce complex, high-dimensional data in the form of images or video streams. This richness of information, however, also presents significant computational challenges for processing and interpretation.

The fundamental principle behind a camera is to capture light reflected or emitted from the environment and form an image on a light-sensitive sensor. This process involves optics, specifically a lens system, which focuses the incoming light rays onto an image plane. The characteristics of the lens, such as its focal length, significantly influence the field of view and the perspective of the captured image. A shorter focal length results in a wider field of view, capturing more of the scene but with less magnification, akin to a wide-angle lens. Conversely, a longer focal length provides a narrower field of view and greater magnification, similar to a telephoto lens.

The image sensor itself is a crucial component, converting the optical information into an electrical signal. Historically, charge-coupled devices (CCDs) were prevalent, but complementary metal-oxide-semiconductor (CMOS) sensors have become dominant due to their lower power consumption, faster readout speeds, and lower manufacturing costs. These sensors are typically comprised of a grid of pixels, where each pixel corresponds to a small area in the captured image. When light strikes a pixel, it generates an electrical charge proportional to the intensity of the light. This charge is then converted into a digital value, forming the digital image.

The resolution of a camera, usually expressed as the number of pixels in the horizontal and vertical dimensions (e.g., 1920x1080 pixels), dictates the level of detail that can be captured. A higher resolution allows for finer discrimination of features within the scene, which is critical for tasks like object recognition or reading small text. However, higher resolution also translates to larger data sizes, increased computational demands for processing, and potentially slower frame rates. The frame rate, measured in frames per second (fps), determines how many images the camera can capture per unit of time. For dynamic scenes or applications requiring real-time tracking, a high frame rate is essential. Low frame rates can lead to motion blur and make it difficult to accurately track fast-moving objects.

Exposure is another critical parameter in image acquisition. It controls the amount of light that reaches the image sensor. This is typically managed by adjusting the aperture (the size of the opening in the lens that allows light to pass through), the shutter speed (the duration for which the sensor is exposed to light), and the sensor's gain (amplification of the signal). Proper exposure is vital for obtaining images with good contrast and dynamic range, ensuring that both bright and dark areas of the scene are captured effectively. Overexposure can lead to washed-out details in bright regions, while underexposure can result in loss of detail in shadows. In robotics, automatic exposure control is often employed to adapt to varying lighting conditions, but manual control or specific exposure settings might be necessary for applications requiring consistent image quality.

Beyond these fundamental aspects, cameras can be broadly categorized based on their imaging capabilities and the type of information they provide. The most basic type is the
monocular camera, which uses a single lens and sensor to capture a 2D image. While widely used for

tasks such as object detection, tracking, and visual odometry (estimating the robot's motion from image sequences), monocular cameras inherently lack direct depth information. Inferring depth from a single 2D image is a challenging problem, often requiring assumptions about the scene or the use of advanced computer vision techniques like structure-from-motion.

To overcome the limitations of monocular vision regarding depth perception,
stereo cameras are employed. A stereo camera system consists of two or more cameras, typically mounted a fixed distance apart, mimicking the human binocular vision system. By comparing the images captured by the two cameras, a disparity map can be computed. The disparity is the difference in the projected position of a 3D point in the two images. By leveraging the known distance between the cameras (the baseline) and the computed disparity, the 3D position of points in the scene can be triangulated, thereby providing depth information. Stereo vision systems can offer accurate depth perception for objects within their field of view and baseline. However, they require careful calibration to establish the precise relative pose between the cameras. Furthermore, accurate disparity computation relies on robust feature matching between the two images, which can be challenging in textureless regions or in the presence of repetitive patterns. The accuracy of depth estimation typically decreases with distance and increases with the camera baseline, but a larger baseline also reduces the overlap in the field of view between the two cameras.

A more direct approach to obtaining depth information is through
depth cameras, also known as structured light or time-of-flight (ToF) cameras. These cameras actively emit light and measure its interaction with the environment to directly compute depth. **RGB-D cameras** (Red, Green, Blue, Depth) are a popular type of depth camera that simultaneously capture color (RGB) images and a per-pixel depth map. These cameras often employ different technologies to acquire depth:

One common technology for RGB-D cameras is **structured light**. This method involves projecting a known pattern of light, such as a grid or a series of infrared dots, onto the scene. A dedicated infrared camera then observes how this pattern is distorted by the geometry of the objects in the scene. By analyzing the deformation of the projected pattern, the depth of each point can be calculated. Structured light systems are generally accurate at short to medium ranges and can provide dense depth information. However, their performance can be affected by ambient light, especially bright sunlight, which can overwhelm the projected pattern. They are also susceptible to issues with highly reflective or transparent surfaces, as these can distort or scatter the projected light in unpredictable ways.

Another prominent technology for depth sensing is **Time-of-Flight (ToF)**. ToF cameras work by emitting pulses of light (often infrared) and measuring the time it takes for the light to travel to an object and return to the sensor. Since the speed of light is constant, the time-of-flight is directly proportional to the distance to the object. Different implementations exist, including direct ToF (dToF) which measures the precise arrival time of individual photons, and indirect ToF (iToF) which measures the phase shift of modulated light. ToF cameras can operate in ambient light conditions and are less affected by surface reflectivity compared to structured light, making them suitable for a wider range of environments. However, their accuracy can be influenced by the material properties of the target surface (e.g., absorption or reflection characteristics) and they can struggle with highly specular surfaces. The effective range of ToF cameras is also a design consideration, with longer-range sensors typically requiring more power and potentially having lower resolution or accuracy.

The data acquired by cameras, particularly images, is typically represented in terms of pixel intensity values. For grayscale images, each pixel has a single value representing its brightness, often ranging from 0 (black) to 255 (white) in an 8-bit representation. Color images, on the other hand, require multiple values per pixel to represent color information. The most common color space for digital cameras is
RGB (Red, Green, Blue), where each pixel is described by the intensity of its red, green, and blue components. Other color spaces, such as HSV (Hue, Saturation, Value) or HSL (Hue, Saturation, Lightness), are sometimes used for specific image processing tasks, as they can decouple color information from brightness, which can be useful for tasks like color-based object segmentation.

The inherent nature of visual data introduces several challenges that must be addressed in robotic perception systems.

Lighting variations are a pervasive problem. Changes in ambient light intensity, shadows, and reflections can drastically alter the appearance of objects and scenes, making it difficult for algorithms to consistently recognize them. For instance, an object that is easily identifiable under bright sunlight might become almost invisible in dim light or when partially obscured by a shadow. Robots operating in dynamic environments often need to employ sophisticated image processing techniques, such as illumination normalization or adaptive thresholding, to mitigate these effects.

Occlusions are another significant challenge. When an object is partially or fully hidden by another object, the robot loses visual information about the occluded part. This can lead to difficulties in object tracking, recognition, and navigation. For example, if a robot is following a person, and the person walks behind a pillar, the robot might lose track of them until they re-emerge. Addressing occlusions often involves using sensor fusion (combining visual data with other sensors like lidar or radar that can see through certain occlusions) or employing predictive models that can infer the likely position or state of the occluded object based on its past motion.

The formation of an image by a camera can be mathematically modeled. A simplified pinhole camera

model assumes a single point of projection and a flat image plane. In this model, a 3D point in the world is projected onto the image plane according to principles of similar triangles. However, real cameras have lenses that introduce distortions.
Lens distortion is a deviation from the ideal pinhole model, primarily categorized into radial distortion and tangential distortion. Radial distortion, common in wide-angle lenses, can cause straight lines to appear curved. Barrel distortion bows lines outward from the center, while pincushion distortion bows them inward. Tangential distortion arises from the lens not being perfectly perpendicular to the image sensor. These distortions can significantly affect the accuracy of measurements made from images, such as distances or angles. Therefore, camera calibration is a critical preprocessing step in many robotic vision applications. Camera calibration involves determining the camera's intrinsic parameters (focal length, principal point, and distortion coefficients) and its extrinsic parameters (its position and orientation in the robot's coordinate frame). This allows for the correction of lens distortions and the accurate mapping between 3D world points and 2D image points.

The data acquired from cameras, whether it's a simple grayscale image, an RGB image, or a depth map, is often preprocessed before being fed into more complex perception algorithms. This preprocessing can include:

Noise Reduction: Applying filters like Gaussian blur or median filters to reduce random noise in the image, which can improve the performance of subsequent algorithms.
Color Space Conversion: Transforming images from RGB to other color spaces like HSV if the application benefits from separating color information from intensity.
Image Resizing or Cropping: Adjusting the image dimensions to match the input requirements of a particular algorithm or to focus on a region of interest.
Geometric Transformations: Correcting for lens distortions or aligning images from multiple cameras.

The choice of camera and its associated parameters is heavily dependent on the specific robotic application. For instance, a mobile robot navigating an indoor environment might benefit from a monocular camera for visual odometry and object detection, potentially augmented by an RGB-D camera for obstacle avoidance and mapping at closer ranges. A surveillance robot might prioritize high-resolution cameras with excellent low-light performance. An industrial robot performing precise assembly tasks might utilize stereo vision or specialized machine vision cameras for accurate 3D inspection.

The output of a camera, typically an image, is a dense representation of visual information. This contrasts with the sparser data provided by many other sensors. For example, a lidar sensor might return a point cloud containing thousands or millions of 3D points, but these points are discrete samples of the environment's geometry. A camera, on the other hand, provides continuous pixel information across its field of view. This density of information is both a strength and a challenge. It allows for rich scene understanding, texture analysis, and the detection of subtle visual cues that might be missed by other sensors. However, processing such a large amount of data in real-time requires significant computational resources and efficient algorithms.

Furthermore, the interpretation of camera data often involves understanding concepts from computer vision and image processing. Techniques like edge detection, feature extraction (e.g., SIFT, SURF, ORB), segmentation, and machine learning-based object recognition are fundamental to unlocking the potential of camera-based sensing. For example, to enable a robot to recognize a specific type of tool, a dataset of images of that tool under various conditions would be used to train a machine learning model. This model would learn to identify distinguishing visual features of the tool, allowing the robot to detect and locate it in new, unseen images.

In summary, vision-based sensing using cameras provides an incredibly rich source of information for robotic systems. From monocular cameras offering 2D perspectives to stereo and depth cameras providing crucial 3D information, these sensors are at the forefront of robotic perception. Understanding the fundamental principles of image acquisition, including resolution, frame rate, and exposure, alongside the properties of image formation and the challenges posed by lighting, occlusions, and lens distortions, is essential for effectively leveraging this powerful sensing modality. The rich data generated by cameras forms the foundation for advanced perception algorithms that enable robots to understand, interact with, and navigate their environments.

In the realm of robotics, understanding the physical layout and the presence of surrounding objects is paramount for safe and effective operation. While vision-based systems excel at interpreting the semantic content of an environment, they often struggle with direct, precise distance measurements, especially in challenging lighting conditions or when dealing with features that lack distinct visual textures. This is where range and proximity sensing technologies become indispensable. These sensors provide the critical quantitative data about the spatial relationships between the robot and its surroundings, forming the bedrock for applications such as obstacle detection, avoidance, mapping, localization, and navigation. Two of the most prominent technologies in this domain are Light Detection and Ranging (LiDAR) and Sonar (Sound Navigation and Ranging).

LiDAR systems, a cornerstone of modern robotic perception, leverage the power of light, specifically lasers, to precisely measure distances to objects. The fundamental principle behind LiDAR is the time-of-flight (ToF) measurement, similar in concept to how active depth cameras operate but typically with far greater range and precision. A LiDAR unit emits pulsed laser beams into the environment and measures the time it takes for these beams to reflect off surfaces and return to a detector. Knowing the speed of light (a constant), the distance to the reflecting object can be calculated with remarkable accuracy using the formula: `distance = (speed of light time of flight) / 2`. The factor of two accounts for the round trip of the laser pulse.

The output of a LiDAR sensor is not a single distance reading but rather a collection of these distance measurements, often forming a detailed three-dimensional representation of the environment. This collection of points, each with an associated XYZ coordinate (or distance and angular position), is commonly referred to as a "point cloud." The density and resolution of this point cloud are determined by the LiDAR's design and scanning mechanism. Early and still prevalent LiDAR systems utilize **mechanical scanning**. These systems typically employ a rotating platform, often with multiple laser emitters and detectors mounted on it. As the platform rotates, the laser beams sweep across the environment, capturing a multitude of distance readings in a full 360-degree horizontal field of view, and often with a significant vertical field of view as well, depending on the number of laser beams and their vertical arrangement. These multi-line scanning LiDARs, with common configurations of 16, 32, or even 64 laser beams stacked vertically, can capture incredibly dense point clouds, providing a rich representation of the environment's geometry. The speed of rotation influences the frame rate at which the point cloud is generated. While offering excellent performance, mechanical scanning LiDARs can be susceptible to wear and tear due to their moving parts, are generally larger and heavier, and can be more expensive.

In response to these limitations, the field has seen significant advancements in **solid-state LiDAR**. These technologies aim to eliminate or minimize moving parts, leading to more robust, compact, and potentially lower-cost sensors. One prominent solid-state approach is **Flash LiDAR**. Similar to how a camera flash illuminates a scene, a Flash LiDAR emits a single, broad pulse of laser light that illuminates an entire area or volume simultaneously. A specialized focal plane array of detectors then captures the reflected light, with each detector element corresponding to a specific point in the scene. This allows for depth measurement across the entire field of view in a single shot, offering very high frame rates and excellent resistance to vibration. However, Flash LiDAR typically has a more limited range compared to mechanical scanning LiDARs and can be more sensitive to ambient light conditions and the reflectivity of surfaces.

Another significant solid-state LiDAR technology is **Optical Phased Arrays (OPAs)** and **MEMS (Micro-Electro-Mechanical Systems) mirrors**. OPAs use arrays of tiny optical antennas to steer the laser beam electronically without any physical movement. By precisely controlling the phase of the light emitted from each antenna, the beam can be directed in any desired direction. MEMS mirrors, on the other hand, use micro-mirrors that can be rapidly tilted to steer the laser beam. These technologies allow for rapid scanning of the laser beam across the field of view, offering flexibility in scan patterns and high angular resolution. They are generally more compact and rugged than traditional mechanical scanners, making them attractive for a variety of robotic applications. However, challenges remain in achieving the same range and field of view as the most advanced mechanical systems while keeping costs down.

The data produced by LiDAR, the point cloud, is a fundamental data structure for robotic perception. Each point typically contains its 3D coordinates (x, y, z) in the sensor's frame of reference. Many LiDAR sensors also provide intensity information, which is a measure of the strength of the returned laser signal. This intensity can be related to the reflectivity of the surface, providing an additional cue for distinguishing different materials or identifying specific features within the point cloud. Some advanced LiDARs also integrate cameras, allowing for the fusion of geometric data from the LiDAR with visual information from the camera, creating what is known as an "RGB-D point cloud" or "colorized point cloud." This fusion greatly enhances the ability to semantically understand the environment by associating color and texture with the precise 3D geometry.

The applications of LiDAR in robotics are vast and

growing. In autonomous vehicles, LiDAR is crucial for creating detailed 3D maps of the environment, detecting other vehicles, pedestrians, and obstacles, and enabling precise localization by matching the sensor's current readings to pre-existing maps. For mobile robots operating in industrial settings, such as warehouses or factories, LiDAR provides reliable obstacle detection for navigation around aisles, machinery, and personnel, ensuring safe operation and efficient path planning. In surveying and mapping, LiDAR is used to create highly accurate digital elevation models and 3D representations of terrain and structures. Even in areas like agriculture and forestry, LiDAR is employed for tasks such as crop monitoring and tree height measurement. The inherent ability of LiDAR to provide accurate depth and geometric information, independent of ambient lighting, makes it a robust choice for many perception tasks where visual systems might falter.

However, LiDAR is not without its limitations. The precision of LiDAR measurements can be affected by the material properties of the reflecting surface. Highly absorptive materials, like dark, matte surfaces, can return weaker signals, reducing the effective range or leading to missed detections. Conversely, highly reflective surfaces, like polished metal or glass, can cause specular reflections, where the laser beam is reflected away from the sensor, leading to inaccurate readings or no reading at all. Weather conditions, particularly fog, heavy rain, or snow, can scatter or absorb the laser light, significantly degrading performance and range. Furthermore, LiDAR sensors can be relatively expensive, especially the high-performance multi-line mechanical scanning units, which can be a significant cost factor for some robotic applications. Interference can also be an issue if multiple LiDAR sensors are operating in close proximity, as their laser pulses might interfere with each other, leading to noisy data.

Complementing LiDAR, and often serving as a more cost-effective alternative for specific tasks, are Sonar sensors. These systems operate by emitting ultrasonic sound waves and measuring the time it takes for these waves to return after reflecting off an object. The principle is analogous to LiDAR's time-of-flight but uses sound instead of light. An ultrasonic transducer acts as both the emitter and receiver. The sensor emits a burst of high-frequency sound waves (typically in the range of 20 kHz to several MHz), and then listens for the echo. The speed of sound in air is significantly slower than the speed of light, and it is also variable, depending on factors such as temperature, humidity, and air pressure. This variability means that sonar-based distance measurements are generally less precise than those from LiDAR, and their accuracy can be affected by environmental conditions. The distance is calculated using a similar time-of-flight formula: `distance = (speed of sound time of flight) / 2`.

Sonar sensors are particularly well-suited for detecting obstacles in close proximity and are widely used in applications where cost is a major consideration or where LiDAR might be overkill or impractical. For example, many domestic robots, such as vacuum cleaners or lawnmowers, use sonar sensors for basic obstacle avoidance. They are also commonly found on parking assist systems in cars, helping drivers detect nearby objects when maneuvering in tight spaces. In some industrial robots, sonar can be used for detecting the presence or absence of objects on a conveyor belt or for proximity sensing during manipulation tasks.

One of the key advantages of sonar is its ability to function in environments where other sensors might struggle. Sound waves can penetrate smoke, fog, and dust that would impede LiDAR and cameras. This makes sonar a valuable option for environments with poor visibility or for detecting objects through obscurants. Furthermore, sonar sensors are generally much less expensive to manufacture than LiDAR systems, making them an attractive choice for mass-produced consumer robots and cost-sensitive industrial applications.

However, sonar sensors also have distinct limitations. The resolution of sonar is typically much lower than that of LiDAR. Because sound waves spread out as they travel, the "beam width" of a sonar sensor is often quite wide, meaning it provides a less precise indication of an object's location and shape. This can lead to "fuzziness" in the perceived environment and can make it difficult to distinguish between closely spaced objects or to detect small obstacles. The angular resolution is also generally poorer than LiDAR. Another significant limitation is the minimum detection range. Sonar sensors cannot reliably detect objects that are too close to the sensor, as the sound pulse may not have enough time to return before the next pulse is emitted or the sensor's electronics may not be able to distinguish the outgoing pulse from a very near echo. This "dead zone" near the sensor means sonar is not suitable for tasks requiring precise close-range sensing.

The detection capabilities of sonar can also be affected by the acoustic properties of the objects in the environment. Soft, sound-absorbing materials, such as foam or fabric, may not reflect sound waves strongly enough to be detected, while very hard, smooth surfaces can cause complex reflections that lead to inaccurate readings. The speed of sound variability is also a concern for precise applications. While some advanced sonar systems employ techniques to compensate for temperature variations, environmental changes can still introduce errors. Finally, the relatively slow speed of sound limits the rate at which sonar can acquire data. Compared to the high frame rates achievable with modern LiDAR and cameras, sonar updates are typically much slower, which can be a bottleneck for robots that need to react quickly to dynamic environments.

The data output from sonar sensors is typically a list of distance measurements corresponding to the direction in which the sensor was pointed or scanned. For simple sonar sensors, this might be a single distance value at a time. More complex sonar systems, such as those used for 3D mapping, might involve arrays of transducers or mechanically steered beams to build up a more comprehensive picture. The output is generally less dense and less geometrically precise than LiDAR point clouds.

In summary, both LiDAR and Sonar sensors play crucial roles in a robot's perception system, providing vital range and proximity information that complements other sensing modalities. LiDAR, with its laser-based time-of-flight measurements, offers high precision, detailed 3D point clouds, and excellent performance in various lighting conditions, making it ideal for applications demanding accurate mapping, localization, and obstacle detection at medium to long ranges. However, its cost and susceptibility to certain environmental factors like fog are considerations. Sonar, on the other hand, utilizes sound waves, providing a more cost-effective solution for obstacle detection, particularly in close-range applications or environments with poor visibility. Its limitations include lower resolution, a minimum detection range, and sensitivity to the acoustic properties of surfaces. Understanding the strengths and weaknesses of each technology is key to selecting the appropriate sensor for a given robotic task and for effectively fusing data from multiple sensors to create a robust and comprehensive understanding of the robot's operating environment.

When a robot embarks on its journey through the physical world, knowing not just *where* obstacles are but also its own movement and orientation is fundamental. This self-awareness of motion and pose is crucial for navigation, control, and maintaining an accurate understanding of its environment over time. While external sensors like LiDAR and sonar provide information about the surroundings, a critical piece of the perception puzzle comes from internal sensors that track the robot's own motion. This subsection delves into two key technologies that empower robots with this proprioceptive sense: Inertial Measurement Units (IMUs) and the broader concept of odometry.

At the heart of a robot's ability to sense its own motion lies the Inertial Measurement Unit (IMU). An IMU is a sophisticated sensor package that typically combines accelerometers and gyroscopes, and sometimes magnetometers, to measure the robot's linear acceleration and angular velocity. These measurements are then used to estimate the robot's orientation, and by extension, its position and velocity over time.

Accelerometers are devices that measure acceleration, which is the rate of change of velocity. They typically work by detecting the inertial force exerted on a proof mass when the sensor is accelerated. Imagine a simple accelerometer: it might consist of a spring-loaded mass. When the sensor accelerates in a particular direction, the mass lags behind due to inertia, and the displacement of the mass from its equilibrium position is proportional to the acceleration. In modern IMUs, these are usually implemented using Micro-Electro-Mechanical Systems (MEMS) technology. These tiny devices use microscopic structures that deflect under acceleration, with their movement often detected capacitively or piezoresistively. Accelerometers are sensitive to both linear motion and the force of gravity. When the IMU is stationary, the accelerometer will still measure a constant acceleration due to gravity, pointing downwards. This gravitational acceleration provides a crucial reference for determining the "down" direction, which is essential for estimating the robot's tilt or pitch and roll angles. By measuring acceleration along its three orthogonal axes (typically x, y, and z), an accelerometer can provide a detailed picture of the robot's translational motion.

Complementing the accelerometers are gyroscopes, which are designed to measure angular velocity, or the rate of rotation around an axis. Traditionally, gyroscopes relied on the principle of conservation of angular momentum, using a spinning rotor. However, modern IMUs almost exclusively employ MEMS gyroscopes. These often utilize a vibrating proof mass. When the sensor rotates, the Coriolis effect causes this vibrating mass to experience a force perpendicular to its motion and the axis of rotation. This force induces a displacement or stress in the MEMS structure, which is then measured. Like accelerometers, MEMS gyroscopes are fabricated on a microscopic scale and are incredibly sensitive. By measuring the angular velocity around each of the three orthogonal axes, gyroscopes provide information about how the robot is tumbling or turning.

The raw data from an IMU—linear acceleration and angular velocity—while informative, needs to be processed to be directly useful for understanding a robot's state. To estimate orientation, the acceleration data can be used to determine the direction of gravity, which establishes the "down" vector. By comparing the measured acceleration vector to the known direction of gravity, algorithms can calculate the robot's tilt (pitch and roll). For the yaw (heading) angle, which is not directly measurable by accelerometers (as gravity acts solely in the vertical direction), gyroscopes are indispensable. Integrating the angular velocity measured by the gyroscopes over time yields the change in angle. For example, if a gyroscope measures a constant angular velocity of 10 degrees per second around the z-axis, after 5 seconds, the yaw angle will have changed by 50 degrees.

Combining the outputs of accelerometers and gyroscopes allows for a more complete estimation of a robot's 3D orientation (roll, pitch, and yaw). This process of fusing sensor data often involves sophisticated algorithms, such as Kalman filters or complementary filters. These filters take the noisy, biased measurements from the individual sensors and combine them in an optimal way to produce a more accurate and stable estimate of the robot's orientation. For instance, gyroscopes are excellent at measuring rapid changes in orientation but tend to drift over time due to integration errors. Accelerometers, on the other hand, are less susceptible to drift in orientation estimation over short periods but are sensitive to vibrations and other linear accelerations that are not related to gravity. By using accelerometers to correct the long-term drift of gyroscopes, and gyroscopes to provide high-frequency updates and track rapid movements, a robust orientation estimate can be achieved.

The data provided by an IMU is crucial for a wide range of robotic applications. In robotics, it is often referred to as estimating the robot's "attitude." This includes its orientation with respect to a global frame of reference (e.g., North, East, Down) or relative to its own previous state. For a flying robot like a drone, maintaining a stable attitude is paramount for flight control. For a ground robot, understanding its pitch and roll is important for navigating uneven terrain and for accurate localization.

However, it is vital to acknowledge the inherent limitations of IMUs. Accelerometers and gyroscopes are susceptible to noise, biases, and temperature variations. More importantly, integrating angular velocity to obtain orientation inherently leads to drift. Over time, small errors in the gyroscope's measurements, when integrated, accumulate, causing the estimated orientation to diverge from the true orientation. This drift is a fundamental challenge in inertial navigation. While accelerometers can help correct for drift in pitch and roll by referencing gravity, they cannot correct for yaw drift in the absence of other sensors. This is why IMUs are often coupled with other sensors, such as magnetometers (which measure the Earth's magnetic field and can provide an absolute heading reference) or GPS (for outdoor navigation), to provide a more globally accurate state estimate.

Beyond measuring self-motion, robots need to estimate their position and trajectory through the environment. This is where the concept of odometry comes into play. Odometry is essentially the process of estimating a robot's change in position and orientation by integrating measurements of its own motion over time. It's analogous to how you might estimate your walking distance by counting your steps and noting your turns.

The most common form of odometry in wheeled robots is wheel odometry. This method relies on sensors called wheel encoders, which are typically attached to the robot's drive wheels. Wheel encoders measure the rotation of each wheel. By knowing the circumference of the wheels and the number of rotations they have made, it is possible to calculate the distance traveled by each wheel. For a differential drive robot with two independently driven wheels, for instance, knowing the distance traveled by each wheel allows for the calculation of the robot's linear and angular displacement. If the wheels rotate by `dR1` and `dR2` respectively, and the robot's wheelbase is `L`, then the change in position (`dx`, `dy`) and change in orientation (`d_theta`) can be calculated. For a simple forward motion, `dx` would be related to the average distance traveled by the wheels. For turning, the difference in distance traveled by the wheels, combined with the wheelbase, indicates the change in heading.

Another form of odometry is visual odometry. This technique uses cameras to estimate the robot's motion. By analyzing sequences of images, visual odometry algorithms can detect features in the environment and track their movement across frames. By understanding how these features have shifted between images, the algorithm can infer the camera's (and thus the robot's) motion in 3D space. This can be done with a single camera (monocular visual odometry), but it often requires additional assumptions or techniques to recover the scale of the motion. Using stereo cameras (two cameras with a known separation) or RGB-D cameras (which provide both color and depth information) can directly provide scale information, making visual odometry more robust.

Inertial odometry is another important technique, leveraging the data from IMUs to estimate motion. By integrating the linear acceleration measurements (after accounting for gravity and noise) and the angular velocity measurements from the IMU, one can estimate the robot's trajectory. As discussed earlier, this is prone to significant drift over time due to the inherent inaccuracies in inertial sensors. However, for short durations or when coupled with other sensors, inertial odometry can provide valuable high-frequency motion estimates.

The primary advantage of odometry, in all its forms, is its ability to provide a continuous estimate of the robot's pose (position and orientation) without requiring external landmarks or signals, making it suitable for indoor navigation or environments where GPS is unavailable. It offers a relatively low-cost and computationally efficient way to track a robot's movement.

However, odometry methods, particularly wheel odometry and inertial odometry, suffer from a critical limitation: the accumulation of errors. Wheel encoders, for example, can be affected by wheel slippage (when the wheel spins without the robot moving), uneven surfaces, and calibration inaccuracies. Any error in the measurement of wheel rotation, no matter how small, will be integrated over time, leading to an increasing divergence between the estimated position and the robot's true position. This phenomenon is known as "drift." The longer the robot moves, the larger the accumulated error becomes. Similarly, as noted with IMUs, integrating angular velocity measurements leads to drift in orientation estimates.

Visual odometry can also be affected by factors such as poor lighting, lack of distinct features in the environment (e.g., a long, featureless corridor), rapid motion that causes motion blur, and dynamic environments where objects are moving. Accumulation of errors is also a concern, though often less severe than in purely inertial or wheel-based odometry, especially if the algorithm can periodically re-localize against stable features.

The consequence of this error accumulation is that pure odometry is not sufficient for long-term, accurate localization. If a robot relies solely on odometry to navigate, its estimated position will gradually become inaccurate, leading to errors in path planning, obstacle avoidance, and mapping. To overcome this limitation, odometry data is almost always fused with information from other sensing modalities and used in conjunction with more sophisticated state estimation techniques.

This is where the concept of "dead reckoning" becomes relevant. Dead reckoning is a general term for estimating a current position based on a previously determined position and subsequent measurements of motion. Odometry is a form of dead reckoning. While useful for short-term estimation and providing high-frequency updates, the inherent drift in dead reckoning necessitates the use of other methods for periodic correction.

To achieve robust and accurate localization and state estimation, robotic systems employ techniques that fuse data from multiple sensors, including odometry.

Sensor fusion algorithms, such as Extended Kalman Filters (EKFs), Unscented Kalman Filters (UKFs), or Particle Filters (also known as Monte Carlo Localization), are designed to combine the complementary strengths of different sensors. For example, an EKF might combine the high-frequency motion updates from wheel odometry and IMUs with occasional absolute position updates from GPS (outdoors) or from matching sensor readings (like LiDAR scans) to a pre-built map (indoors). This fusion process allows the system to correct the drift introduced by dead reckoning and maintain a more accurate estimate of the robot's true state (position, orientation, velocity).

In essence, IMUs provide the raw ingredients of self-motion (acceleration and angular velocity), which, when processed, allow for the estimation of orientation and velocity. Odometry, in its various forms (wheel, visual, inertial), builds upon these motion measurements to estimate the robot's trajectory. However, the inherent drift in these dead reckoning methods means they are best used as a component within a larger state estimation framework. The continuous estimation of motion provided by IMUs and odometry is indispensable for real-time control and for providing a high-frequency baseline for more robust localization techniques that rely on external sensing and map matching. The interplay between sensing self-motion and sensing the environment forms the foundation of a robot's ability to navigate and operate intelligently within its world.

The preceding discussion has established the critical role of sensors in enabling robots to perceive their surroundings and to understand their own motion and orientation. While vision, depth sensing, and inertial measurement units provide a rich tapestry of environmental awareness and proprioception, a complete robotic system often necessitates additional sensory modalities to achieve sophisticated interaction with the physical world. These modalities can provide detailed feedback on physical contact, enable precise localization in global frames, and are indispensable for tasks ranging from delicate manipulation to autonomous navigation across vast distances. This section will explore three such essential sensor types: tactile sensors, force/torque sensors, and Global Navigation Satellite Systems (GNSS), highlighting their unique contributions to a robot's operational capabilities.

Tactile sensing, in its broadest definition, refers to a robot's ability to perceive physical contact and pressure distribution across its surfaces, particularly on manipulators such as grippers or end-effectors. This capability is fundamental for tasks requiring interaction with objects, where simply knowing an object's position is insufficient; understanding the nature of the contact is paramount. Imagine a robot tasked with picking up a fragile egg. While vision might confirm the egg's location and shape, it cannot convey the force being applied. A tactile sensor, integrated into the robot's gripper, can detect the pressure exerted, allowing the robot to adjust its grip to avoid crushing the delicate object. Conversely, a robot tasked with opening a stuck jar would need to apply a certain level of force, which tactile feedback would help it gauge and control.

The underlying technology for tactile sensing is diverse, reflecting the complexity of replicating human touch. Many tactile sensors operate on principles of capacitance, resistance, or optical changes triggered by deformation. Capacitive tactile sensors, for instance, often employ an array of electrodes separated by a dielectric material. When pressure is applied, the dielectric material deforms, changing the capacitance between the electrodes. By measuring these capacitance variations across the sensor array, a detailed pressure map can be generated, providing information about the location, magnitude, and even the shape of the contact. Resistive tactile sensors, on the other hand, utilize materials whose electrical resistance changes with applied pressure. As pressure increases, the contact area between conductive elements increases, or the material itself compresses, leading to a decrease in resistance. Piezoelectric materials, which generate an electric charge when subjected to mechanical stress, are also employed in tactile sensing.

Beyond simple pressure mapping, advanced tactile sensors can also detect texture, slip, and temperature. Texture sensing, for example, involves arrays of very fine elements that can resolve minute surface irregularities, enabling a robot to distinguish between different materials or identify surface features. Slip detection is critical for grasping and manipulation. By sensing tangential forces or minute changes in normal force due to sliding, a robot can detect when an object is slipping from its grasp and initiate corrective actions, such as increasing grip force. Some tactile sensors can even provide information about the temperature of an object, which can be crucial for handling sensitive materials or for inferring properties like material composition.

The integration of tactile sensors into robotic end-effectors dramatically enhances their dexterity and robustness. For instance, in robotic surgery, tactile feedback allows surgeons to "feel" the tissues they are manipulating, enabling more precise and safer procedures. In manufacturing, tactile sensing can be used for automated assembly, inspection, and sorting of components. In domestic robotics, it could allow a robot to gently pick up laundry or prepare food without damaging ingredients. The rich data provided by tactile sensors allows for more intelligent and adaptive interactions with the physical environment, moving beyond purely visual or geometric understanding. However, developing truly human-like tactile sensing remains a significant research challenge, particularly in achieving the spatial resolution, sensitivity, and responsiveness of biological systems.

Closely related to tactile sensing, but often focused on larger-scale forces and torques, are force and torque sensors. While tactile sensors might measure pressure distribution across a surface, force and torque sensors are typically designed to measure the resultant forces and moments exerted at a specific point, often at the wrist of a robotic manipulator or at the base of a mobile robot's joints. These sensors are crucial for controlling the interaction forces between the robot and its environment, enabling compliant motion and safe human-robot collaboration.

Force and torque sensors are commonly based on strain gauges. Strain gauges are devices that change their electrical resistance when they are stretched or compressed. In a force/torque sensor, a precisely machined elastic element (often made of metal) is instrumented with multiple strain gauges. When an external force or torque is applied to this elastic element, it deforms slightly. This deformation causes the strain gauges attached to it to change their resistance. By strategically placing and wiring these strain gauges, it's possible to measure forces along three orthogonal axes (Fx, Fy, Fz) and torques (or moments) around those same axes (Tx, Ty, Tz), often referred to as a 6-axis force/torque sensor. The electrical signals from the strain gauges are then processed to provide a quantitative measurement of the applied forces and torques.

The applications of force and torque sensing are extensive. In industrial robotics, they enable "force control" or "compliance control." Instead of moving to a predefined position, a robot can be programmed to apply a specific force, or to react compliantly to external forces. For example, when inserting a peg into a hole, slight misalignments can prevent smooth insertion. A robot equipped with a force/torque sensor can detect the resistance and make small adjustments to its position and orientation to guide the peg into the hole, effectively "feeling" its way in. This is far more robust than relying solely on precise kinematic models, which may not account for manufacturing tolerances or environmental variations.

Human-robot collaboration, or "cobotics," heavily relies on force and torque sensors for safety. In environments where robots and humans work in close proximity, the robot must be able to detect unexpected collisions and react instantly to prevent injury. Force sensors allow the robot to limit the forces it exerts, and to stop or retract upon sensing contact with a human or other unexpected obstacle. This is essential for creating robots that can work alongside people in shared workspaces, enhancing productivity without compromising safety.

Furthermore, force and torque sensing plays a vital role in advanced manipulation tasks. When performing tasks like polishing, grinding, or deburring, maintaining a consistent contact force is critical for achieving the desired surface finish. A robot can use its force sensor to regulate the pressure applied to the workpiece, ensuring uniform results. In robotic assembly, it can detect if a part is correctly seated or if excessive force is being applied during a connection.

The successful deployment of force and torque sensors requires careful consideration of their limitations. They are sensitive to vibration, temperature changes, and overloads, which can lead to inaccurate readings or even damage to the sensor. Calibration is also crucial, as the relationship between applied force and the sensor's output needs to be precisely known. Nevertheless, their ability to provide direct feedback on physical interaction makes them indispensable for robots operating in dynamic and contact-rich environments.

While IMUs and odometry provide a robot with a sense of its own motion and position relative to its starting point, a critical requirement for many robotic applications, particularly those involving navigation over extended distances or in areas where no prior map exists, is the ability to determine one's absolute position within a global coordinate system. This is the domain of Global Navigation Satellite Systems (GNSS), with the most well-known example being the Global Positioning System (GPS).

GNSS technologies, which also include systems like GLONASS (Russia), Galileo (Europe), and BeiDou (China), enable devices to determine their precise location anywhere on Earth by receiving signals from a constellation of satellites. The fundamental principle involves measuring the time it takes for signals from multiple satellites to reach the receiver. Each satellite broadcasts its precise orbital position and the time the signal was sent. By knowing the distance to at least four satellites (calculated from the time difference), a receiver can trilaterate its position in three dimensions (latitude, longitude, and altitude) and also determine its precise time.

For mobile robots, GNSS offers a powerful solution for outdoor localization. A robot can be programmed to follow a predetermined path, or to navigate to specific waypoints, by continuously receiving its global coordinates. This is invaluable for applications such as autonomous agricultural machinery, delivery robots operating in urban environments, drones for aerial surveying and inspection, and exploration robots in remote areas. The ability to acquire an absolute global position means that robots are not confined by the limitations of internal sensors or the need for pre-existing maps, allowing for operation in previously unvisited territories.

However, GNSS, particularly GPS, is not without its challenges and limitations. The accuracy of standard civilian GPS receivers typically ranges from a few meters to tens of meters, which may be insufficient for many robotic applications requiring centimeter-level precision. This accuracy is affected by several factors. "Multipath interference" is a significant problem, where satellite signals bounce off buildings, terrain, or other objects before reaching the receiver. This creates longer signal paths, leading to errors in distance calculation and, consequently, position estimation.

"Signal blockage" is another common issue, especially in urban canyons (areas with tall buildings), dense forests, or even indoors, where the line of sight to enough satellites can be obstructed. When a receiver cannot "see" at least four satellites, it cannot compute a 3D position. "Atmospheric effects," such as ionospheric and tropospheric delays, can also introduce errors as satellite signals travel through different layers of Earth's atmosphere. The "satellite geometry" (the spatial arrangement of the visible satellites) also plays a role; a poor geometric distribution can lead to less accurate position fixes.

To overcome these limitations and achieve higher accuracy, various augmentation techniques are employed. "Differential GPS" (DGPS) uses a fixed ground-based reference station that broadcasts its known position and any calculated errors to nearby mobile receivers. By comparing the receiver's calculated position with the reference station's true position, corrections can be applied. "Real-Time Kinematic" (RTK) GPS takes this a step further, using carrier phase measurements of the GNSS signals, in addition to code measurements, to achieve centimeter-level accuracy in near real-time. This requires a base station and a rover unit (on the robot), both equipped with RTK-capable receivers.

Another approach is to use "inertial navigation systems" (INS) in conjunction with GNSS. As previously discussed, IMUs provide high-frequency estimates of motion. When combined with GNSS, the IMU can bridge the gaps during periods of GNSS signal loss or inaccuracy, while the GNSS can periodically correct the inherent drift of the IMU. This fusion, often implemented using Kalman filters, creates a robust navigation system that leverages the strengths of both sensor types. For instance, a drone might use IMUs for immediate flight control and attitude stabilization, while relying on RTK GPS for accurate waypoint navigation and absolute position correction.

In essence, GNSS sensors provide a robot with an external, absolute reference for its position on Earth. While standard GNSS accuracy might be suitable for general navigation, achieving the precision required for many robotic tasks necessitates specialized GNSS receivers and augmentation techniques, or their fusion with other sensor modalities. The ability to accurately determine and maintain a global position is a cornerstone of autonomous systems operating in the open world, enabling them to navigate complex environments, perform coordinated tasks over large areas, and return to designated locations with confidence.

The discussion of tactile, force, and GNSS sensors

underscores a fundamental principle in modern robotics: the power of sensor fusion. Each sensor modality has its strengths and weaknesses. IMUs and wheel odometry provide excellent short-term motion estimates but suffer from drift. Vision systems offer rich environmental perception but can be hampered by lighting conditions and textureless surfaces. GNSS provides global positioning but can be inaccurate or unavailable in certain environments. Tactile and force sensors excel at interaction but provide limited information about the robot's overall state or environment.

By integrating data from multiple, diverse sensors, robotic systems can achieve a level of robustness and accuracy that would be impossible with any single sensor alone. Complex algorithms, such as Extended Kalman Filters, Unscented Kalman Filters, or Particle Filters, are employed to fuse these disparate data streams. These algorithms weigh the information from each sensor based on its reliability and characteristics, producing a more accurate and complete estimation of the robot's state, including its position, orientation, velocity, and interaction forces. This integrated approach allows robots to overcome individual sensor limitations, adapt to dynamic environments, and perform increasingly sophisticated tasks with greater confidence and reliability. The careful selection and intelligent fusion of these essential sensor types are therefore critical for unlocking the full potential of robotic systems across a vast spectrum of applications.

Chapter 3: Sensor Calibration and Data Preprocessing

The preceding discussion has illuminated the diverse array of sensors that endow robots with the ability to perceive and interact with their environments. From the subtle nuances of touch conveyed by tactile sensors to the global positioning capabilities of GNSS, each sensor modality contributes a unique facet to a robot's understanding of its state and surroundings. However, the mere presence of these sophisticated sensing devices does not inherently guarantee their utility. A fundamental prerequisite for extracting meaningful and trustworthy information from any sensor, regardless of its complexity or technological sophistication, is the process of calibration. This section delves into the critical necessity of sensor calibration, emphasizing why it is an indispensable step in the development of any robust and reliable robotic system. Without it, even the most advanced hardware can become a source of significant error, undermining the very purpose for which it was integrated.

The inherent nature of physical sensors means they are susceptible to a range of imperfections, both in their construction and their operation. These imperfections, often subtle, can introduce systematic deviations between the sensor's reported measurement and the true physical quantity it is intended to represent. These deviations are not random; they follow predictable patterns or biases. For instance, a camera lens might have a slight distortion, causing straight lines to appear curved, particularly towards the edges of the image. An accelerometer, even when perfectly stationary, might report a non-zero acceleration due to internal imbalances. Similarly, the electronic components that process the raw sensor signals can introduce offsets or scaling errors. These imperfections are a direct consequence of manufacturing tolerances, material properties, environmental influences (such as temperature variations), and the physical principles upon which the sensor operates. Consequently, a sensor that has not undergone calibration is akin to a measuring tape that has been stretched or shrunk; its readings, while seemingly precise, are fundamentally inaccurate and will lead to erroneous conclusions about the physical world.

The necessity of calibration stems directly from its role in rectifying these systematic errors. Through a well-defined calibration procedure, the relationship between the sensor's output and the actual physical phenomenon is established. This process typically involves exposing the sensor to known stimuli or reference values and observing the corresponding sensor readings. For example, a camera might be calibrated by capturing images of a precisely patterned checkerboard from various angles. By analyzing how the checkerboard squares appear in the image, the camera's intrinsic parameters (like focal length and principal point) and distortion coefficients can be mathematically determined. For an inertial measurement unit (IMU), calibration involves measuring its output when subjected to known accelerations and rotations, and also under static conditions to identify biases and scale factors for each axis. For force/torque sensors, calibration involves applying known forces and torques and recording the sensor's response, allowing for the creation of a mapping from sensor voltage to force/torque values.

The impact of uncalibrated sensors reverberates throughout the entire robotic system. Consider the field of perception. If a camera is not calibrated, its measurements of object distances or sizes will be skewed by lens distortion. This can lead to misinterpretations of the environment, affecting tasks like object recognition, collision avoidance, and grasp planning. For example, a robot attempting to pick up an object might misjudge its size and orientation due to uncorrected camera distortions, leading to a failed grasp. Similarly, if a depth sensor suffers from uncalibrated spatial offsets or scale errors, the perceived shape and distance of objects will be inaccurate, potentially causing the robot to collide with obstacles or misjudge the traversability of terrain.

Localization and mapping are perhaps even more critically dependent on sensor calibration. Odometry systems, whether based on wheel encoders or inertial measurements, are notoriously prone to accumulating errors over time. Wheel encoders can be affected by wheel slip, uneven surfaces, and variations in tire pressure. IMUs, as previously mentioned, suffer from drift, where small errors in acceleration and angular velocity measurements integrate over time to produce significant errors in estimated position and orientation. Without proper calibration to account for biases, scale factors, and cross-axis sensitivities, these odometric estimates will quickly diverge from the robot's true position. When these uncalibrated odometry readings are used to build a map of the environment or to localize the robot within an existing map, the inaccuracies are compounded. A poorly calibrated odometry system can lead to a map that is distorted and inconsistent, making subsequent localization attempts futile. Imagine a robot trying to navigate a building using a map built with uncalibrated odometry; the robot might appear to be in a hallway when it is actually in a room, or vice-versa, leading to a complete breakdown of navigation.

Control algorithms also suffer immensely from the lack of sensor calibration. Many robotic control strategies rely on accurate measurements of joint angles, velocities, forces, and external disturbances. If a joint encoder is uncalibrated, the robot might not know its actual limb configuration, leading to incorrect kinematic calculations and flawed trajectory execution. For robots that interact physically with their environment, such as those performing assembly tasks or human-robot collaboration, accurate force and torque sensing is paramount. An uncalibrated force/torque sensor could provide misleading readings, causing the robot to apply too much or too little force, potentially damaging itself, the workpiece, or injuring a human collaborator. For example, in a pick-and-place operation, if the force sensor is not calibrated, the robot might exert excessive pressure and crush the object it is trying to manipulate.

Furthermore, the fusion of data from multiple sensors, a cornerstone of modern robotics for achieving robustness and accuracy, becomes profoundly problematic without proper calibration. When sensor data is fused, algorithms like Kalman filters or particle filters attempt to combine information from different sources, weighting them according to their estimated reliability. If one or more of these sensors are uncalibrated, their reported data will be systematically biased. The fusion algorithm will then incorporate these biased measurements, potentially leading to an even worse overall state estimate than if only the more reliable sensor had been used. For instance, fusing uncalibrated camera data with uncalibrated IMU data for visual-inertial odometry could result in a combined estimate that is significantly more erroneous than either sensor's reading alone. The calibration process ensures that all sensors are speaking a common, accurate language, allowing for their effective integration into a coherent understanding of the robot's state.

The consequences of neglecting sensor calibration extend beyond mere performance degradation; they can have significant safety implications. A robot operating with inaccurate sensor data may perceive a clear path when an obstacle is present, leading to collisions. It might misjudge the forces it is exerting, leading to mechanical failure or harm to humans. In critical applications like autonomous driving or medical robotics, the requirement for highly accurate and reliable sensor data, achievable only through meticulous calibration, is non-negotiable. The integrity of the entire robotic system hinges on the trustworthiness of its sensory inputs.

In essence, sensor calibration is not an optional or a minor tuning step; it is a foundational requirement. It transforms raw sensor outputs from potentially misleading signals into reliable data that can be meaningfully interpreted and utilized. It underpins the accuracy of perception, the reliability of localization, the precision of control, and the effectiveness of sensor fusion. Without it, even the most sophisticated robotic hardware is rendered fundamentally compromised, incapable of performing its intended tasks safely and effectively. Therefore, understanding the principles of calibration and implementing rigorous calibration procedures for all onboard sensors is an absolutely indispensable part of building any competent robotic system.

The ability of a robot to perceive and interact with its environment relies heavily on its vision system. Cameras, the primary sensors for this visual perception, provide a rich stream of data that, when processed correctly, can unlock a wide range of functionalities, from object recognition and navigation to manipulation and human-robot interaction. However, raw image data from a camera is not directly usable for precise geometric tasks. This is due to inherent characteristics of the camera itself, including its internal optical properties and its physical placement on the robot. To bridge this gap between raw pixels and a meaningful understanding of the 3D world, a fundamental process known as camera calibration is indispensable. This process establishes a precise mathematical relationship between the 3D points in the real world and their corresponding 2D projections in the image plane.

Camera calibration can be broadly categorized into two main types: intrinsic and extrinsic calibration. These two processes address different aspects of the camera's relationship with the world and the robot. Intrinsic calibration deals with the internal parameters of the camera – essentially, how the camera "sees" the world through its lens and sensor. Extrinsic calibration, on the other hand, determines the camera's pose, i.e., its position and orientation, relative to a reference coordinate system, typically the robot's own base frame. Both are critical for accurate 3D reconstruction and geometric reasoning.

Intrinsic Camera Calibration

Intrinsic calibration aims to quantify the internal optical characteristics of the camera system. These characteristics dictate how light rays from the 3D world are projected onto the 2D image sensor. The primary intrinsic parameters are:

1.

Focal Length (fx, fy): This parameter relates to the distance between the optical center of the lens and the image sensor. In most modern cameras, the optical axis is perpendicular to the image sensor. However, in simplified pinhole camera models, the focal length is often represented by two values, fx *and* fy, corresponding to the focal length along the horizontal and vertical axes of the image sensor. These values are typically measured in pixels. If the sensor pixels are square, then fx *and* fy will be equal.

2.
Principal Point (cx, cy): This is the projection of the optical center of the lens onto the image plane. Ideally, it should lie at the center of the image sensor. However, due to manufacturing imperfections, the principal point is often slightly offset from the exact center. (cx, cy) are also typically expressed in pixels.

3.
Skew Coefficient (α): This parameter accounts for non-orthogonality between the image sensor's horizontal and vertical axes. In an ideal scenario, these axes are perfectly perpendicular, leading to a skew coefficient of zero. However, mechanical misalignments during sensor assembly can introduce a slight skew. While often assumed to be zero in many applications, it can be a relevant parameter for high-precision systems.

4.
Lens Distortion Coefficients ($k1, k2, k3, p1, p_2$): Lenses, particularly those with wide fields of view or simple construction, are not perfect. They introduce

distortions that cause straight lines in the 3D world to appear curved in the 2D image. These distortions are typically modeled as radial and tangential components.
Radial Distortion: This is the most common type of lens distortion and causes features to appear further from or closer to the center of the image than they should be. It is primarily characterized by radial distance from the image center. Commonly, it's modeled using polynomial terms:
k_1: This is the first-order radial distortion coefficient, which typically accounts for barrel distortion (lines bowing outwards) or pincushion distortion (lines bowing inwards).
k_2: This second-order coefficient refines the radial distortion model.
k_3: A third-order radial distortion coefficient, often used for more complex distortion profiles.
Tangential Distortion: This arises from imperfections in the lens mounting, where the lens planes are not perfectly parallel to the image sensor. It causes objects to appear shifted tangentially to the radial direction. It is modeled by two coefficients:
p_1: Controls tangential distortion along the x-axis.
p_2: Controls tangential distortion along the y-axis.

These intrinsic parameters are often consolidated into the
intrinsic camera matrix, denoted by K. For a simplified model (assuming zero skew), this matrix is:

$$

K = \begin{bmatrix}

fx & 0 & cx \\
0 & fy & cy \\
0 & 0 & 1
\end{bmatrix}
$$

This matrix is fundamental in transforming 3D points in camera coordinates to 2D image coordinates. When lens distortion is considered, the projection of a 3D point (X, Y, Z) in camera coordinates is first computed using the pinhole model, resulting in normalized image coordinates (x', y'):

$$
x' = X/Z
$$
$$
y' = Y/Z
$$

These normalized coordinates are then distorted using the distortion coefficients:

$$
xd = x'(1 + k1 r^2 + k2 r^4 + k3 r^6) + 2p1 x' y' + p2 (r^2 + 2x'^2)
$$
$$

$$
y_d = y'(1 + k_1 r^2 + k_2 r^4 + k_3 r^6) + p_1 (r^2 + 2y'^2) + 2p_2 x' y'
$$

where $r^2 = x'^2 + y'^2$ is the squared radial distance from the principal point. The distorted normalized coordinates (x_d, y_d) are then mapped to pixel coordinates (u, v) using the intrinsic matrix:

$$
u = f_x x_d + c_x
$$

$$
v = f_y y_d + c_y
$$

Alternatively, the distortion model can be applied in the pixel domain. The process of estimating these intrinsic parameters is typically performed by capturing images of a known calibration pattern, such as a checkerboard or a chessboard, from multiple viewpoints. The algorithm then analyzes the observed positions of the pattern's features (e.g., the corners of the squares) in the images and uses optimization techniques to find the set of intrinsic parameters that best explain the observed feature locations. The more viewpoints and the more diverse the viewing angles, the more robust and accurate the calibration will be.

Extrinsic Camera Calibration

While intrinsic calibration describes how the camera maps 3D points to its image plane, extrinsic calibration defines the camera's spatial relationship with respect to a world or robot coordinate frame. This involves determining the camera's
pose, which is composed of its 3D **rotation** and 3D **translation**.

1.
Rotation: This describes the orientation of the camera's coordinate frame relative to the reference frame. It is commonly represented by a 3x3 rotation matrix (R) or a rotation vector (e.g., Euler angles or axis-angle representation).
2. **Translation:** This describes the position of the camera's optical center relative to the origin of the reference frame. It is represented by a 3D vector (t).

Together, the rotation matrix R and translation vector t form the
extrinsic camera matrix, often denoted as $[R|t]$. This matrix transforms points from the world coordinate frame into the camera's coordinate frame:

$$

\mathbf{P}{camera} = R \mathbf{P}{world} + t
$$

where $\mathbf{P}

*{world}$ is a point in the world frame and $\mathbf{P}*{camera}$ is the same point expressed in the camera's frame.

The goal of extrinsic calibration is to find the R and t that accurately map known 3D points in the robot's environment (or on the robot itself) to their observed 2D projections in the camera's image. This is crucial for tasks such as:

3D Reconstruction: Combining information from multiple views or sensor modalities (like depth sensors) requires knowing the relative poses of the sensors.
Robot Navigation: If the camera is used for localization or mapping, its pose relative to the robot's base frame is essential.
Object Grasping and Manipulation: For a robot arm to grasp an object detected by a camera, the object's position in the robot's coordinate system must be known, which requires the camera's extrinsic parameters.
Sensor Fusion: Integrating camera data with other sensors, like LiDAR or IMUs, necessitates a common coordinate frame and thus requires accurate extrinsic calibration between all sensors.

Similar to intrinsic calibration, extrinsic calibration often involves using a known calibration pattern. However, the emphasis shifts to knowing the precise 3D coordinates of the pattern's features in the robot's reference frame. For example, if a checkerboard is mounted on the robot's base, its 3D coordinates are known relative to that base. By detecting the checkerboard's corners in the camera image, we can establish correspondences between 3D points in the robot frame and their 2D projections in the image. Using these correspondences, along with the already computed intrinsic parameters, algorithms can solve for the rotation R and translation t that describe the camera's pose relative to the robot.

A common scenario is calibrating a camera mounted on a robot arm. In this case, the desired extrinsic parameters are the transformation from the robot's tool center point (or end-effector frame) to the camera frame, or vice-versa. This is often achieved by moving the robot arm to various known poses and capturing images of a stationary calibration pattern, or by moving the calibration pattern while the robot arm remains in a fixed pose.

Calibration Techniques: The Checkerboard Method

One of the most widely adopted and effective methods for camera calibration, encompassing both intrinsic and extrinsic parameters, is the checkerboard method. This technique relies on a planar pattern with a grid of alternating black and white squares. The key advantage of the checkerboard is that the corners of the squares are easily and reliably detectable.

Algorithm Overview:

1.
Pattern Generation: Create a checkerboard pattern with a known size of squares (e.g., 2.5 cm x 2.5 cm) and a specific number of internal corners (e.g., 6x7 internal corners, meaning 7x8 squares). The exact dimensions are crucial for accurate metric calibration.

2.
Image Acquisition: Capture multiple images of the checkerboard from various viewpoints and orientations. It's important that the checkerboard is visible in each image and that the viewpoints are diverse enough to constrain all the camera parameters effectively. For intrinsic calibration, views where the checkerboard is nearly parallel to the image plane are beneficial. For extrinsic calibration, views from different distances and angles are helpful.

3.
Corner Detection: For each captured image, an algorithm is used to detect the precise pixel coordinates of the internal corners of the checkerboard. This is typically a sub-pixel precise operation to achieve high accuracy. Libraries like OpenCV provide robust functions for this purpose (e.g., `findChessboardCorners`). The output of this stage is a

set of 2D pixel coordinates for the detected corners in each image.

4.
3D Object Point Generation: For each image, the corresponding 3D coordinates of the checkerboard corners in a chosen object coordinate system are generated. If the checkerboard is considered to lie on the $z=0$ plane, and its top-left corner is at the origin of the object coordinate system, then the 3D coordinates of the corners (i, j) can be represented as $(j \times \text{squaresize}, i \times \text{squaresize}, 0)$, where i and j are the row and column indices of the corner, and `square_size` is the physical dimension of one square.

5.
Initial Parameter Estimation (Intrinsic): Using a subset of the detected 2D corner points and their corresponding 3D object points, an initial estimation of the intrinsic camera parameters ($fx, fy, cx, cy, k1, k2, \dots$) is performed. This can be done using methods like the Direct Linear Transformation (DLT) algorithm, which provides a linear solution for the projection matrix, followed by decomposing this matrix to extract the intrinsic and extrinsic parameters. Alternatively, more sophisticated iterative optimization methods can be employed.

6.
Refinement using Optimization (Bundle Adjustment): The initial estimates are then refined using a non-linear optimization technique, commonly

known as Bundle Adjustment. This is an iterative process that simultaneously optimizes the 3D structure of the scene (the 3D points) and the camera parameters (both intrinsic and extrinsic) to minimize the reprojection error. The reprojection error is the difference between the observed 2D pixel coordinates of a feature and the 2D coordinates predicted by projecting the estimated 3D point using the current camera parameters. By minimizing this error across all detected points in all images, the most accurate estimates for both intrinsic and extrinsic parameters are obtained.

Mathematically, Bundle Adjustment seeks to minimize the sum of squared reprojections errors:

$$
\sum_{i=1}^{N} \sum_{j=1}^{M_i} || \mathbf{m}_{ij} - \pi(\mathbf{K}, \mathbf{R}_j, \mathbf{t}_j, \mathbf{X}_i) ||^2
$$

where:

N is the number of 3D points (e.g., checkerboard corners).
M_i is the number of images where the i-th 3D point is visible.
$\mathbf{m}_{ij}$ is the observed 2D pixel coordinate of the i-th point in the j-th image.

$\mathbf{K}$ are the intrinsic camera parameters (assumed constant for all images if calibrating intrinsics).
$\mathbf{R}_j, \mathbf{t}_j$ are the rotation and translation (extrinsic parameters) for the j-th image.
$\mathbf{X}_i$ are the 3D coordinates of the i-th point.
$\pi(\cdot)$ is the projection function that maps a 3D point to a 2D image point, including the effects of lens distortion.

The optimization variables typically include the intrinsic camera matrix K (if not fixed), the distortion coefficients, and for each image, the pose (rotation R_j and translation t_j) of the camera relative to the object's coordinate frame.

7.
Output: The result of the calibration process is a set of intrinsic parameters (focal length, principal point, distortion coefficients) and, for each image captured during calibration, the extrinsic parameters (rotation and translation) that define the camera's pose relative to the checkerboard's coordinate frame. If the checkerboard's coordinate frame is aligned with the robot's coordinate frame, then these extrinsic parameters directly provide the camera's pose with respect to the robot.

The Camera Projection Matrix

The culmination of camera calibration, particularly intrinsic and extrinsic calibration combined, is the **camera projection matrix**, often denoted as P. This 3x4 matrix encapsulates the entire transformation from 3D world coordinates to 2D image coordinates. It is formed by combining the intrinsic camera matrix K and the extrinsic parameters $[R|t]$:

$$
P = K [R|t]
$$

In homogeneous coordinates, a 3D point $\mathbf{X}{world} = [X, Y, Z, 1]^T$ *in the world frame is projected onto the image plane to give homogeneous image coordinates* $\mathbf{x}{image} = [u, v, 1]^T$ via:

$$
s \begin{bmatrix} u \\ v \\ 1 \end{bmatrix} = P \begin{bmatrix} X \\ Y \\ Z \\ 1 \end{bmatrix}
$$

where s is a scaling factor, typically representing the depth of the point in the camera's coordinate system. After the matrix multiplication, the homogeneous image coordinates are de-homogenized by dividing the first two components by the third:

$$
u = (s \cdot u') / s
$$

$$
v = (s \cdot v') / s
$$

The projection matrix P is fundamental for numerous computer vision tasks, including:

3D Reconstruction: It allows for the triangulation of 3D points from multiple 2D views. By knowing the projection matrices for two or more cameras observing the same point, one can triangulate its 3D location.
Geometric Vision: It is used in various geometric operations such as homography estimation, fundamental matrix computation, and essential matrix computation, which are all crucial for understanding scene geometry and camera motion.
Augmented Reality: Overlaying virtual objects onto real-world camera feeds requires accurate knowledge of the camera's pose and intrinsic parameters to render virtual objects correctly in perspective.
Robotics: For tasks like visual servoing, where robot motion is controlled based on visual feedback, the projection matrix is essential for relating image features to robot motion commands.

The accuracy of the projection matrix directly dictates

the accuracy of any subsequent geometric computations. Therefore, robust and precise camera calibration is a non-negotiable first step for any application relying on the geometric interpretation of camera images. The checkerboard method, due to its reliability, ease of implementation, and ability to estimate both intrinsic and extrinsic parameters simultaneously, has become a de facto standard in the field. However, it's important to note that this method assumes a planar calibration target. For applications requiring calibration of cameras with extreme distortions or those used in non-planar environments, alternative calibration methods might be considered. Nonetheless, the fundamental concepts of intrinsic and extrinsic parameters, and the goal of obtaining an accurate projection matrix, remain central to all camera calibration procedures.

The preceding discussion underscored the critical role of accurate sensor data, particularly from cameras, in enabling robots to perceive and interact with their environment. However, a robot's perception system often comprises a suite of sensors, each with its own characteristics and calibration requirements. Among these, Light Detection and Ranging (LiDAR) and Inertial Measurement Units (IMUs) are paramount for tasks involving precise spatial understanding, navigation, and state estimation. This section delves into the calibration techniques essential for extracting reliable data from these vital sensors.

LiDAR Calibration

LiDAR sensors, which use pulsed laser light to measure distances to objects, provide rich 3D point cloud data representing the environment. For this data to be geometrically accurate and useful, the LiDAR sensor itself must be calibrated. LiDAR calibration generally addresses two main aspects: the accuracy of range measurements and the precise angular positioning of the laser beams.

Range Accuracy Calibration:
LiDAR systems determine the distance to an object by measuring the time of flight of a laser pulse or by analyzing the phase shift of a modulated laser beam. Imperfections in the sensor's electronics, detector, or the laser itself can lead to systematic errors in these time-of-flight or phase measurements, which translate directly into range errors.

A common approach to calibrate range accuracy involves measuring the LiDAR's distance to a highly reflective, planar target at various distances. Ideally, the target should be constructed from a material that reflects the laser light back with minimal scattering or absorption, and it should be precisely manufactured to be perfectly flat. The LiDAR sensor is then positioned at a known distance from this target, or the target is moved to known distances from the sensor. By comparing the measured distances from the LiDAR to the actual, known distances to the target, a calibration curve or correction factor can be determined. This process often reveals a relationship that is not strictly linear, especially at very short or very long ranges. For instance, some LiDARs exhibit a slight bias at close range due to limitations in their time-of-flight measurement circuitry, while others might show a decrease in accuracy at the edge of their operational range due to signal degradation.

Multiple measurements at different distances are essential to characterize this error profile. These measurements can then be used to fit a polynomial function or a lookup table that corrects the raw range readings from the LiDAR. For example, if the LiDAR consistently reports a distance that is 0.05 meters shorter than the actual distance for ranges between 1 and 10 meters, a constant offset of +0.05 meters can be applied. If the error is more complex, a second-order polynomial might be fitted: $R_{corrected} = a \cdot R_{measured}^2 + b \cdot R_{measured} + c$. The coefficients a, b, and c are determined through the fitting process. This refined range data is crucial for applications like 3D mapping and object detection, where precise distance measurements are foundational.

Angular Calibration:
A spinning LiDAR sensor consists of an array of laser beams, each emitted at a specific vertical angle and swept horizontally by the rotation of the sensor head. For accurate 3D point cloud generation, it is imperative

to know the precise horizontal and vertical angle of each individual laser beam at the moment it strikes a target.

Vertical Angle Calibration: Each laser beam within the sensor array is designed to be at a specific vertical angle relative to the sensor's horizontal plane. Due to manufacturing tolerances, these angles may not be perfectly precise. Calibrating the vertical angles involves determining the actual vertical angle for each laser channel. One method is to scan a vertical wall or a target with known vertical features. By analyzing the point cloud data, particularly the points that should lie on a horizontal plane, differences in the vertical position of points generated by different laser channels can reveal angular offsets. Alternatively, one can use a precisely angled jig to align the LiDAR and measure the angles.

Horizontal Angle Calibration (Angular Resolution and Offset): The horizontal angle is determined by the rotation speed of the LiDAR and the timing of each laser pulse. While the rotation speed is often assumed to be constant, variations can occur. More critically, the precise horizontal angle at which each beam is fired relative to a reference point on the sensor's rotation can have an offset or an incorrect angular resolution.

To calibrate this, a common technique involves scanning a very narrow vertical target, such as a thin pole or a sheet of paper held vertically. The LiDAR scans this target as it rotates. The resulting data will show a distinct pattern of points hitting the target. By analyzing the angular separation between consecutive hits on the

target and the overall shape of the scanned object, corrections can be applied. If the LiDAR's horizontal angular resolution is slightly off (e.g., instead of 0.2 degrees per step, it's 0.21 degrees), the point cloud will appear slightly stretched or compressed horizontally. Similarly, an angular offset will cause the entire scanned object to appear rotated by a small, constant angle.
A more robust method is to use a known geometric structure, such as a corner formed by two perpendicular walls or a series of precisely spaced vertical posts. By observing how these structures are represented in the point cloud, the angular calibration parameters can be refined. For example, if the LiDAR scans a corner, the transition from hitting one wall to the other should occur at a precise horizontal angle change. Deviations from this expected change can be used to correct the angular encoder readings or the timing of the laser firings.

The output of LiDAR calibration is a refined set of parameters that can be applied to the raw sensor readings to produce accurate range and angular measurements. This leads to a more faithful representation of the environment in the 3D point cloud, which is crucial for subsequent processing steps such as SLAM (Simultaneous Localization and Mapping), object recognition, and obstacle avoidance.

Inertial Measurement Unit (IMU) Calibration

An Inertial Measurement Unit (IMU) typically comprises accelerometers and gyroscopes, which measure linear acceleration and angular velocity, respectively. These sensors are fundamental for estimating a robot's motion, orientation, and position over time, especially when combined with other sensors like cameras or GPS. However, raw IMU data is susceptible to various errors that must be calibrated out to ensure reliable state estimation.

Accelerometer Calibration:

Accelerometers measure the acceleration experienced along their sensitive axes. The primary calibration parameters for accelerometers are:

1.
Bias (Offset): Even when stationary, an accelerometer can report a non-zero acceleration due to inherent electronic offsets. This bias can drift with temperature. Static calibration aims to determine this bias when the sensor is at rest.
2. **Scale Factor:** The accelerometer's output reading might not be exactly proportional to the actual acceleration. A scale factor corrects for this linear relationship. For example, if the accelerometer reports 1.01 g when subjected to 1 g of acceleration, its scale factor is approximately 1/1.01.
3. **Misalignment:** The sensitive axes of the accelerometer might not be perfectly orthogonal to

each other or aligned with the sensor's housing. Misalignment introduces cross-axis sensitivities, meaning acceleration along one axis can be partially measured by another.

Accelerometer Calibration Procedure (Static):
A common and effective method for static calibration involves placing the IMU in multiple known orientations relative to gravity. Since gravity provides a constant acceleration of approximately 9.81 m/s^2 along the direction of gravity, it acts as a known external force.

The IMU is placed in at least six distinct static orientations. These orientations are typically achieved by placing the IMU on a level surface, then rotating it 90 degrees around each of its three axes in both directions. For example:

Orientation 1: IMU lies flat, sensitive axes aligned with x, y, z. Gravity acts along -z.
Orientation 2: IMU rotated 180 degrees around the x-axis. Gravity acts along +z.
Orientation 3: IMU rotated 90 degrees around the x-axis. Gravity acts along +y.
Orientation 4: IMU rotated -90 degrees around the x-axis. Gravity acts along -y.
Orientation 5: IMU rotated 90 degrees around the y-axis. Gravity acts along -x.
Orientation 6: IMU rotated -90 degrees around the y-axis. Gravity acts along +x.

In each orientation, the IMU is allowed to settle for a short period, and its readings from the accelerometers are recorded. Let the gravity vector be $\mathbf{g} = [0, 0, -g]^T$ in the world frame (where $g \approx 9.81 \, \text{m/s}^2$). If the IMU's coordinate frame in a particular orientation is represented by a rotation matrix $R
*i$, then the accelerometer readings $\mathbf{a}*i$ in this orientation should ideally correspond to $R_i \mathbf{g}$.

The calibration problem then becomes finding the bias vector $\mathbf{b}
*a$, the scale factor matrix $S*a$ (often a diagonal matrix), and the misalignment matrix M_a (which captures non-orthogonality) that best satisfy the relationship:

$\mathbf{a}
*i = S*a M*a R*i \mathbf{g} + \mathbf{b}_a$

Alternatively, a more comprehensive model can be expressed as:

$\mathbf{a}
*{raw, i} = S*a \mathbf{a}*{ideal, i} + \mathbf{b}*a$

where $\mathbf{a}
{ideal, i}$ is the acceleration measured in the sensor's body frame, accounting for gravity and any external accelerations. In static tests, external accelerations are

zero, so $\mathbf{a}{ideal, i} = Ri \mathbf{g}$. The calibration parameters Sa and $\mathbf{b}_a$ can be estimated using least-squares optimization. For a full characterization, including misalignment, a slightly more complex linear system can be formulated. For instance, if we assume a model where the measured acceleration is a linear combination of ideal accelerations along orthogonal axes, plus bias, the six orientations provide enough linear constraints to solve for the unknown parameters.

Gyroscope Calibration:
Gyroscopes measure angular velocity. Similar to accelerometers, gyroscopes suffer from bias and scale factor errors, and their sensitive axes might not be perfectly orthogonal.

1.
Bias (Offset): Even when stationary, a gyroscope can report a non-zero angular velocity. This bias is highly temperature-dependent.
2. **Scale Factor:** The output of the gyroscope might not be perfectly proportional to the actual angular velocity.
3. **Misalignment:** The axes of the gyroscopes can be misaligned with each other and with the sensor's housing.

Gyroscope Calibration Procedure (Static and Dynamic):
Static calibration for gyroscopes is challenging because gravity does not induce angular velocity. Therefore,

static calibration primarily focuses on estimating the bias.

Static Bias Estimation: The IMU is kept perfectly still for an extended period, and the gyroscope readings are averaged. This average value is taken as the bias. However, this bias can drift, so it's often continuously estimated and compensated for during operation.

Dynamic Calibration: To estimate scale factors and misalignment, the IMU needs to experience known angular velocities. This can be achieved by mounting the IMU on a precision turntable that can rotate at precisely controlled angular velocities. By commanding the turntable to rotate at known rates around specific axes, the gyroscope's response can be measured. The data collected from these controlled rotations allows for the estimation of scale factors and misalignment matrices, similar to how it was done for accelerometers.

A more sophisticated approach, often used in practice, is

dynamic multi-position calibration. This method utilizes a series of precisely defined motions and orientations. For instance, the IMU might be rotated in a controlled manner around a fixed axis, then switched to another axis, and perhaps moved through a trajectory By collecting data across these diverse motions, a more robust estimation of all calibration parameters (bias, scale factor, and misalignment) can be performed using optimization techniques that minimize the error between the measured angular velocities and the expected ones.

Temperature Compensation:
Both accelerometer and gyroscope biases are highly sensitive to temperature. Therefore, an important part of IMU calibration is to characterize the bias drift with temperature. This is typically done by placing the IMU in a temperature-controlled environment and varying the temperature across its operational range while measuring the bias of each sensor. A look-up table or a polynomial function relating temperature to bias can then be used to compensate for temperature-induced drifts during operation.

The result of IMU calibration is a set of parameters (biases, scale factors, misalignment matrices) that are used to transform the raw sensor readings into accurate measurements of linear acceleration and angular velocity. This significantly improves the accuracy of dead reckoning, which is the process of estimating position and orientation based solely on IMU data, and enhances the performance of sensor fusion algorithms that integrate IMU data with other sensors.

Sensor Fusion and Extrinsic Calibration

While intrinsic calibration focuses on the internal characteristics of individual sensors, **extrinsic calibration** is crucial for determining the spatial relationship between different sensors mounted on the robot. This includes the transformation (rotation and translation) between a camera's coordinate frame and a LiDAR's coordinate frame, or between an IMU's frame and the robot's base frame. This process is often referred to as **sensor-to-sensor calibration** or **hand-eye calibration** when a camera is involved and is being calibrated with respect to a robot's end-effector.

The accurate extrinsic calibration of multiple sensors is fundamental for
sensor fusion, where data from different modalities is combined to achieve a more comprehensive and robust perception of the environment. For example:

LiDAR-Camera Calibration: To project 3D LiDAR points onto a 2D camera image, or to overlay camera features onto a 3D point cloud, the relative pose between the LiDAR and the camera must be known. This involves finding the rotation matrix R_{LC} and translation vector t_{LC} that transform points from the LiDAR frame to the camera frame: $\mathbf{P}_{camera} = R_{LC} \mathbf{P}_{LiDAR} + t_{LC}$. Calibration typically involves observing common features or objects in both the LiDAR point cloud and the camera images. For instance, a checkerboard pattern can be detected in camera images, and its corresponding 3D points can be identified in the LiDAR scan. By establishing correspondences between these features, the extrinsic transformation can be estimated.
IMU-LiDAR/Camera Calibration: For applications like visual-inertial odometry (VIO) or LiDAR-inertial odometry (LIO), knowing the fixed spatial relationship between the IMU and the LiDAR or camera is essential.

This allows for the fusion of motion estimates from the IMU with the environmental perception from the LiDAR or camera. Calibration methods often involve capturing synchronized data from the IMU and the other sensor while the robot performs a series of motions. Algorithms then estimate the transformation that best aligns the estimated motion or observed features from the different sensors. For example, if the IMU provides estimates of rotation and the camera detects features, the calibration can involve finding the IMU-to-camera transform that minimizes the visual reprojection error of features when using IMU-derived motion for tracking.

Common techniques for extrinsic calibration include:

Pattern-Based Calibration: Similar to camera calibration, known patterns (e.g., checkerboards, spheres, ArUco markers) are placed in the environment and observed by all sensors simultaneously. The 3D positions of these patterns in a common reference frame are known, and correspondences between sensor readings are established to solve for the relative transformation.
Manual Alignment and Refinement: For a coarse initial alignment, sensors might be physically mounted using jigs that approximate the correct relative poses. This initial estimate is then refined using iterative optimization techniques that minimize a cost function based on synchronized sensor data.

Motion-Based Calibration: The robot performs controlled motions, and data from all sensors are logged. Algorithms then estimate the extrinsic parameters by optimizing for consistency across the sensor data. For example, if the IMU indicates rotation and the LiDAR observes a static environment, the estimated rotation from the IMU should be consistent with the lack of significant relative motion of environmental features in the LiDAR scans, after accounting for the extrinsic transform.

The accuracy of extrinsic calibration directly impacts the effectiveness of sensor fusion. Misalignments can lead to erroneous fusion results, causing significant errors in state estimation, mapping, and control. For instance, a slight error in the extrinsic calibration between a LiDAR and a camera can cause 3D points projected from LiDAR onto the camera image to appear misaligned with actual image features, leading to poor performance in tasks like visual odometry or object detection.

In summary, the calibration of LiDAR and IMU sensors, along with the extrinsic calibration between multiple sensors, forms a crucial layer of preprocessing for robotic perception systems. It ensures that the raw sensor measurements are transformed into accurate, reliable, and geometrically consistent data, which is the bedrock for sophisticated robot behaviors and intelligent decision-making. Without these meticulous calibration procedures, the rich information provided by these advanced sensors would be fundamentally compromised, limiting the robot's ability to navigate, understand, and interact with its world safely and effectively.

Before delving into the complex algorithms for sensor fusion, it is imperative to prepare the raw data streams from disparate sensors into a coherent and consistent format. This preprocessing phase is often overlooked but is absolutely critical for the success of any multi-sensor perception system. Without meticulous preprocessing, even the most sophisticated fusion algorithms will produce erroneous results, undermining the robot's ability to accurately perceive and interact with its environment. This section will focus on the essential steps: ensuring temporal coherence through time synchronization, establishing spatial coherence via coordinate transformations, and enhancing data quality through noise reduction.

Time Synchronization: Aligning the Temporal Fabric

Robots are dynamic systems operating in a constantly evolving world. Different sensors, due to their internal mechanisms, processing speeds, and communication protocols, capture information at slightly different times. For effective sensor fusion, it is paramount that data points from different sensors, when combined, represent the same instant in time. Imagine a robot moving forward; if the camera image and the LiDAR scan used for fusion are not perfectly synchronized, the robot might perceive an obstacle that is no longer there or fail to detect an approaching one. This temporal misalignment can lead to significant errors in state estimation, localization, and obstacle avoidance.

The fundamental goal of time synchronization is to align the timestamps of data packets originating from various sensors. This can be achieved through several mechanisms, broadly categorized into hardware-based and software-based synchronization.

Hardware-Based Synchronization: This is the most accurate and preferred method. It involves using a common clock signal or a trigger pulse to timestamp data from all sensors simultaneously.

Global Positioning System (GPS) Time: Many sophisticated robotic systems, especially those operating outdoors, utilize GPS signals. GPS receivers provide highly accurate, synchronized time stamps. By connecting sensors to a system that receives GPS time, or by using a GPS-disciplined oscillator, all sensor data can be timestamped with a common, precise reference. This is particularly effective when dealing with sensors spread across a large robot or even multiple robots. For example, a system might have a central time server that receives GPS time and then distributes it to individual sensor processing units via a network protocol like Network Time Protocol (NTP) or Precision Time Protocol (PTP).

Hardware Triggering/Pulsing: In many sensor systems, particularly those designed for integrated robotics, there is a capability for hardware triggering. A central synchronization unit can emit a hardware pulse (e.g., a rising or falling edge on a digital line). Each sensor is configured to timestamp its data acquisition or processing completion event based on this pulse. For instance, a camera might capture a frame when it receives a trigger, and a LiDAR might record a scan's start time upon receiving the same trigger. This ensures that all sensor events are initiated or marked by a common temporal reference. This method is often implemented using GPIO pins or dedicated synchronization interfaces on the sensors themselves. The precision here can be in the order of microseconds.
Time-Division Multiplexing (TDM): In some specialized communication buses, TDM can be employed. Each sensor is allocated a specific time slot within a frame. During its allocated slot, the sensor transmits its data, and the timestamp is implicitly known by the receiver based on the known slot timing. While effective for certain communication architectures, it can be less flexible than event-based triggering.

Software-Based Synchronization: When hardware synchronization is not feasible or not precise enough, software-based methods are employed. These methods rely on network protocols and algorithms to adjust timestamps after data has been acquired.

Network Time Protocol (NTP): NTP is a widely used protocol for synchronizing computer clocks over a network. While effective for synchronizing the clocks of different processing units, it typically offers accuracy in the range of milliseconds, which might be insufficient for high-frequency sensor fusion tasks requiring sub-millisecond precision. However, it can serve as a baseline for synchronizing system clocks before more fine-grained synchronization is attempted.
Precision Time Protocol (PTP) / IEEE 1588: PTP is a more advanced protocol designed for higher precision time synchronization, often down to the nanosecond level in specialized implementations. It is particularly useful in industrial automation and robotics where precise coordination is critical. PTP establishes a master clock and synchronizes slave clocks to it, taking into account network latencies and delays. This makes it a strong candidate for synchronizing sensor data streams when hardware triggering is not an option.
Timestamp Interpolation/Extrapolation: Even with synchronization, there can be small discrepancies or gaps in data arrival. A common software technique is to use interpolation or extrapolation based on the timestamps of acquired data. For example, if a camera frame arrives slightly late, its content can be analyzed to estimate the robot's state at an earlier, synchronized timestamp. Conversely, if a LiDAR scan is slightly delayed, its points can be transformed into the expected pose of the robot at the synchronized time. A more sophisticated approach involves using an underlying state estimator (e.g., an Extended Kalman Filter) that maintains a continuous estimate of the robot's state. When data from a sensor arrives, its timestamp is used

to query the state estimator for the robot's pose at that exact moment. This ensures that sensor data is always associated with the correct robot pose, even if there are minor timing variations.

Challenges in Time Synchronization:

Clock Drift: Even synchronized clocks can drift over time. Periodic resynchronization is often necessary.
Network Latency and Jitter: Network communication introduces variable delays, which can complicate software-based synchronization.
Sensor Latency: The time it takes for a sensor to acquire, process, and transmit data is not always constant or known. This internal sensor latency needs to be accounted for. Some sensors provide internal timestamps upon data acquisition, while others only provide timestamps upon transmission, introducing an additional unknown delay.
Data Buffering: Sensors may buffer data, further complicating the exact timing of events. Understanding how each sensor buffers data is crucial.

For robust sensor fusion, it is essential to understand the timestamping mechanism of each sensor and implement a synchronization strategy that achieves the required precision for the application. This often involves a combination of hardware and software techniques, with continuous monitoring and recalibration to account for drift and changing network conditions.

Coordinate Transformation: Aligning Spatial References

In a typical robotic system, multiple sensors and components are mounted at different locations and orientations relative to each other and to the robot's base. Each sensor operates within its own local coordinate frame (e.g., the camera frame, the LiDAR frame, the IMU frame). For data from these sensors to be meaningfully fused, it must be represented in a common, unified coordinate frame. This is achieved through coordinate transformations, specifically by defining and applying rigid body transformations (rotations and translations) between these different frames.

The primary transformations involved are:

Sensor-to-Sensor Transformations (Extrinsic Calibration): As discussed in the previous section, extrinsic calibration determines the fixed relative pose (rotation and translation) between any two sensors. For example, the transformation from the LiDAR frame (L) to the camera frame (C), denoted as $T{CL}$, *is crucial for projecting LiDAR points into the camera image or vice-versa. This transformation is typically*

*represented by a rotation matrix $R*{CL}$ and a translation vector t_{CL}:
$\mathbf{p}*C = R*{CL} \mathbf{p}*L + t*{CL}$
where $\mathbf{p}*L$ is a point in the LiDAR frame and $\mathbf{p}*C$ is the corresponding point in the camera frame. This transformation is static, meaning it does not change during operation, and is determined once through extrinsic calibration.

Sensor-to-Robot Base Transformations: Each sensor's pose relative to the robot's main reference frame, often the robot's base frame (B), must also be known. For instance, the transformation from the camera frame (C) to the robot base frame (B), T_{BC}, is needed to understand where the camera is located on the robot. This can be obtained either by directly measuring the sensor's position and orientation on the robot or by calibrating it relative to a known object in the robot's environment.
$\mathbf{p}*B = R*{BC} \mathbf{p}*C + t*{BC}$

World Frame: For global localization and mapping, a consistent world frame (W) is established. This frame might be fixed at the robot's starting position, tied to a global navigation system like GPS, or dynamically defined by a SLAM system. Transformations from the robot base frame to the world frame (T_{WB}) are then used to map all sensor data into this global reference.

The Transformation Chain:
The ability to transform data between any two frames forms a transformation chain. If we have data in the LiDAR frame and want to represent it in the camera frame, and we know $T{CL}$ and $T{BC}$ (transformation from Camera to Base), we can express LiDAR points in the Base frame first, then in the Camera frame. However, it's more common to know $T{LC}$ (LiDAR to Camera). If we have $T{LC}$ and $T{CB}$ (Camera to Base), then a point in the LiDAR frame $\mathbf{p}L$ can be transformed to the Base frame $\mathbf{p}_B$ as:
$\mathbf{p}B = T{CB} \mathbf{p}C = T{CB} (T{LC} \mathbf{p}L)$
$T{LB} = T{CB} T_{LC}$
where T_{LB} is the transformation from LiDAR to Base.

In practice, transformations are often represented using homogeneous coordinates and 4x4 matrices for easier composition:

$\begin{bmatrix} \mathbf{p}' \\ 1 \end{bmatrix} = \begin{bmatrix} R & t \\ \mathbf{0}^T & 1 \end{bmatrix} \begin{bmatrix} \mathbf{p} \\ 1 \end{bmatrix}$
where $\mathbf{p}'$ is the transformed point, $\mathbf{p}$ is the original point, R is the 3x3 rotation matrix, t is the 3x1 translation vector, $\mathbf{0}^T$ is a 1x3 zero vector, and 1 is a scalar.

Applying Transformations in Practice:

LiDAR to Camera: To visualize LiDAR points on a camera image, each 3D point $(\text{x}L, \text{y}L, \text{z}L)$ from the LiDAR is first transformed into the camera frame using $T{LC}$: $(\text{x}C, \text{y}C, \text{z}C) = R{LC} [\text{x}L, \text{y}L, \text{z}L]^T + t{CL}$. This 3D point in the camera frame can then be projected onto the 2D image plane using the camera's intrinsic calibration parameters (focal length, principal point) via the pinhole camera model:
$u = fx \frac{\text{x}C}{\text{z}C} + cx$
$v = fy \frac{\text{y}C}{\text{z}C} + cy$
The resulting (u, v) coordinates indicate where the LiDAR point appears in the image.

IMU to Robot Base: An IMU, often located at a different position and orientation than the robot's base, provides measurements of linear acceleration and angular velocity in its own frame. To use these measurements for state estimation relative to the robot's base, they must be transformed. If the IMU frame is I and the robot base frame is B, with transformation $T{BI}$, then an angular velocity $\boldsymbol{\omega}I$ measured by the IMU is transformed to the base frame as:
$\boldsymbol{\omega}B = R{BI} \boldsymbol{\omega}_I$
Similarly, linear acceleration $\mathbf{a}_I$ is transformed:
$\mathbf{a}B = R{BI} \mathbf{a}I + \mathbf{v}{BI} \times \boldsymbol{\omega}_{BI}$

where $\mathbf{v}{BI}$ and $\boldsymbol{\omega}{BI}$ are the translational and angular velocities of the IMU frame with respect to the Base frame, and $\mathbf{v}{BI} \times \boldsymbol{\omega}{BI}$ is the Coriolis acceleration term. If the IMU is rigidly attached to the base, then $\mathbf{v}{BI}$ and $\boldsymbol{\omega}{BI}$ are zero, and the transformation simplifies to just the rotation: $\mathbf{a}B = R{BI} \mathbf{a}_I$.

Dynamic Transformations:
It's important to distinguish between static and dynamic transformations. The extrinsic calibration parameters ($T{LC}$, $T{BI}$, etc.) represent static relationships between sensors. However, the transformation from the robot's base frame to the world frame (T_{WB}) is dynamic, as it represents the robot's pose in the environment, which changes over time. This dynamic transformation is what state estimation and localization algorithms aim to estimate.

The process of coordinate transformation is ubiquitous in multi-sensor fusion. It is the mechanism by which disparate data sources are unified, allowing for coherent perception. Errors in extrinsic calibration, which define these transformations, directly translate into spatial misalignments in fused data, leading to significant performance degradation.

Noise Reduction: Enhancing Data Quality

Raw sensor data is inherently noisy. This noise can arise from various sources, including sensor electronics, environmental conditions, physical phenomena, and quantization errors. If left unaddressed, this noise can propagate through fusion algorithms, leading to unreliable state estimates and poor decision-making. Therefore, noise reduction is a crucial preprocessing step. Various filtering techniques are employed, with the choice depending on the sensor characteristics, the type of noise, and the desired trade-off between noise reduction and data responsiveness.

Types of Noise:

Additive Noise: This is noise that is simply added to the true signal, often modeled as a zero-mean random process (e.g., Gaussian noise). Accelerometer and gyroscope readings often exhibit this type of noise.
Multiplicative Noise: This noise scales the true signal, often proportional to the signal's magnitude.
Outliers/Spikes: These are sudden, large errors that deviate significantly from the expected signal. They can be caused by transient environmental factors or sensor malfunctions.
Quantization Noise: This arises from the discretization of continuous sensor readings into finite digital values.

Common Noise Reduction Techniques:

Moving Average Filter: This is one of the simplest filters. It computes the average of the last N data points.
$y[n] = \frac{1}{N} \sum_{i=0}^{N-1} x[n-i]$
where $x[n]$ is the input signal and $y[n]$ is the output. Moving averages are effective at smoothing out random fluctuations but introduce a lag (phase shift) proportional to N. They are particularly useful for reducing high-frequency noise in sensor readings like velocity or position estimates derived from slower-changing phenomena. For example, if a robot's speed is being estimated from noisy wheel encoder data, a moving average can smooth out the speed fluctuations.

Low-Pass Filters: These filters allow low-frequency components of a signal to pass through while attenuating high-frequency components.
Butterworth Filter: This is a maximally flat magnitude response filter. It is monotonic in the passband and stopband.
Chebyshev Filter: These filters achieve a steeper transition band than Butterworth filters but introduce ripple in either the passband (Type I) or stopband (Type II).
Bessel Filter: These filters are optimized for linear phase response, meaning they introduce minimal phase distortion. This is crucial for applications where the timing of signal features is important.

Low-pass filters can be implemented in the time domain (e.g., as Finite Impulse Response (FIR) or Infinite Impulse Response (IIR) filters) or the frequency domain using Fourier Transforms. For real-time applications, IIR filters are often preferred due to their computational efficiency. They are widely used to filter sensor data such as camera images (to reduce sensor noise before feature extraction) or IMU readings.

Kalman Filter (and its variants): The Kalman filter is a powerful recursive algorithm for estimating the state of a linear dynamic system from a series of noisy measurements. It provides an optimal estimate in the mean-squared error sense for linear systems with Gaussian noise.
The Kalman filter operates in two steps:
1. **Prediction:** Based on a system model, it predicts the next state and its associated uncertainty.
2. **Update:** It incorporates the latest measurement to refine the predicted state, reducing the uncertainty.

For non-linear systems, such as those commonly encountered in robotics (e.g., robot motion dynamics, sensor models), extensions of the Kalman filter are used:

Extended Kalman Filter (EKF): Linearizes the non-linear system and measurement models around the current state estimate. This can be computationally efficient but may diverge if the linearization is inaccurate or the system is highly non-linear. **Unscented Kalman Filter (UKF):** Uses a deterministic sampling technique called the unscented transform to capture the mean and covariance of the state distribution. It is generally more robust than the EKF for highly non-linear systems and does not require explicit Jacobian calculations.

Kalman filters are indispensable for sensor fusion. They can fuse data from multiple sensors (e.g., IMU, wheel encoders, GPS, cameras, LiDAR) by treating measurements from different sensors as inputs to the filter's update step. By explicitly modeling the system dynamics and measurement noise, the Kalman filter can produce a smoother and more accurate estimate of the robot's state (position, velocity, orientation) than using any single sensor alone. For instance, in visual-inertial odometry, an EKF or UKF is used to fuse noisy IMU readings with visual odometry estimates from a camera, resulting in a more stable and accurate trajectory estimation.

Median Filter: This is a non-linear filter that replaces each data point with the median of its neighboring data points. It is particularly effective at removing salt-and-pepper noise (outliers/spikes) without significantly blurring edges or sharp features in the data. This makes it suitable for processing image data or discrete sensor readings where occasional erroneous spikes might occur.

Gaussian Filter: A special case of the low-pass filter that uses a Gaussian kernel for smoothing. It provides a good balance between smoothing and preserving image features and is often used in image processing.

Applying Noise Reduction to Different Sensors:

IMUs: Gyroscope and accelerometer data are notoriously noisy. Kalman filtering is commonly used to fuse IMU data with other sensors and simultaneously smooth the IMU readings. Simple moving averages or low-pass filters can be applied to the raw IMU output for initial smoothing, but this must be done cautiously to avoid excessive lag.
Cameras: Image noise can be reduced using Gaussian or median filters, especially in low-light conditions. However, aggressive filtering can blur important visual features needed for tasks like object recognition or visual odometry. Denoising algorithms are often applied as an early step in the computer vision pipeline.
LiDAR: LiDAR point clouds can be noisy due to atmospheric conditions (e.g., fog, rain), reflective surfaces, and sensor limitations. Techniques like statistical outlier removal (SOR) can be used to identify and remove points that are statistically unlikely to belong to the main surface, effectively cleaning up spurious points. Voxel grid downsampling, while primarily for reducing data density, also has a slight smoothing effect.

The selection of a noise reduction technique involves a trade-off. Aggressive filtering can effectively remove noise but may introduce lag, distort signal characteristics, or discard valuable high-frequency information. Conversely, less aggressive filtering preserves more detail but leaves more noise in the data. The optimal choice depends on the specific application requirements and the characteristics of the sensors involved. Often, a combination of techniques is employed in a multi-stage processing pipeline to achieve the desired level of data quality for subsequent fusion operations.

Data Augmentation and Cleaning for Perception

As we transition from general sensor data preprocessing to the specific demands of robotic perception, especially those powered by machine learning, the nature of data preparation takes on new dimensions. Perception systems, tasked with interpreting the robot's environment to identify objects, estimate distances, and understand spatial relationships, often rely on vast and diverse datasets. Machine learning models, in particular, thrive on this data abundance; the more varied the examples they are exposed to, the more robust and generalizable their learned representations become. However, acquiring sufficiently large and diverse real-world datasets can be prohibitively expensive and time-consuming. This is where

data augmentation emerges as a powerful technique to artificially expand training datasets. Simultaneously, even meticulously collected data can contain errors, noise, or irrelevant information that can degrade perception performance. Hence, **data cleaning** remains a critical, albeit often more nuanced, step for perception-specific data.

Data Augmentation: Expanding the Perceptual Horizon

Data augmentation is a set of strategies used to increase the amount and diversity of training data by applying realistic transformations to existing data. The core idea is that if a model can perform well on slightly altered versions of the same data, it will be more resilient to variations encountered in the real world. For perception tasks involving visual data (e.g., cameras) or 3D data (e.g., LiDAR), this can involve a wide array of transformations that simulate changes in viewpoint, lighting, scale, and other environmental factors without requiring new data collection.

1. Geometric Transformations: These transformations alter the spatial arrangement of the data, mimicking changes in camera pose, object orientation, or perspective.

Rotation: Rotating an image or a point cloud by a certain angle. For a camera image, this simulates the robot or the object being at a different orientation. For a LiDAR point cloud, rotating it can represent the sensor scanning from a slightly different angle or the robot itself being tilted. The rotation must be applied consistently across all dimensions and be accompanied by appropriate transformations of associated labels (e.g., bounding boxes for object detection would need to be re-projected).
Example: An autonomous vehicle's camera sees a stop sign. If the training data only contains images of the stop sign directly in front, the model might struggle to recognize it when it appears at a 30-degree angle due to the vehicle turning. Augmenting the dataset with rotated versions of the stop sign image helps the model learn to identify it from various viewpoints.

Translation: Shifting the image or point cloud horizontally or vertically. This is analogous to an object appearing in a different part of the sensor's field of view. For bounding box labels, this requires adjusting the coordinates of the box to match the shifted image.
Example: A robot's vision system is trained to detect a charging station. If the training data always shows the station in the center of the image, the model might fail to detect it when it appears on the periphery. Translating the image, and the bounding box, allows the model to learn detection across the entire sensor frame.

Scaling (Zooming): Resizing the image or point cloud. This simulates objects appearing closer or further away from the robot. Scaling of bounding boxes must also be performed, and care must be taken to ensure that scaled objects remain within the sensor's view. Zooming in can highlight fine details, while zooming out tests the model's ability to recognize objects at a distance.
Example: A drone's perception system needs to identify landing pads from various altitudes. Scaling augmentation allows the model to recognize the same landing pad whether it is viewed from 10 meters or 100 meters above, by simulating different distances.

Flipping: Horizontally or vertically flipping the data. Horizontal flipping is particularly common and effective for objects with left-right symmetry (e.g., cars, people), as it effectively doubles the dataset size. Vertical flipping is less common for ground-based robots but can be useful for aerial or underwater robots, or for objects with vertical symmetry.
Example: A robot arm learning to grasp objects. If the training data shows an object being approached from the left, augmenting with a horizontally flipped version of the image helps the robot learn to approach and grasp it from the right, which is often a mirrored action.

Shearing: Skewing the image or point cloud along one of the axes. This can simulate perspective distortions or slight tilts of the sensor.

Example: A robot navigating a warehouse with narrow aisles. Shearing augmentation can help its perception system cope with the slight visual distortions that occur when viewing objects from extreme angles within confined spaces.

2. Photometric/Color Transformations: These transformations alter the pixel values or intensity levels, simulating changes in lighting conditions, camera settings, or sensor characteristics. These are primarily applied to camera data.

Brightness Adjustment: Increasing or decreasing the overall brightness of an image. This simulates different times of day or varying ambient light.
Example: A robot operating both indoors under fluorescent lights and outdoors under sunlight. Augmenting with adjusted brightness helps its object recognition system perform reliably in both scenarios.

Contrast Adjustment: Modifying the difference between the darkest and brightest parts of an image. This simulates different weather conditions (e.g., hazy versus clear days) or camera sensor responses.
Example: A self-driving car needs to detect pedestrians in both foggy and clear conditions. Adjusting contrast during training can expose the model to a wider range of visual clarity.

Saturation Adjustment: Altering the intensity of colors in an image. This helps the model become invariant to the vibrancy of colors, which can vary significantly with lighting.
Example: A robot used for agricultural monitoring might encounter plants with different color saturations due to nutrient levels or watering. Saturation augmentation can help it consistently identify plant types regardless of these color variations.

Hue Adjustment: Shifting the dominant colors in an image. This is useful for objects whose color might be perceived differently under different lighting or due to sensor biases.
Example: Identifying colored markers used for robot navigation. Hue augmentation can help the system recognize a specific shade of red even if it appears slightly more orange or pink under certain lighting conditions.

Adding Noise: Introducing random noise (e.g., Gaussian noise, salt-and-pepper noise) to the image. This makes the model more robust to sensor noise, transmission artifacts, or low-light imaging imperfections.
Example: A security robot using a camera in a dimly lit environment. Adding noise to training images prepares its recognition system to function even when the acquired images are inherently noisy.

3. Advanced Augmentation Techniques:

Cutout/Random Erasing: Randomly masking out square regions of an image. This forces the model to learn from incomplete information and focus on other salient features, improving its robustness to occlusions.
Example: A robot navigating a cluttered environment. If a critical object (like a sign) is partially occluded by another object in the training image, Random Erasing can simulate this and train the model to still identify the object based on its visible parts.

Mixup: Linearly interpolating two images and their corresponding labels. This creates synthetic training samples that lie between existing data points, encouraging smoother decision boundaries and improving generalization.
Example: Training a robot to distinguish between two similar objects, say a cup and a mug. Mixup could create an intermediate image that is a blend of a cup and a mug, forcing the model to learn a more nuanced distinction.

CutMix: Similar to Mixup, but instead of blending pixel values, a patch from one image is cut and pasted onto another, with the label being a weighted combination based on the area of the patches. This encourages the model to utilize a wider range of features.

Simulations and Domain Randomization: For tasks where direct data collection is difficult or dangerous (e.g., robotic manipulation in hazardous environments), synthetic data generated in simulators (like Gazebo, CoppeliaSim, or Unity) can be used. Domain randomization involves systematically varying parameters within the simulator (e.g., textures, lighting, object poses, physics properties) to create a diverse set of synthetic data. The hope is that the learned model will generalize to the real world, a phenomenon known as "sim-to-real transfer."
Example: Training a robot arm to pick and place objects. Generating thousands of variations of object poses, lighting, and textures in a simulator, and then fine-tuning the resulting model on a smaller set of real-world data, can significantly reduce the need for extensive real-world training.

The efficacy of data augmentation lies in applying transformations that are
realistic for the target environment and task. Over-augmentation or applying irrelevant transformations can degrade performance. Therefore, careful consideration of the sensor modalities, the robot's operating domain, and the specific perception task is crucial when designing an augmentation strategy.

Data Cleaning: Refining the Perception Input

While data augmentation expands the dataset, data cleaning focuses on removing or correcting erroneous, noisy, or irrelevant information from the existing data. For perception systems, this is vital because raw sensor readings can be corrupted by various factors, and machine learning models can learn and amplify these errors if not addressed.

1. Outlier Detection and Removal: Outliers are data points that significantly deviate from the expected distribution of the majority of the data.

For Camera Images: Outliers can manifest as corrupted pixels, sensor artifacts (e.g., dead pixels, blooming), or transient environmental effects (e.g., lens flare, brief reflections).
Techniques:
Statistical Methods: Calculating Z-scores or using interquartile range (IQR) to identify pixels or regions with unusually high or low intensity values compared to their neighbors.
Median Filtering: As mentioned previously, a median filter is excellent at removing salt-and-pepper noise (isolated spikes) while preserving edges.
Image Inpainting/Denoising Algorithms: Advanced algorithms can detect and attempt to reconstruct corrupted or anomalous regions based on surrounding valid information.

For LiDAR Point Clouds: Outliers in LiDAR data can arise from several sources:

Atmospheric Effects: Fog, rain, or dust can cause spurious points to be detected.
Reflective Surfaces: Highly reflective or specular surfaces (like polished metal or mirrors) can lead to erroneous reflections.
Multi-path Reflections: Laser beams bouncing off multiple surfaces before returning to the sensor.
Sensor Noise: Random errors in distance or angle measurements.
Techniques:
Statistical Outlier Removal (SOR): This is a widely used technique. For each point, it calculates the mean distance to its k nearest neighbors. If the mean distance is above a certain threshold (or if a certain percentage of neighbors are too far away), the point is considered an outlier and removed. This is highly effective for removing isolated spurious points.
Radius Outlier Removal: Similar to SOR, but instead of using nearest neighbors, it considers all points within a specified radius. Points with fewer than a minimum number of neighbors within that radius are removed. This is useful for removing points in sparse regions or small clusters of noise.
RANSAC (Random Sample Consensus): While often used for fitting models (e.g., planes), RANSAC can be adapted for outlier removal. It iteratively selects random subsets of points to fit a model (e.g., a ground plane). Points that are far from the best-fit model are considered outliers. This is particularly effective for removing noise points that do not conform to expected geometric structures like floors or walls.
Intensity Thresholding: Some LiDAR sensors provide reflectivity information. Points with unusually low or

high reflectivity might indicate noise or unusual surface properties and can be filtered out.

2. Removing Irrelevant Information: Perception systems often need to focus on specific aspects of the scene and ignore others.

Ground Plane Segmentation: For many ground-based robots (e.g., autonomous vehicles, warehouse robots), the ground plane itself is not of primary interest for tasks like object detection or obstacle avoidance, beyond its role in supporting objects. Removing the ground plane from LiDAR scans significantly reduces computational load and can simplify subsequent processing.
Techniques: Algorithms like iterative RANSAC or grid-based segmentation are commonly used to identify and remove the ground plane.

Region of Interest (ROI) Filtering: Limiting processing to a specific spatial area within the sensor's field of view. This is useful when the robot knows that relevant objects will only appear in certain locations.
Example: A robot arm designed to work on a specific assembly line might only need to process camera data within a predefined rectangular region where parts are presented.

Occupancy Grid Filtering (for 3D data): In some cases, point clouds can be voxelized into an occupancy grid. Filtering operations can then be applied to remove sparse voxels or small connected components that might represent spurious detections or sensor noise.

3. Handling Sensor-Specific Artifacts:

Camera Distortion Correction: While often part of intrinsic calibration, it's crucial to ensure that lens distortion (e.g., barrel or pincushion distortion) is corrected for all images used in perception tasks. Uncorrected distortion can lead to misinterpretation of object shapes and relative positions.

LiDAR Ring Artifacts: Some LiDAR sensors can exhibit "ring artifacts" where points from specific sensor "rings" (individual laser channels) might be systematically noisy or misaligned. Specialized filtering or calibration procedures might be needed to mitigate these.

4. Data Imbalance Handling (Related to Cleaning/Augmentation):
In perception tasks, datasets often suffer from class imbalance, where certain objects or scenarios appear far more frequently than others. While not strictly "cleaning," addressing this imbalance is a form of data preparation to prevent biased models.

Techniques:
Undersampling: Removing data from majority classes.
Oversampling: Duplicating data from minority classes.
Synthetic Data Generation (SMOTE - Synthetic Minority Over-sampling Technique): Creating synthetic samples for minority classes.
Weighted Loss Functions: During training, assigning higher weights to misclassifications of minority classes.
Data augmentation, as described earlier, is also a powerful tool for addressing class imbalance by artificially increasing the representation of minority classes.

The process of cleaning perception data is iterative and domain-specific. What constitutes an "outlier" or "irrelevant information" depends heavily on the application. For instance, while ground plane removal is standard for autonomous driving, it might be undesirable for a robot tasked with inspecting subterranean environments. Thorough understanding of the sensor characteristics, the robot's operational context, and the goals of the perception system are paramount when devising effective data cleaning strategies. By combining robust data augmentation with meticulous data cleaning, the foundational data layer for sophisticated perception algorithms is built, paving the way for more accurate, reliable, and generalized robotic intelligence.

Chapter 4: Perception and Machine Learning Foundations

Perception forms the bedrock upon which all sophisticated robotic capabilities are built. Without the ability to sense and interpret the external world, a robot is effectively blind and deaf, incapable of interacting meaningfully with its environment or executing any intelligent behavior. At its core, perception in robotics is the process of translating raw, often noisy, sensor data into a coherent and actionable understanding of the robot's surroundings, its own internal state, and the relationships between them. This transformation is not merely about data processing; it is about making sense of the complex, dynamic, and often unpredictable reality that robots are designed to navigate.

The spectrum of raw sensor data is vast, encompassing inputs from cameras, LiDAR, radar, ultrasonic sensors, tactile sensors, inertial measurement units (IMUs), and more. Each sensor modality provides a unique window into the world, with its own strengths, limitations, and characteristic types of noise and artifacts. Perception, therefore, involves not only processing individual sensor streams but also fusing information from multiple sources to create a more robust and comprehensive picture. This multisensory integration is crucial for overcoming the inherent weaknesses of any single sensor and for achieving a richer, more accurate representation of the environment.

The ultimate goal of perception is to extract meaningful information that can inform a robot's subsequent

actions. This information can take many forms, leading to a diverse set of perceptual tasks.
Object recognition is a fundamental task, enabling the robot to identify and classify specific entities within its environment, such as distinguishing a chair from a table, a pedestrian from a vehicle, or a specific tool from a collection of items. This often involves sophisticated machine learning models trained on large datasets, as discussed in previous sections. Beyond mere classification, **object detection** aims to not only identify objects but also to localize them within the sensor data, typically by providing bounding boxes or masks.

Another critical perceptual task is **scene understanding**. This goes beyond identifying individual objects to comprehending the overall context and structure of the environment. It involves recognizing the spatial relationships between objects, understanding the layout of the scene (e.g., identifying rooms, corridors, or open spaces), and inferring the semantic meaning of the scene (e.g., recognizing that a scene depicts a kitchen, a factory floor, or a park). This level of understanding allows robots to anticipate how the environment might behave and to plan their actions accordingly. For instance, understanding that a door leads to another room enables a robot to navigate through it, while recognizing a cluttered area might necessitate a slower, more cautious approach.

Obstacle detection and avoidance are paramount for safe and effective robot operation, especially in dynamic environments. This task involves identifying any entity that could impede the robot's movement or pose a collision risk, and then determining its trajectory and speed. This requires not only detecting static obstacles but also dynamic ones, such as moving people or other robots, and predicting their future positions. The information gathered from obstacle detection directly feeds into the robot's motion planning and control systems, ensuring it can navigate without collisions.

Beyond the external environment, perception also plays a vital role in estimating the robot's own **state**. This includes determining its precise location and orientation within a known or unknown map (**localization**), understanding how fast it is moving (**velocity estimation**), and sensing the forces acting upon it (**force and torque sensing**). For mobile robots, accurate localization is essential for navigation and path planning. For robotic manipulators, understanding joint positions, end-effector forces, and torques is crucial for precise control and safe interaction with objects.

The outputs of these perceptual tasks are not ends in themselves but rather the essential inputs for higher-level functions, primarily **decision-making** and **control**. A robot's ability to perceive an object of interest (e.g., a package to be picked up) directly triggers a decision-making process that might involve planning a grasping strategy. Similarly, detecting an obstacle initiates a control sequence to alter the robot's trajectory. The perception system acts as the robot's sensory interface to the world, providing the critical intelligence that allows it to move, interact, and achieve its objectives.

The efficacy of a robot's perception system directly dictates its overall performance. A perception system that is prone to errors, misses important objects, or misinterprets the environment will inevitably lead to flawed decision-making and potentially unsafe or ineffective actions. Conversely, a robust and accurate perception system empowers a robot with the confidence to operate autonomously, to adapt to unforeseen circumstances, and to perform complex tasks with a high degree of reliability. This underscores why perception is not just one component among many but a foundational pillar upon which all other robotic intelligence rests. The richness and accuracy of the robot's internal model of the world, derived from perception, are directly correlated with its ability to exhibit intelligent behavior.

The process of perception is inherently tied to the sensor modalities available to the robot and the specific tasks it is designed to perform. For instance, a robot operating in a structured indoor environment, like a warehouse, might rely heavily on LiDAR for precise distance measurements and mapping, complemented by cameras for object identification. An autonomous vehicle, on the other hand, requires a more complex suite of sensors, including cameras for reading traffic signs and detecting lane markings, radar for all-weather object detection and velocity estimation, and LiDAR for high-resolution 3D mapping and obstacle detection.

Each sensor provides a different perspective and type of data, and the perception system must be adept at harmonizing these disparate inputs.

Consider the task of navigating a cluttered environment. A simple ultrasonic sensor might detect the presence of an obstacle but provide little information about its shape, size, or type. A camera, however, could provide rich visual information, allowing for object recognition and estimation of its exact boundaries. LiDAR can offer precise depth information, creating a detailed 3D point cloud of the obstacle, which is invaluable for path planning. The perception system's role is to fuse these different pieces of information. It might use LiDAR to identify a large, solid object, then use camera data to determine if it's a wall, a piece of furniture, or a person, and use this combined understanding to decide whether to go around it, stop, or wait.

The complexity of perception is further amplified by the dynamic nature of the real world. Environments are rarely static. Objects move, lighting conditions change, and unexpected events occur. A perception system must be able to cope with this variability. For a robot designed to work alongside humans, like a collaborative robot (cobot) in a factory, the ability to accurately track human movement and predict their intentions is critical for safety. This involves not just detecting the human but also understanding their pose, gestures, and velocity to anticipate potential collisions or collaborative actions.

Furthermore, perception is an iterative process. A robot doesn't just perceive its environment once. It constantly updates its understanding as new sensor data arrives. This continuous feedback loop allows the robot to refine its internal model of the world, correct initial misinterpretations, and adapt to changes. For example, if a robot is navigating towards a door and initially perceives it as closed, but as it approaches, the door opens, the perception system must detect this change and adjust the robot's planned path accordingly. This requires the perception system to be not only accurate but also responsive and capable of real-time updates.

The reliance on machine learning, particularly deep learning, has revolutionized robotic perception. Convolutional Neural Networks (CNNs) have become ubiquitous for image analysis, enabling highly accurate object detection, segmentation, and classification. Similarly, point cloud processing techniques, often inspired by CNN architectures, are employed for analyzing LiDAR data. These learning-based approaches can discover complex patterns and features directly from data, often surpassing traditional hand-crafted feature extraction methods. However, as previously discussed, the effectiveness of these methods is heavily dependent on the quality and quantity of the training data, highlighting the intertwined nature of perception, machine learning, and data management.

The "understanding" that a robot derives from perception is, of course, different from human consciousness or subjective experience. It is a computational representation, a model of the world that the robot can use to reason and act. This model typically includes information about the geometry of the environment, the identity and properties of objects, and the robot's own spatial and temporal state. The sophistication of this internal model directly influences the robot's ability to perform complex tasks. A robot that only knows "there is an object in front" is far less capable than one that knows "there is a red, cylindrical can approximately 10 cm away, which is a beverage container."

The challenge in robotic perception extends to scenarios where sensor data is inherently ambiguous or incomplete. For example, inferring the 3D shape of an object from a single 2D image is an ill-posed problem, as depth information is lost during the projection. Similarly, distinguishing between objects with similar visual appearances but different functionalities requires a deeper semantic understanding. These challenges drive the development of more advanced perception algorithms, including those that leverage temporal information (e.g., tracking objects over time), incorporate physical models of the world, or employ sophisticated reasoning mechanisms.

In essence, perception is the robot's primary means of

sensing its reality. It is the process through which raw sensory signals are transformed into knowledge about the world. This knowledge is then the indispensable fuel for all subsequent intelligent actions. Without a capable perception system, a robot remains a mere mechanical automaton; with it, it gains the capacity for awareness, interaction, and ultimately, intelligence. The continuous advancements in sensor technology and machine learning are constantly pushing the boundaries of what is possible in robotic perception, enabling robots to perform increasingly complex tasks in ever more challenging environments. The journey from raw sensor data to actionable understanding is the defining characteristic of a perceptive robot, and it is a journey that underpins the entire field of robotics.

The ability of a robot to perceive its environment, as discussed previously, is the first critical step towards intelligent behavior. However, raw sensor data, even after processing, often contains complexities and nuances that are exceptionally challenging, if not impossible, to explicitly program using traditional rule-based methods. Imagine trying to write a comprehensive set of rules to account for every possible variation in lighting, texture, and viewpoint that a robot's camera might encounter when trying to identify a specific object. This is where the power of Machine Learning (ML) comes into play. ML offers a paradigm shift, enabling robots to learn from data, adapt to novel situations, and improve their performance over time without explicit reprogramming for every conceivable scenario. It transforms robots from pre-programmed automata into systems that can exhibit a form of "learning" and "intelligence" by extracting patterns and making predictions from experience.

At its core, Machine Learning is about developing algorithms that allow computer systems to learn from data without being explicitly programmed for every specific task. Instead of an engineer meticulously crafting every conditional statement and logical operation, ML algorithms are fed large amounts of data, and they learn to identify patterns, relationships, and underlying structures within that data. This learned knowledge can then be applied to new, unseen data to make predictions, classifications, or decisions. For robotics, this means that a robot can learn to recognize objects, navigate complex terrains, or predict the behavior of its environment by analyzing data collected from its sensors during various operations or through simulated experiences. The more data the ML model is exposed to, and the more varied that data is, the more robust and generalizable its learned capabilities become.

There are several primary paradigms within Machine Learning, each suited for different types of problems and data. The most prominent among these are supervised learning, unsupervised learning, and reinforcement learning. Understanding these fundamental paradigms is crucial for appreciating how ML is applied to various robotic tasks.

Supervised Learning: This is perhaps the most widely used and understood form of ML. In supervised learning, the algorithm is trained on a dataset where each data point is paired with a corresponding correct output or "label." Think of it like a student learning with a teacher who provides the answers. The goal of the algorithm is to learn a mapping function from the input data to the output labels. Once trained, the model can then predict the label for new, unseen input data.

In the context of robotics, supervised learning is instrumental for tasks such as object recognition and classification. For instance, to train a robot to identify different types of tools, engineers would provide a dataset of images, where each image is meticulously labeled with the name of the tool it contains (e.g., "wrench," "screwdriver," "hammer"). The ML model, often a type of neural network like a Convolutional Neural Network (CNN), analyzes these labeled images and learns to associate visual features—shapes, textures, colors, patterns—with specific tool categories. After sufficient training, when the robot encounters a new image from its camera, the trained model can predict which tool is present with a high degree of accuracy.

Another common application of supervised learning in robotics is in
semantic segmentation, where the goal is not just to identify an object but to classify every pixel in an image

according to the object it belongs to. For example, in an autonomous driving scenario, a supervised learning model would be trained on images where pixels are labeled as "road," "car," "pedestrian," "sky," "building," and so on. This detailed pixel-level understanding is vital for safe navigation, allowing the vehicle to precisely distinguish drivable areas from obstacles or other road users.

Feature Extraction plays a vital role in supervised learning, especially when dealing with raw sensor data. Raw data, like pixel values from an image or raw points from a LiDAR scan, can be high-dimensional and noisy. Feature extraction involves transforming this raw data into a more informative and concise representation, often called "features." These features highlight the essential characteristics of the data that are relevant for the learning task. Traditionally, engineers would manually design feature extraction methods (e.g., edge detection, corner detection for images). However, modern deep learning models, particularly CNNs, can learn to perform feature extraction automatically as part of the training process. They learn hierarchical representations, where early layers detect simple features like edges and corners, and later layers combine these to detect more complex patterns and object parts.

Model Training is the process of feeding the data to the ML algorithm and adjusting its internal parameters to minimize errors or achieve a desired outcome. For supervised learning, this typically involves an iterative process where the model makes predictions on the training data, compares its predictions to the true labels, calculates an error or "loss," and then uses an optimization algorithm (like gradient descent) to adjust its parameters to reduce that error. This process is repeated for many cycles (epochs) until the model's performance on the training data reaches a satisfactory level.

Evaluation Metrics are essential for assessing how well a trained ML model performs. For classification tasks, common metrics include **accuracy** (the proportion of correct predictions), **precision** (the proportion of positive identifications that were actually correct), and **recall** (the proportion of actual positives that were identified correctly). For tasks like object detection or segmentation, more complex metrics like Intersection over Union (IoU) are used. These metrics help researchers understand the strengths and weaknesses of a model and guide further improvements.

Unsupervised Learning: In contrast to supervised learning, unsupervised learning algorithms are given data without explicit labels. The goal here is for the algorithm to discover inherent patterns, structures, or relationships within the data on its own. It's like a student being asked to find groupings or anomalies in a set of objects without being told what those groupings or anomalies should be.

Unsupervised learning is incredibly valuable in robotics for tasks where obtaining labeled data is difficult or impossible. One common application is **clustering**, where the algorithm groups similar data points together. For example, a robot exploring an unknown environment might use unsupervised clustering on its sensor readings to identify distinct types of terrain or surfaces. It could cluster LiDAR points or camera features to differentiate between grass, gravel, pavement, or water without prior knowledge of these categories.

Another key unsupervised learning technique is **dimensionality reduction**. Real-world sensor data can be extremely high-dimensional, making it computationally expensive to process and store. Dimensionality reduction techniques aim to represent the data in a lower-dimensional space while preserving as much of the important information as possible. Principal Component Analysis (PCA) is a classic example, finding the directions (principal components) in the data that capture the most variance. Autoencoders, a type of neural network, are also widely used for learning compact, low-dimensional representations of data. This can be useful for tasks like compressing sensor readings or for creating efficient representations of the environment for navigation.

Unsupervised learning can also be used for **anomaly detection**. A robot can learn what "normal" operational data or environmental states look like. If it then encounters data that deviates significantly from this learned norm, it can flag it as an anomaly. This could be crucial for detecting sensor malfunctions, unexpected environmental changes, or potential hazards that weren't anticipated during supervised training. For instance, a robot arm might learn the typical force profiles associated with successful grasping. If it suddenly experiences a much higher or lower force than usual, it could signal a problem with the object or the grasping process.

Reinforcement Learning (RL): This paradigm is perhaps the most exciting and promising for enabling robots to learn complex behaviors through interaction. In reinforcement learning, an "agent" (the robot) learns to make a sequence of decisions by trying to maximize a cumulative "reward" signal it receives from its environment. The agent learns through trial and error, receiving positive rewards for actions that lead to desirable outcomes and negative rewards (or penalties) for undesirable ones.

The core components of an RL system are the agent, the environment, states, actions, and rewards. The robot is the agent. The physical world or a simulated version of it is the environment. A "state" is a representation of the current situation of the agent and the environment (e.g., the robot's position, sensor readings, object locations). An "action" is a decision the agent makes, such as moving forward, turning, or grasping an object. The "reward" is a scalar value provided by the environment after an action is taken, indicating how good or bad that action was in the context of the current state and the overall goal.

Reinforcement learning is particularly well-suited for sequential decision-making tasks where the optimal strategy is not immediately obvious and may depend on a long series of actions. For robotics, this includes tasks like locomotion, manipulation, and navigation in dynamic environments. For example, training a quadrupedal robot to walk might involve defining states based on its joint angles and velocities, actions as commands to motors, and rewards for moving forward and staying upright, with penalties for falling. Through thousands or millions of simulated (or real-world) trials, the RL algorithm can learn a policy—a mapping from states to actions—that allows the robot to walk effectively.

One of the significant advantages of RL is its ability to discover emergent behaviors that might be difficult for humans to engineer explicitly. Consider a robot tasked with stacking blocks. A supervised approach might require meticulously defining how to pick up, move, and place each block. An RL agent, however, can learn this entire sequence through rewards for successfully placing blocks and penalties for dropping them or creating unstable structures. This allows RL to tackle problems with very complex state and action spaces.

However, RL also presents significant challenges.

Exploration vs. Exploitation is a fundamental dilemma: should the agent explore new actions to potentially find better rewards, or should it exploit its current knowledge to maximize immediate rewards? Finding the right balance is critical for efficient learning. Furthermore, **sample efficiency** can be an issue; RL often requires a vast number of interactions with the environment to learn effectively, which can be time-consuming and costly in the real world. This has led to significant research in areas like **model-based RL**, where the agent learns a model of the environment's dynamics and uses it to plan more efficiently, and **sim-to-real transfer**, where models are trained in simulation and then adapted for use on physical robots.

Bridging the Paradigms: It's important to note that these ML paradigms are not mutually exclusive and are often combined in sophisticated robotic systems. For example, a robot might use supervised learning to accurately detect objects in its camera feed, then use reinforcement learning to decide how to best grasp and manipulate those detected objects. Unsupervised learning could be used initially to discover meaningful features or representations from raw sensor data, which are then fed into a supervised or reinforcement learning model. This hybrid approach leverages the strengths of each paradigm to address the multifaceted challenges of robotic intelligence.

The integration of Machine Learning into robotics is not merely about replacing traditional algorithms; it's about augmenting them with the capacity to learn, adapt, and generalize. As robots are increasingly deployed in unstructured and dynamic environments, the need for systems that can cope with unforeseen situations and learn from experience becomes paramount. ML provides the foundational tools and techniques that enable robots to move beyond pre-programmed behaviors and towards true autonomy and adaptability. The ongoing research in ML continues to push the boundaries of what robots can achieve, making them more capable, versatile, and integrated partners in our lives.

The process of applying ML to robotics typically involves a well-defined workflow. First, the **problem is defined**: what specific task should the robot learn to perform? This could range from simple tasks like recognizing a specific object to highly complex ones like performing delicate surgery or navigating through a crowded city. Next, the **data is collected**. This is a crucial step, as the quality and quantity of data directly influence the performance of the ML model. For supervised learning, this involves collecting and meticulously labeling vast datasets. For reinforcement learning, it might involve setting up environments for extensive simulation or carefully designed real-world interaction protocols.

Following data collection, the next stage is **feature engineering and selection**, though as mentioned, deep learning models can automate much of this. However, understanding the relevant features—what aspects of the sensor data are most indicative of the desired outcome—is still critical. For instance, when training a robot to detect slippery surfaces, features related to texture, visual reflectivity, and possibly even acoustic properties (if using microphones) might be important. For object detection, features related to edges, corners, color histograms, and specific shape descriptors are relevant.

Then comes
model selection and training. This involves choosing the appropriate ML algorithm or architecture (e.g., a CNN for image processing, a Recurrent Neural Network (RNN) for sequential data, a Q-learning algorithm for RL) and then training it on the collected and preprocessed data. This training phase is computationally intensive and often requires significant processing power, sometimes leveraging GPUs or TPUs to accelerate the calculations. During training, the model's parameters are iteratively adjusted to minimize errors or maximize rewards.

Once the model is trained, it needs to be
evaluated. This is where the carefully selected evaluation metrics come into play. The model's performance is assessed on a separate dataset (the validation or test set) that it has not seen during training. This provides an unbiased estimate of how well the model will perform on new, unseen data in the real world. If the performance is not satisfactory, the process may loop back to earlier stages: collecting more data, trying different feature engineering techniques, selecting a different model architecture, or fine-tuning the training process.

Finally, the trained and validated **model is deployed** onto the robot. This involves integrating the ML model into the robot's software architecture, ensuring it can receive real-time sensor data and output its predictions or actions efficiently. The deployed model is then put to the test in real-world scenarios. Even after deployment, the learning process doesn't necessarily stop. Many modern robotic systems employ **online learning** or **continuous learning**, where the robot continues to gather data and update its models as it operates, allowing it to adapt to changing conditions and improve its performance over time without needing to be taken offline for retraining.

The successful application of ML in robotics hinges on a deep understanding of both the ML techniques and the specific robotic domain. For instance, knowledge of robot kinematics and dynamics is crucial when designing actions and reward functions for reinforcement learning tasks involving manipulation or locomotion. Similarly, an understanding of sensor characteristics and limitations is vital for effective feature engineering and for interpreting the output of ML models.

Key Concepts in ML for Robotics:

Data Representation: How sensor data is converted into a format that ML algorithms can understand. This

can range from raw pixel values and point clouds to hand-crafted features or learned embeddings.
Model Architectures: The specific structure of the ML algorithm used. For example, Convolutional Neural Networks (CNNs) are excellent for image data, capturing spatial hierarchies. Recurrent Neural Networks (RNNs) and Transformers are adept at handling sequential data like time-series sensor readings or language.
Loss Functions: Mathematical functions that quantify the error between the model's predictions and the ground truth in supervised learning, or measure the discrepancy from a desired state.
Optimization Algorithms: Methods used to adjust the model's parameters to minimize the loss function (e.g., Gradient Descent, Adam optimizer).
Overfitting and Underfitting: Common pitfalls in model training. Overfitting occurs when a model learns the training data too well, including its noise, and fails to generalize to new data. Underfitting occurs when a model is too simple to capture the underlying patterns in the data. Techniques like regularization and cross-validation are used to combat these issues.
Transfer Learning: A powerful technique where a model pre-trained on a large dataset (e.g., ImageNet for image recognition) is adapted for a new, related task. This can significantly reduce the amount of data and training time required for a new robotic application. For example, a robot learning to identify specific industrial components might benefit from a model initially trained on general object recognition.
Sim-to-Real Transfer: As mentioned earlier, training ML models in simulation offers significant advantages

in terms of speed, safety, and data generation. However, bridging the gap between simulated and real-world performance is a major research area. Techniques like domain randomization (varying simulation parameters like lighting, textures, physics) and domain adaptation are employed to improve the transferability of learned models.

In essence, Machine Learning provides robots with the indispensable capability to learn from experience, adapt to the unpredictable nature of the real world, and perform tasks that would be prohibitively complex to program explicitly. It is not a magical solution but a powerful set of tools that, when applied thoughtfully and in conjunction with a deep understanding of robotics principles, can unlock unprecedented levels of autonomy and intelligence. As we delve deeper into specific robotic applications, the role and implementation of these ML concepts will become even more apparent. The journey from raw sensor data to intelligent action is now more profoundly shaped by the principles of Machine Learning than ever before.

The remarkable advancements in robotic perception and decision-making over the past decade are inextricably linked to the rise of Deep Learning (DL). While traditional Machine Learning techniques provide a powerful framework for robots to learn from data, Deep Learning represents a significant evolution within this field, particularly in its ability to automatically learn complex representations directly from raw sensory input. This capability has been a game-changer, especially for tasks involving perception, where the sheer variability and richness of real-world data pose immense challenges for handcrafted feature extraction.

At the heart of Deep Learning lies the concept of Artificial Neural Networks (ANNs), often simply referred to as neural networks. These are computational models inspired by the structure and function of the biological neural networks found in the human brain. A neural network is composed of interconnected processing units called "neurons" or "nodes," organized in layers. These neurons receive input signals, perform a simple computation, and then transmit an output signal to other neurons.

The fundamental building block of an ANN is a single neuron. This neuron receives multiple input signals, each multiplied by a corresponding "weight." These weighted inputs are then summed together, and a "bias" term is added. This sum represents the neuron's activation level before it is passed through an "activation function." The activation function introduces non-linearity into the network, which is crucial for learning complex patterns. Without non-linearity, a multi-layered neural network would essentially behave like a single-layer network, severely limiting its representational power. Common activation functions include the Rectified Linear Unit (ReLU), sigmoid, and tanh functions, each offering different properties that can influence the learning process.

Neural networks are typically organized into layers: an

input layer, one or more hidden layers, and an output layer. The input layer receives the raw data from the robot's sensors—for example, pixel values from an image, readings from a LiDAR sensor, or joint encoder values from a robotic arm. Each neuron in the input layer corresponds to a feature in the input data. The hidden layers are where the bulk of the computation and feature extraction occurs. Neurons in a hidden layer receive inputs from the neurons in the preceding layer, apply weights and biases, and pass the result through an activation function. The output layer produces the final result of the network, such as a classification label (e.g., "object A is a cup"), a probability distribution over possible classes, or a set of continuous values (e.g., coordinates for object detection).

The power of deep neural networks stems from their depth—the presence of multiple hidden layers. Each layer in a deep network learns to represent the data at a different level of abstraction. For instance, in a neural network designed for image recognition, the first few layers might learn to detect simple features like edges, corners, and color blobs. Subsequent layers build upon these basic features to recognize more complex patterns, such as textures, shapes, and object parts. The deepest layers can then combine these learned representations to identify entire objects. This hierarchical learning process allows deep networks to automatically discover intricate patterns and structures in data that would be extremely difficult to define manually.

The process of training a neural network, that is, teaching it to perform a specific task, involves adjusting the weights and biases of its neurons. The most common and effective training algorithm for deep neural networks is called **backpropagation**. Backpropagation is an iterative process that leverages gradient descent optimization. It works in two main phases: a forward pass and a backward pass.

In the **forward pass**, an input sample is fed through the network, layer by layer, until an output is produced at the output layer. This output is then compared to the desired or "ground truth" output for that input. The difference between the predicted output and the ground truth is quantified by a **loss function**, which essentially measures the error or "cost" of the network's prediction.

In the **backward pass**, the error calculated by the loss function is propagated backward through the network. Using the chain rule of calculus, backpropagation computes the gradient of the loss function with respect to each weight and bias in the network. This gradient indicates the direction and magnitude of change needed for each parameter to reduce the overall error. Once these gradients are computed, an optimization algorithm, such as stochastic gradient descent (SGD) or its variants like Adam or RMSprop, uses these gradients to update the weights and biases. This update aims to nudge the parameters in a direction that minimizes the loss function. This entire cycle—forward pass, loss calculation, backward pass, and parameter update—is repeated for many input samples, often grouped into "batches," and for multiple passes over the entire training dataset (known as "epochs"). Through this iterative process, the neural network gradually learns to make more accurate predictions by minimizing the loss.

The effectiveness of backpropagation is heavily reliant on the choice of activation functions and the loss function. Differentiable activation functions are essential for the chain rule to be applied during the backward pass. The loss function must also be differentiable with respect to the network's outputs. For classification tasks, common loss functions include cross-entropy, which is particularly well-suited for probability distributions. For regression tasks, mean squared error (MSE) is frequently used.

While the core principles of ANNs and backpropagation are universal, different architectures have been developed to excel at specific types of data and tasks. For robotics, two architectures have proven particularly influential: Convolutional Neural Networks (CNNs) and Recurrent Neural Networks (RNNs).

Convolutional Neural Networks (CNNs) are specifically designed to process data with a grid-like topology, such as images. Their architecture is inspired by the visual cortex of animals, and they are exceptionally good at detecting spatial hierarchies of features. A key component of CNNs is the **convolutional layer**. Instead of fully connecting every neuron to every neuron in the previous layer, convolutional layers use small filters (also called kernels) that slide over the input data (e.g., an image). Each filter is designed to detect a specific pattern, such as a horizontal edge, a vertical edge, or a specific color gradient. As the filter slides, it performs a dot product with the input region it covers, producing a "feature map" that highlights where the specific pattern was detected in the input.

Multiple filters are used in a convolutional layer, each learning to detect a different pattern. This allows the layer to capture a rich set of low-level features. Following the convolutional layers, **pooling layers** are often employed. Pooling layers, such as max pooling or average pooling, reduce the spatial dimensions (width and height) of the feature maps while retaining the most important information. This helps to make the network more robust to small variations in the position of features and also reduces computational complexity.

As data passes through successive convolutional and pooling layers in a CNN, the network learns increasingly complex and abstract features. The early layers might detect edges, while deeper layers can combine these to recognize shapes, textures, and eventually parts of objects. The final layers of a CNN are typically fully connected layers, similar to those in a standard feedforward neural network, which take the high-level features learned by the convolutional layers and use them for tasks like classification or object detection.

For robotics, CNNs are foundational for visual perception tasks. A robot equipped with cameras can use a CNN to:

Object Recognition and Classification: Identify and categorize objects in its field of view (e.g., distinguishing between a tool, a person, or a wall).
Semantic Segmentation: Assign a class label to every pixel in an image, allowing the robot to understand which regions correspond to drivable surfaces, obstacles, or sky. This is crucial for autonomous navigation.
Pose Estimation: Determine the position and orientation of objects in 3D space based on 2D image data.
Scene Understanding: Comprehend the overall context of an environment, which aids in planning and navigation.

Consider a robotic arm tasked with picking up specific items from a cluttered bin. A CNN can process camera images to accurately locate and identify the target item, even if it's partially obscured or viewed from an unusual angle. This visual understanding is a prerequisite for the robot to then plan and execute the grasping maneuver.

Recurrent Neural Networks (RNNs), on the other hand, are designed to handle sequential data, where the order of information matters. Unlike feedforward networks where information flows in one direction, RNNs have connections that loop back on themselves, allowing them to maintain an internal "memory" of previous inputs. This "memory" is essentially a hidden state that is updated at each time step based on the current input and the previous hidden state.

The recurrent nature of RNNs makes them suitable for tasks involving time-series data, such as:

Speech Recognition: Processing sequences of audio signals to transcribe spoken words.
Natural Language Processing (NLP): Understanding and generating human language, where the meaning of words depends heavily on their context within a sentence or document.
Time-Series Prediction: Forecasting future values in sequences like sensor readings over time.

In robotics, RNNs are particularly useful for tasks that involve understanding temporal dynamics or making decisions over time. For example:

Robot Trajectory Prediction: Predicting the future path of dynamic objects (e.g., a person walking) based on their past movements. This is vital for collision avoidance and safe navigation.

Sequential Task Execution: Learning to perform a series of actions in a specific order, where the decision at each step depends on what has happened previously.
Sensor Fusion over Time: Integrating data from multiple sensors over extended periods, where the temporal correlation between sensor readings is important for a comprehensive understanding of the environment.
Human-Robot Interaction: Understanding and responding to human commands or intentions expressed through speech or gesture sequences.

A common limitation of simple RNNs is their difficulty in learning long-term dependencies; they tend to "forget" information from many time steps ago. To address this, more advanced architectures like **Long Short-Term Memory (LSTM)** and **Gated Recurrent Unit (GRU)** networks have been developed. LSTMs and GRUs incorporate sophisticated gating mechanisms that allow them to selectively remember or forget information, making them much more effective at capturing dependencies over longer sequences.

For a robot navigating a busy environment, an RNN (or LSTM/GRU) could process a sequence of LiDAR scans and camera frames to build a dynamic map of moving obstacles and predict their future trajectories. This temporal awareness allows the robot to plan its movements more effectively and avoid collisions. Similarly, in human-robot collaboration, an RNN could process a sequence of a human's gestures and speech to infer their overall intent for a collaborative task.

The development of these deep learning architectures, particularly CNNs and RNNs, has significantly boosted the capabilities of robots in perception and understanding. They have moved us closer to robots that can not only sense their environment but also interpret it in a meaningful way, paving the path for more intelligent and autonomous systems. The ability of these networks to learn directly from raw data, without the need for laborious manual feature engineering, has democratized advanced perception capabilities and opened up a vast array of new possibilities for robotic applications. This transition from handcrafted features to learned representations is a hallmark of the deep learning revolution and a critical foundation for the future of robotics.

The remarkable advancements in robotic perception and decision-making over the past decade are inextricably linked to the rise of Deep Learning (DL). While traditional Machine Learning techniques provide a powerful framework for robots to learn from data, Deep Learning represents a significant evolution within this field, particularly in its ability to automatically learn complex representations directly from raw sensory input. This capability has been a game-changer, especially for tasks involving perception, where the sheer variability and richness of real-world data pose immense challenges for handcrafted feature extraction.

At the heart of Deep Learning lies the concept of Artificial Neural Networks (ANNs), often simply referred to as neural networks. These are computational models inspired by the structure and function of the biological neural networks found in the human brain. A neural network is composed of interconnected processing units called "neurons" or "nodes," organized in layers. These neurons receive input signals, perform a simple computation, and then transmit an output signal to other neurons.

The fundamental building block of an ANN is a single neuron. This neuron receives multiple input signals, each multiplied by a corresponding "weight." These weighted inputs are then summed together, and a "bias" term is added. This sum represents the neuron's activation level before it is passed through an "activation function." The activation function introduces non-linearity into the network, which is crucial for learning complex patterns. Without non-linearity, a multi-layered neural network would essentially behave like a single-layer network, severely limiting its representational power. Common activation functions include the Rectified Linear Unit (ReLU), sigmoid, and tanh functions, each offering different properties that can influence the learning process.

Neural networks are typically organized into layers: an input layer, one or more hidden layers, and an output layer. The input layer receives the raw data from the robot's sensors—for example, pixel values from an image, readings from a LiDAR sensor, or joint encoder values from a robotic arm. Each neuron in the input layer corresponds to a feature in the input data. The hidden layers are where the bulk of the computation and feature extraction occurs. Neurons in a hidden layer receive inputs from the neurons in the preceding layer, apply weights and biases, and pass the result through an activation function. The output layer produces the final result of the network, such as a classification label (e.g., "object A is a cup"), a probability distribution over possible classes, or a set of continuous values (e.g., coordinates for object detection).

The power of deep neural networks stems from their depth—the presence of multiple hidden layers. Each layer in a deep network learns to represent the data at a different level of abstraction. For instance, in a neural network designed for image recognition, the first few layers might learn to detect simple features like edges, corners, and color blobs. Subsequent layers build upon these basic features to recognize more complex patterns, such as textures, shapes, and object parts. The deepest layers can then combine these learned representations to identify entire objects. This hierarchical learning process allows deep networks to automatically discover intricate patterns and structures in data that would be extremely difficult to define manually.

The process of training a neural network, that is, teaching it to perform a specific task, involves adjusting the weights and biases of its neurons. The most common and effective training algorithm for deep neural networks is called

backpropagation. Backpropagation is an iterative process that leverages gradient descent optimization. It works in two main phases: a forward pass and a backward pass.

In the
forward pass, an input sample is fed through the network, layer by layer, until an output is produced at the output layer. This output is then compared to the desired or "ground truth" output for that input. The difference between the predicted output and the ground truth is quantified by a **loss function**, which essentially measures the error or "cost" of the network's prediction.

In the
backward pass, the error calculated by the loss function is propagated backward through the network. Using the chain rule of calculus, backpropagation computes the gradient of the loss function with respect to each weight and bias in the network. This gradient indicates the direction and magnitude of change needed for each parameter to reduce the overall error. Once these gradients are computed, an optimization algorithm, such as stochastic gradient descent (SGD) or its variants like Adam or RMSprop, uses these gradients to update the weights and biases. This update aims to nudge the parameters in a direction that minimizes the loss function. This entire cycle—forward pass, loss calculation, backward pass, and parameter update—is repeated for many input samples, often grouped into "batches," and for multiple passes over the entire training dataset (known as "epochs"). Through this

iterative process, the neural network gradually learns to make more accurate predictions by minimizing the loss.

The effectiveness of backpropagation is heavily reliant on the choice of activation functions and the loss function. Differentiable activation functions are essential for the chain rule to be applied during the backward pass. The loss function must also be differentiable with respect to the network's outputs. For classification tasks, common loss functions include cross-entropy, which is particularly well-suited for probability distributions. For regression tasks, mean squared error (MSE) is frequently used.

While the core principles of ANNs and backpropagation are universal, different architectures have been developed to excel at specific types of data and tasks. For robotics, two architectures have proven particularly influential: Convolutional Neural Networks (CNNs) and Recurrent Neural Networks (RNNs).

Convolutional Neural Networks (CNNs) for Vision

Convolutional Neural Networks (CNNs) represent a profound breakthrough in the field of computer vision and, by extension, robotic perception. Their architecture is meticulously designed to process data with a grid-like topology, with images being the most prominent example. The inspiration for CNNs can be traced back to the biological visual cortex of animals, which exhibits a hierarchical processing of visual information. This neurobiological parallel is key to understanding why CNNs are so exceptionally adept at detecting and learning spatial hierarchies of features within images. Unlike traditional neural networks that treat input data as a flat vector, CNNs leverage the spatial structure of images, processing local regions and building up an understanding of the image content progressively.

The cornerstone of a CNN's architecture is the **convolutional layer**. This layer deviates significantly from the fully connected layers found in standard feedforward neural networks. Instead of connecting every neuron in a layer to every neuron in the previous layer, convolutional layers employ small, learnable filters, often referred to as kernels. These filters are designed to slide, or "convolve," across the input data, such as an image. Each filter is trained to detect a specific, localized pattern within the input. For instance, one filter might specialize in identifying horizontal edges, another in vertical edges, and yet another might be sensitive to a particular color gradient or texture. As a filter moves across the image, it performs a dot product with the small patch of the input it currently covers. The result of this operation is a new matrix called a "feature map." This feature map essentially highlights the locations in the original input where the specific pattern detected by the filter was present.

A single convolutional layer typically employs multiple filters, each learning to recognize a different visual feature. This ensemble of filters allows the layer to extract a diverse set of low-level features from the input. For example, in the initial layers of a CNN processing images, these filters might learn to detect simple geometric primitives like lines, curves, corners, and basic color patterns. As the data progresses through deeper convolutional layers, the filters in these layers operate on the feature maps generated by the preceding layers. This means that deeper filters are not looking at the raw image pixels but at combinations of the simpler features detected earlier. Consequently, deeper convolutional layers learn to recognize more complex and abstract patterns, such as textures (e.g., the weave of fabric, the roughness of bark), recurring shapes (e.g., circles, squares), and even parts of objects (e.g., an eye, a wheel, a handle). This hierarchical learning process, where simple features are combined to form progressively more complex representations, is the essence of how CNNs achieve their remarkable visual understanding capabilities.

Following the convolutional layers, **pooling layers** are frequently integrated into CNN architectures. The primary role of pooling layers is to progressively reduce the spatial dimensions (width and height) of the feature maps. This reduction serves two critical purposes. Firstly, it helps to make the network more robust to small variations in the position of features within the image. For example, if a specific edge feature shifts slightly in the image, a pooling operation ensures that the overall activation in that region remains significant, thereby preventing the network from being overly sensitive to precise pixel locations. Secondly, pooling significantly reduces the computational complexity of the network. By downsampling the feature maps, fewer parameters and computations are required in subsequent layers, leading to faster training and inference times, and also mitigating the risk of overfitting. Common pooling operations include **max pooling**, which takes the maximum value within a local window of the feature map, and **average pooling**, which computes the average value within that window. Max pooling is often preferred as it tends to preserve the most salient features.

After a series of convolutional and pooling layers have extracted and refined hierarchical features, the output typically consists of high-dimensional feature maps that encapsulate abstract representations of the image content. These feature maps are then usually fed into one or more
fully connected layers, which are the traditional layers found in standard feedforward neural networks. In this context, the fully connected layers act as a classifier or regressor. They take the learned high-level features as input and perform the final task, such as assigning a class label to the entire image (e.g., "cat," "dog," "car") or predicting bounding box coordinates for object detection. The dense connections in these final layers allow for a global combination of the detected features to make the ultimate decision.

The effectiveness of CNNs in visual perception tasks for robotics is vast and transformative. A robot equipped with cameras can leverage CNNs to perform a wide array of critical functions:

Object Recognition and Classification: This is perhaps the most well-known application. A CNN can be trained to identify and categorize objects present in the robot's field of view. For a robot operating in an industrial setting, this could mean distinguishing between different types of tools, components, or finished products. In a domestic environment, it could be recognizing household items like cups, plates, books, or furniture. This capability is fundamental for robots to interact meaningfully with their surroundings, for example, to grasp a specific object or to identify potential hazards. The ability to differentiate between a "mug" and a "glass," even if they share similar shapes, relies on the CNN's learned ability to discern subtle textural or pattern differences.

Image Classification: While related to object recognition, image classification typically refers to assigning a single label to an entire image. For a mobile robot navigating an outdoor environment, this could mean classifying an image as "road," "sidewalk," "building," or "vegetation." This high-level understanding of the scene is crucial for navigation and task planning. For instance, a robot might need to classify an area as "walkable" before attempting to traverse it.

Semantic Segmentation: This is a more granular form of image understanding. Instead of just identifying objects or classifying the whole scene, semantic segmentation assigns a class label to every single pixel in an image. This allows the robot to create a detailed map of its environment, understanding not just that there is an object, but which specific pixels belong to that object, and what its category is. For example, in autonomous driving, semantic segmentation is vital for distinguishing drivable areas (road surface) from non-drivable areas (curbs, buildings, other vehicles, pedestrians). A robot arm in a warehouse might use semantic segmentation to identify the precise boundaries of a pallet to avoid collisions while maneuvering. For a service robot in a home, it could delineate "floor," "wall," "table," and "chair," enabling it to navigate without bumping into furniture or getting lost.

Object Detection: Going a step further than classification, object detection not only identifies what objects are present but also precisely locates them within the image by drawing bounding boxes around them. This is essential for tasks where the robot needs to interact with specific instances of objects. For example, a warehouse robot might need to detect individual boxes on a shelf, even if multiple boxes are close together. A medical robot assisting in surgery might need to detect the precise location of a tumor or a vital organ. Object detection models often output a list of detected objects, each with a class label and the coordinates of its bounding box. This information is directly usable for path planning and manipulation.

Pose Estimation: This task involves determining the 3D position and orientation (pose) of an object relative to the robot or a global coordinate frame. For a robot tasked with picking up an object, knowing its exact pose in 3D space is paramount for planning the trajectory of its manipulator. CNNs can be trained to infer this 3D pose from 2D image data, often by predicting keypoints on the object or by directly estimating the transformation parameters. This is critical for tasks like industrial assembly, where components must be fitted together with high precision, or for a robot that needs to insert a plug into a socket.

Scene Understanding: This is a broader concept that encompasses the interpretation of the overall context and relationships within a visual scene. A CNN can help a robot infer not just individual objects but also how they relate to each other and to the environment. For instance, understanding that a "cup" is "on" a "table" provides richer contextual information than just knowing that both objects exist. This deeper understanding aids in more sophisticated planning, such as a robot deciding to clear items from a table before setting it for a meal, or understanding that a person is "facing" a specific direction, which might indicate their intention.

Consider a practical example of a robotic arm tasked with picking specific items from a cluttered bin. A traditional approach would require meticulously designing algorithms to identify edges, corners, and shapes, then trying to assemble these into a model of the target object. This is brittle; a slight change in lighting, object orientation, or the presence of occlusions could easily break the system. With a CNN, the robot can feed images from its camera directly into the network. The CNN, through its learned hierarchy of features, can robustly identify the target item amidst the clutter, even if it is partially hidden or at an unusual angle. It can simultaneously provide the object's location and potentially its orientation, enabling the robot to plan a precise grasping motion.

Another scenario involves a mobile robot navigating a dynamic environment, such as a factory floor with moving forklifts and pedestrians. While a simple occupancy grid might detect obstacles, a CNN processing camera feeds can go further. It can classify detected objects as "person," "forklift," or "cart," and even predict their future motion based on their current trajectory. This allows for more intelligent path planning, such as anticipating a pedestrian's movement to avoid a potential collision rather than just reacting to an obstacle that has already entered its path. This temporal aspect, combined with sophisticated spatial perception, is what elevates CNNs to a powerful tool for robotic autonomy.

The development and widespread availability of large datasets, such as ImageNet, coupled with significant advancements in computational power (especially GPUs), have been instrumental in the success of CNNs. These networks, trained on millions of images, learn incredibly rich and generalizable visual representations. Transfer learning, a technique where a CNN pre-trained on a large dataset is fine-tuned for a specific robotics task, has further accelerated the deployment of CNNs, as it significantly reduces the amount of data and training time required for a new application.

In summary, Convolutional Neural Networks have revolutionized robotic vision by providing an end-to-end solution for learning complex visual features directly from raw pixel data. Their layered architecture, incorporating convolutional, pooling, and fully connected layers, allows for the hierarchical extraction of information, progressing from simple edges to intricate object recognition. This capability has unlocked a new era of perception for robots, enabling them to perform sophisticated tasks like object detection, semantic segmentation, and pose estimation with unprecedented accuracy and robustness, paving the way for more intelligent, autonomous, and capable robotic systems across diverse applications.

The ability of a robot to perceive and interact with its environment hinges critically on how it represents the incoming sensory information and, consequently, how it extracts meaningful characteristics - or *features* - from this data. This process is foundational to both traditional machine learning paradigms and the more recent, powerful deep learning architectures that have reshaped robotic capabilities. Understanding data representation and feature engineering is not merely an academic exercise; it is the crucial bridge that translates raw sensor streams into actionable knowledge for intelligent robotic systems.

Traditional Machine Learning: The Art of Feature Engineering

Before the advent of deep learning's automatic feature extraction, the efficacy of traditional machine learning algorithms in robotics was heavily dependent on the manual craft of *feature engineering*. This involved domain experts and researchers meticulously designing algorithms to identify and extract salient information from sensor data. The goal was to transform raw, often high-dimensional, sensor readings into a lower-dimensional, more informative set of features that a machine learning model could readily process and learn from. The quality and relevance of these engineered features directly dictated the performance of the subsequent classification, regression, or clustering tasks.

For visual perception, particularly from camera inputs, this was a rich area of development. **Scale-Invariant Feature Transform (SIFT)**, for instance, became a cornerstone for robust feature detection and description. SIFT features are designed to be invariant to image scale, rotation, and illumination changes, making them ideal for tasks like object recognition, image stitching, and localization. The SIFT algorithm first identifies keypoints in an image that are stable across different scales and orientations. These keypoints are typically located at areas with significant local variations in intensity, such as corners or distinctive texture regions. Around each keypoint, a descriptor is computed by analyzing the gradient orientations within a local neighborhood. This descriptor is a histogram of gradient orientations, smoothed and normalized, which captures the local texture and structure around the keypoint. The invariance properties are achieved through careful selection of sampling regions and orientation assignments. For example, by assigning a dominant orientation to each keypoint based on the local image gradients, the descriptors become rotationally invariant.

Another prominent technique for image feature extraction in traditional machine learning was the **Histogram of Oriented Gradients (HOG)**. HOG features represent the distribution of gradient orientations within localized regions of an image. The process involves computing gradients (both magnitude and direction) for each pixel, then dividing the image into small cells and computing a histogram of gradient orientations for each cell. These cell histograms are then normalized over larger blocks to achieve greater robustness to illumination and shadowing variations. The resulting feature vector, a concatenation of these block-normalized histograms, can effectively characterize the shape and appearance of objects, particularly for pedestrian detection and general object recognition. The success of HOG lies in its ability to capture local shape information without requiring explicit knowledge of the object's geometry.

Beyond images, robotic perception often involves processing data from sensors like LiDAR, which generate dense point clouds representing the 3D structure of the environment. Extracting meaningful features from point clouds presented a unique set of challenges due to their unordered nature and the inherent sparsity or density variations. Traditional approaches often involved geometric feature extraction. For example,
Point Feature Histograms (PFH) and its more robust successor, **Fast Point Feature Histograms (FP FH)**, were developed to describe the local geometric properties around a point. These methods compute histograms of surface normals and point relative positions within a local neighborhood. By analyzing the distribution of these geometric properties, one can characterize the shape of the surface at that point, allowing for the identification of corners, planes, or curved surfaces.

Statistical features were also heavily utilized. From

point cloud data, one could compute features like the local point density, the variance of point positions within a neighborhood, or measures of surface curvature estimated from local point distributions. For object recognition from 3D data, features like the centroid, bounding box dimensions, and moments of inertia of a segmented object were common. These low-level geometric and statistical descriptors provided a compact and informative representation that could be fed into classifiers like Support Vector Machines (SVMs) or Random Forests.

The manual nature of feature engineering meant that it was often highly domain-specific and required significant effort to adapt to new sensors or tasks. For instance, features optimized for detecting rigid objects in a structured environment might perform poorly when dealing with deformable objects or in highly dynamic scenarios. Furthermore, the performance ceiling of traditional methods was often limited by the ingenuity of the human designer; it was difficult to discover complex, high-level abstract features that might be implicitly present in the data but not easily described by simple geometric or statistical properties.

Deep Learning: The Era of Automatic Feature Learning

Deep learning architectures, as discussed previously, fundamentally shifted this paradigm by automating the

process of feature extraction. Instead of relying on human experts to handcraft features, deep neural networks, particularly Convolutional Neural Networks (CNNs) for vision and point cloud processing, learn these features directly from the raw sensor data during the training process. This capability is often referred to as
representation learning.

The core idea is that through the layered structure of a deep network, progressively more abstract and useful features are learned. In the context of images, the early layers of a CNN learn low-level features such as edges, corners, and color blobs, much like early visual processing in biological systems. As data flows through deeper layers, these basic features are combined to form more complex representations: textures, simple shapes, then parts of objects (e.g., a wheel, an eye), and eventually, entire objects. The network learns which features are important for a given task through the backpropagation algorithm, which adjusts the network's parameters (weights and biases) to minimize a defined loss function.

This end-to-end learning approach has several profound advantages for robotics. Firstly, it dramatically reduces the need for manual feature engineering, saving significant development time and expertise. Secondly, deep networks can discover complex, non-linear relationships and subtle patterns in the data that might be missed by human intuition. This allows them to achieve significantly higher performance on challenging perception tasks, especially those involving high variability in appearance, lighting, and pose.

The choice of data representation remains crucial even in deep learning, though the focus shifts from designing specific feature extractors to preparing data in a format that the network can efficiently process.

Data Representation for Different Sensor Modalities

The type of sensor dictates the most appropriate data representation for input into a machine learning model, whether traditional or deep.

Images: As discussed extensively, images are typically represented as multi-dimensional arrays (tensors) of pixel values. For a grayscale image, this is a 2D array (height x width), where each element is a single intensity value. For a color image (e.g., RGB), it becomes a 3D array (height x width x channels), where each pixel has three values corresponding to the red, green, and blue intensity components. CNNs are inherently well-suited for this grid-like data structure. The convolutional operations naturally exploit the spatial locality and hierarchical structure present in images. Even for tasks like object detection or segmentation, which require localization information, the image is often processed directly, with the network learning to output coordinates or masks.

Point Clouds: Data from LiDAR, depth cameras, or stereo vision systems is often represented as a collection of 3D points, where each point has coordinates (x, y, z). Additionally, each point might carry other attributes such as intensity, color (if from a depth camera with RGB), or surface normal information. Representing point clouds for deep learning models requires special consideration due to their unordered nature. Unlike images, where the order of pixels has inherent spatial meaning, the order of points in a raw point cloud is arbitrary.

Several deep learning approaches have been developed to handle this:

Voxelization: The 3D space occupied by the point cloud can be discretized into a grid of voxels (3D pixels). Each voxel can then be represented as a feature vector indicating whether it contains points and what their properties are (e.g., average color, density). This voxel representation can then be processed by 3D CNNs, which are analogous to 2D CNNs but operate on volumetric data. The primary trade-off here is the resolution; higher resolution requires more memory and computation.

Point-based Networks: Architectures like PointNet and PointNet++ directly process the raw point cloud without explicit discretization. PointNet operates by learning a permutation-invariant function that maps each point to a global feature vector. It uses shared multi-layer perceptrons (MLPs) to process each point independently and then employs a symmetric function (like max pooling) to aggregate these features into a global representation, ensuring permutation invariance. PointNet++ builds upon this by introducing hierarchical feature learning, capturing local structures at different scales. These networks treat each point as an independent data sample, applying MLPs to extract features from individual points and then aggregating them.

Graph-based Networks: Point clouds can be naturally represented as graphs, where points are nodes and edges connect neighboring points. Graph Neural Networks (GNNs) can then be applied to process this graph structure, learning features that leverage the connectivity and local geometry between points.

Projection-based Methods: Some methods project the 3D point cloud onto 2D planes (e.g., spherical projection, cylindrical projection, or multiple orthographic views) to create image-like representations. These projections can then be processed by standard 2D CNNs. While simpler to implement and leverage existing CNN architectures, these methods can suffer from information loss and distortion due to the projection process.

Other Sensor Modalities: Robotic systems often incorporate a variety of other sensors, such as:

IMU (Inertial Measurement Unit): Provides acceleration and angular velocity data. This is typically represented as time-series data, often in the form of vectors or matrices representing sensor readings over time. Recurrent Neural Networks (RNNs) or Transformers are well-suited for processing such sequential data.

Joint Encoders: Measure the angles or positions of robot joints. This is also time-series data, usually represented as vectors of joint values at each time step.

Force/Torque Sensors: Measure forces and torques at robot joints or end-effectors. These are typically represented as time-series vectors.

GPS/Odometry: Provide positional and velocity information, often as coordinate values or velocity vectors over time.

For these modalities, the data is often represented as

sequences of vectors. Feature engineering might involve calculating moving averages, differences between consecutive readings, or statistical summaries (mean, variance) over time windows. In deep learning, RNNs (like LSTMs and GRUs) and Transformer networks are commonly used to learn temporal dependencies and extract features from these time-series sensor streams.

The Synergy of Data Representation and Feature Learning

The power of deep learning in robotics lies in its ability to learn optimal representations directly from raw data. This means that the network can, in principle, discover the most discriminative features for a given task, without human intervention. However, the initial data representation still plays a crucial role in facilitating this learning process.

For instance, if a robot needs to perform object recognition from a depth camera, representing the data as a raw point cloud might allow a point-based network to learn powerful 3D geometric features. Alternatively, if the object's texture is also important, combining the depth data with color information from an RGB camera and processing it as a multi-channel image through a CNN might be more effective. If the task involves understanding the motion of objects over time, fusing the point cloud or image data with IMU readings and processing the combined sequence with a recurrent or attention-based model could yield superior results.

The process of designing a robotic perception system often involves an iterative interplay between choosing the appropriate sensor modalities, deciding on the best data representation for those modalities, and selecting a suitable machine learning architecture (often deep learning-based) to learn features from that representation. For example, a robot designed for navigating complex, cluttered indoor environments might utilize LiDAR for precise 3D mapping and obstacle avoidance, and cameras for object recognition and semantic understanding. The LiDAR data might be represented as voxels or directly processed by point-based networks, while camera images would be fed into CNNs. Features learned from each modality could then be fused at a later stage in the network or as post-processing steps to achieve a comprehensive understanding of the robot's surroundings.

The trend in modern robotics is towards increasingly sophisticated end-to-end learning systems where the entire pipeline, from raw sensor input to final decision or action, is handled by a single, large neural network or a modular system of interconnected networks. In such systems, the concept of "feature engineering" as it was traditionally understood becomes largely obsolete, replaced by the network's inherent ability to learn hierarchical representations. This has opened up new frontiers for robotic autonomy, enabling robots to tackle tasks that were previously intractable due to the complexity of manual feature design, and paving the way for more adaptable, robust, and intelligent robotic systems in diverse and challenging environments. The ongoing research in this area focuses on developing more efficient and powerful network architectures, exploring novel data representations, and improving the interpretability of the learned features.

Chapter 5: Advanced Perception Techniques

Building upon the foundation of learned representations, the subsequent challenge for a robot is to not only understand the general characteristics of its environment but also to identify and classify specific objects within it. This process, broadly termed *object detection and recognition*, is a cornerstone of advanced robotic perception, enabling a robot to interact meaningfully with the world. While early computer vision systems relied on hand-crafted features and template matching, modern robotics predominantly leverages sophisticated deep learning architectures that excel at this task.

The previous discussion highlighted how Convolutional Neural Networks (CNNs) learn hierarchical features from raw sensor data. This capability forms the bedrock for modern object detection systems. Instead of simply classifying an entire image, object detection aims to not only identify
what objects are present but also *where* they are located within the scene, typically by drawing bounding boxes around them. This localization aspect is critical for robots, as it informs their actions – a robot needs to know *where* to reach for an object, or *where* an obstacle is to navigate around it.

A significant leap in object detection performance came with the development of end-to-end trainable deep learning frameworks. These models can process an entire image and output both the class of detected objects and their spatial coordinates simultaneously, often eliminating the need for separate feature extraction and candidate region proposal stages that characterized earlier methods.

One of the most influential families of object detection algorithms is based on the concept of *region proposals*. These methods first identify potential regions in an image that are likely to contain objects and then classify the objects within these proposed regions.

Faster R-CNN (Region-based Convolutional Neural Network) exemplifies this approach. It significantly improves upon its predecessors, R-CNN and Fast R-CNN, by integrating the region proposal generation into the neural network itself. The process begins with a backbone CNN (e.g., VGG, ResNet) that extracts a rich feature map from the input image. Overlaid onto this feature map is a **Region Proposal Network (RPN)**. The RPN slides a small network over the feature map and, at each location, predicts the probability of an object being present and refines the coordinates of potential bounding boxes. These proposed regions, along with their associated scores, are then passed to a second stage. This second stage also utilizes the same backbone CNN features and, for each proposed region, performs classification (identifying the object class) and further refines the bounding box coordinates. The key innovation of Faster R-CNN is that the RPN is trained jointly with the detection network, making the entire system end-to-end trainable and significantly faster than previous region-proposal-based methods. This allows robots to achieve high accuracy in identifying objects across a wide range of categories.

While region-proposal-based methods like Faster R-CNN offer high accuracy, they can sometimes be computationally intensive, which is a critical consideration for real-time robotic applications. To address this, a new class of object detectors emerged: **single-shot detectors**. These models perform localization and classification in a single forward pass of the network, drastically reducing computation time.

You Only Look Once (YOLO) is a prime example of a single-shot detector that has become extremely popular in robotics due to its speed. The YOLO architecture divides the input image into a grid. Each grid cell is responsible for predicting bounding boxes for objects whose center falls within that cell, as well as the class probabilities for those objects. Crucially, each grid cell predicts a fixed number of bounding boxes and associated confidence scores, which reflect the likelihood of a box containing an object and the accuracy of the box. The network then refines these predictions and outputs a list of detected objects. YOLO's strength lies in its unified architecture, which processes the entire image at once, leading to exceptionally fast inference times suitable for real-time perception tasks. Subsequent versions of YOLO (e.g., YOLOv3, YOLOv4, YOLOv5, YOLOv7, YOLOv8) have introduced further improvements in accuracy and speed through architectural enhancements, new loss functions, and better training strategies.

Another prominent single-shot detector is **Single Shot MultiBox Detector (SSD)**. Similar to YOLO, SSD performs detection in a single pass. However, SSD employs a different strategy for feature extraction and prediction. It uses a base network (like VGG) and adds several convolutional layers to create a multi-scale feature hierarchy. Object detection is then performed at multiple feature maps at different resolutions. This allows SSD to detect objects of various sizes more effectively. For smaller objects, predictions are made on higher-resolution feature maps, while larger objects are detected on lower-resolution maps. SSD also uses a set of default bounding boxes (similar to anchor boxes in Faster R-CNN) at each feature map location, which are then adjusted to better match the object's shape and location. The speed of SSD is generally comparable to YOLO, making it another strong contender for real-time robotic perception.

The choice between these different object detection frameworks often involves a trade-off between accuracy and speed. For applications requiring the utmost precision, Faster R-CNN-like architectures might be preferred. For tasks where real-time performance is paramount, such as autonomous navigation or robotic manipulation in dynamic environments, YOLO or SSD are often the go-to choices. Researchers have also developed numerous hybrid approaches and specialized models tailored for specific robotic contexts, such as detecting small objects in aerial imagery or recognizing industrial parts on an assembly line.

Beyond simply drawing bounding boxes, many robotic tasks benefit from a more granular understanding of object boundaries. This is where **instance segmentation** comes into play. While object detection identifies the presence and location of objects with bounding boxes, and semantic segmentation assigns a class label to each pixel in an image (e.g., all pixels belonging to "car" are labeled as such, regardless of individual car instances), instance segmentation goes a step further. It aims to identify and delineate *each individual instance* of an object, even if they are of the same class. For a robot, this means not only knowing that there are three chairs in a room but also being able to distinguish between chair A, chair B, and chair C, and precisely outlining their contours.

A pioneering instance segmentation framework is **Mask R-CNN**. Building upon Faster R-CNN, Mask R-CNN extends its architecture to predict an object mask in parallel with the existing bounding box and class predictions. For each proposed region, in addition to predicting the object class and refining the bounding box, a small convolutional network generates a binary mask that precisely outlines the object. This mask is then refined and resized to match the detected object. Mask R-CNN achieves high accuracy in both object detection and instance segmentation by effectively combining object localization with pixel-level prediction. This level of detail is invaluable for complex robotic manipulation tasks. For example, when a robot arm needs to grasp a specific object from a pile of similar objects, precise segmentation of each object is essential to avoid collisions and ensure a successful grasp.

Other notable instance segmentation methods include **YOLACT (You Only Look At CoefficienTs)** and **SOLO (Segmenting Objects by Locations)**. YOLACT, for instance, is designed for real-time instance segmentation by generating a set of prototype masks and predicting per-instance coefficients to combine these prototypes, effectively creating instance masks at high speed. SOLO, on the other hand, segments objects based on their location, directly predicting instance masks without relying on object proposals or anchor boxes, offering a simpler and often faster approach.

The ability of instance segmentation to provide pixel-level masks offers several advantages for robotics. In navigation, precise object boundaries can lead to more informed path planning, allowing robots to navigate closer to objects without collision. In manipulation, it enables robots to grasp objects with greater precision, avoiding interference with adjacent items. For human-robot interaction, understanding the exact shape and extent of objects can facilitate more natural and intuitive interactions. For instance, a robot could understand that a human is pointing to a specific cup, not just the general area where cups are located.

A critical aspect of object detection and recognition in robotics, regardless of the chosen framework, is the ability to perform reliably under varying environmental conditions. Robots often operate in dynamic and unpredictable settings, where objects might be:

Occluded: Partially or fully hidden by other objects.
Illuminated differently: Subject to shadows, bright light, or low-light conditions.
Scaled variably: Appearing at different distances from the robot's sensors.
In different poses: Viewed from various angles.
Deformable: Not rigid, changing shape (e.g., cloth, soft robots).

Deep learning models, especially those trained on large and diverse datasets, have shown remarkable robustness to many of these challenges. Techniques like data augmentation during training are crucial. This involves artificially modifying the training data to simulate real-world variations. For example, randomly cropping images to simulate occlusion, adjusting brightness and contrast to mimic different lighting, or rotating and scaling images to represent different viewpoints and distances.

For handling occlusion, models like Faster R-CNN and Mask R-CNN, which predict bounding boxes or masks for each detected object, can often infer the presence of an occluded object based on visible parts. More advanced techniques involve part-based models that can recognize objects even when only a few key parts are visible, or using contextual information from the surrounding scene to infer the presence and identity of occluded objects.

Scale variation is typically handled by employing multi-scale feature extraction, as seen in SSD and some versions of YOLO, or by using feature pyramids, where image features are analyzed at multiple resolutions. This allows the detector to identify both small, distant objects and large, nearby objects effectively.

The ability to generalize to unseen poses and variations is also a product of training on diverse datasets that capture a wide range of object appearances. Furthermore, incorporating 3D information from sensors like LiDAR or depth cameras can significantly improve robustness to pose variations, as 3D shape is inherently more invariant to viewpoint changes than 2D appearance.

The robustness of object detection and recognition systems is directly tied to the quality and diversity of the data they are trained on. For robots operating in specific environments, such as a factory floor, a warehouse, or a domestic setting, fine-tuning general-purpose models on datasets collected from those specific environments can lead to substantial performance improvements. This is often referred to as *domain adaptation*.

The ongoing research in object detection and

recognition for robotics is pushing the boundaries of what is possible. This includes developing more efficient architectures for real-time deployment on resource-constrained robotic platforms, improving performance in challenging conditions like adverse weather or extreme lighting, and enabling robots to recognize and understand novel objects they have not encountered during training (few-shot or zero-shot learning). The ultimate goal is to equip robots with a perception system that is as comprehensive and adaptable as human vision, allowing them to navigate, interact with, and understand the complexities of the physical world with a high degree of autonomy and intelligence. This continuous evolution of object detection and recognition techniques is fundamental to unlocking the full potential of robotics in a wide array of applications, from industrial automation and logistics to healthcare and autonomous exploration.

Semantic segmentation represents a significant advancement beyond object detection by moving from bounding boxes to pixel-level classification. Instead of simply identifying the presence and location of objects, semantic segmentation aims to assign a class label to every single pixel in an image. This provides a much richer, more detailed understanding of the scene's composition. For instance, in an autonomous driving context, semantic segmentation can distinguish not only that a "car" is present but also delineate the exact pixels belonging to the road, the sky, buildings, other vehicles, and pedestrians. This granular understanding is crucial for tasks requiring precise navigation and interaction with the environment.

The core idea behind semantic segmentation is to take an input image and output a segmentation map of the same spatial dimensions, where each pixel is assigned a class label. This requires models that can capture both high-level semantic information (what is the object?) and low-level spatial information (where exactly is it located?). Early approaches often relied on traditional computer vision techniques combined with machine learning classifiers, but the advent of deep learning has revolutionized this field.

Convolutional Neural Networks (CNNs) have proven to be exceptionally well-suited for semantic segmentation. However, standard CNNs, which typically downsample the input image to extract features, lose spatial resolution, making them unsuitable for direct pixel-wise prediction. To address this, specialized CNN architectures have been developed. A seminal architecture in this regard is the **Fully Convolutional Network (FCN)**. FCNs replace the fully connected layers in traditional CNNs with convolutional layers, allowing them to accept arbitrary input image sizes and produce output maps of corresponding spatial resolution.

FCNs typically employ an encoder-decoder structure. The encoder part, often a pre-trained classification network like VGG or ResNet, progressively downsamples the input image, extracting increasingly abstract and semantic features. The decoder part then upsamples these features, gradually recovering the spatial resolution and producing a dense prediction map. To improve the accuracy of the upsampling process and preserve finer details, FCNs utilize techniques such as skip connections, which combine high-level semantic features from the encoder with low-level spatial features from earlier layers. This allows the decoder to leverage both the "what" and the "where" information more effectively. For example, when segmenting a road, the low-level features might highlight edges and textures indicative of asphalt, while the high-level features confirm that this region is indeed part of the drivable surface.

Building upon the foundation laid by FCNs, numerous other semantic segmentation architectures have emerged, each with its own innovations. One particularly influential architecture is **U-Net**. Originally developed for biomedical image segmentation, U-Net shares the encoder-decoder structure with FCNs but features a more symmetric architecture with a greater number of skip connections. These connections are more densely integrated, allowing for a more precise localization of segmented objects. The U-Net's contracting path (encoder) captures context, while the expanding path (decoder) enables precise localization. The skip connections concatenate feature maps from the encoder to the corresponding layers in the decoder, enabling the network to combine information from different scales. This is highly beneficial for segmenting objects with intricate shapes and fine details. In robotics, U-Net's ability to produce highly accurate segmentation masks can be invaluable for tasks such as identifying surgical instruments or precisely mapping complex indoor environments.

Another important class of semantic segmentation models are those based on **Dilated Convolutions (also known as Atrous Convolutions)**. Standard convolutional layers with small receptive fields struggle to capture sufficient context in a single layer, especially at higher resolutions. Dilated convolutions introduce gaps in the kernel, allowing it to cover a larger area of the input feature map without increasing the number of parameters or computational cost. This effectively increases the receptive field of the convolutional filters, enabling the network to capture more contextual information while maintaining the spatial resolution. Architectures like **DeepLab** leverage dilated convolutions to achieve state-of-the-art performance in semantic segmentation. DeepLab models typically use a backbone network like ResNet and apply dilated convolutions in the later stages to expand the receptive field without losing resolution. They also incorporate techniques like Conditional Random Fields (CRFs) to refine the segmentation boundaries, making them smoother and more accurate.

The output of semantic segmentation is a dense per-pixel classification. This provides a comprehensive understanding of the scene's layout, which is critical for many robotic applications. For example, in autonomous navigation, a robot needs to know which areas are traversable (road, sidewalk) and which are not (buildings, cars, pedestrians). Semantic segmentation provides this information directly. In manipulation tasks, understanding the semantic categories of surrounding objects can help a robot anticipate interactions and plan its movements more effectively. For instance, if a robot is tasked with picking up a tool from a workbench, semantic segmentation can identify the workbench, the tools, and any other objects that might obstruct its path or be sensitive to disturbance.

However, semantic segmentation has a limitation: it treats all objects of the same class identically. If there are multiple cars in a scene, semantic segmentation will label all pixels belonging to any car as "car," but it cannot distinguish between individual car instances. This is where
instance segmentation comes into play, offering a more refined level of perception.

Instance segmentation not only classifies each pixel into a category but also distinguishes between different instances of the same category. This means that if there are three cars in an image, instance segmentation will generate three distinct masks, each corresponding to a single car. This ability to differentiate individual objects is crucial for robots operating in complex and dynamic environments where precise object manipulation and interaction are required.

A pioneering and highly effective framework for instance segmentation is

Mask R-CNN. As its name suggests, Mask R-CNN extends the Faster R-CNN object detection framework. Faster R-CNN, as discussed previously, first proposes regions of interest (RoIs) and then classifies objects and refines their bounding boxes within these regions. Mask R-CNN adds a third parallel branch to this process: a mask prediction head. For each RoI that is identified as an object, Mask R-CNN predicts a binary mask that precisely outlines the object's contour. This mask is generated at a higher resolution than the bounding box, allowing for very fine-grained segmentation.

The Mask R-CNN architecture works by first using a backbone network (like ResNet) to extract features from the input image. Then, a Region Proposal Network (RPN) generates candidate object regions. These RoIs are then passed through a RoIAlign layer, which accurately aligns the extracted features with the RoIs, crucial for precise mask prediction. Finally, for each RoI, Mask R-CNN predicts the object's class, refines its bounding box, and simultaneously generates a pixel-wise segmentation mask. The key innovation is the parallel prediction of masks alongside class labels and bounding boxes, making it an end-to-end trainable system for instance segmentation.

For robotics, Mask R-CNN's capabilities are immense. Consider a robot arm tasked with sorting a bin of mixed objects. Instance segmentation allows the robot to identify each individual object, even if they are of the same type, and precisely delineate their boundaries. This enables the robot to grasp each object without colliding with its neighbors. In collaborative robotics, where a robot works alongside humans, precise instance segmentation can help the robot understand which specific objects humans are interacting with, facilitating safer and more intuitive collaboration. For example, if a human is holding a specific wrench, instance segmentation can pinpoint that exact wrench, allowing the robot to avoid interfering with the human's task.

While Mask R-CNN has achieved remarkable success, its two-stage approach (region proposal followed by classification and mask prediction) can still be computationally intensive. This has spurred research into
real-time instance segmentation methods. One notable example is **YOLACT (You Only Look At CoefficienTs)**. YOLACT aims to achieve high-speed instance segmentation by decoupling the mask generation process. It first generates a set of prototype masks that represent different spatial structures. Then, for each detected object, it predicts a set of coefficients. These coefficients are then used to linearly combine the prototype masks, producing the final instance-specific masks. This parallel processing approach allows YOLACT to achieve near real-time frame rates, making it suitable for applications where rapid perception is critical.

Another interesting approach is
SOLO (Segmenting Objects by Locations). Unlike Mask R-CNN, which relies on region proposals, SOLO

directly predicts instance masks based on the location of objects within the image. It divides the image into a grid, and each grid cell is responsible for predicting masks for objects whose centers fall within that cell. SOLO segments objects by their categories and their spatial locations, effectively assigning pixels to object instances without explicit object proposals. This direct prediction mechanism can lead to simpler and faster inference times.

The ability to perform both semantic and instance segmentation is foundational for robots to develop a rich and nuanced understanding of their operational environment. Semantic segmentation provides the "what" and "where" at a scene-wide level, defining the types of surfaces, navigable areas, and static background elements. Instance segmentation then refines this understanding by identifying and separating individual objects, enabling precise interaction and manipulation.

Consider a robot tasked with performing complex assembly operations in a manufacturing setting. Semantic segmentation would first identify the areas of the workspace as "table," "conveyor belt," and "parts bin." Then, instance segmentation would be employed to identify each individual screw, bolt, or component within the parts bin, even if they are of the same type. This allows the robot's manipulator to precisely grasp and place each component according to the assembly plan. Without instance segmentation, the robot might struggle to differentiate between closely packed components, leading to errors or dropped parts.

In the realm of autonomous driving, the distinction between semantic and instance segmentation is critical. Semantic segmentation identifies the road, sidewalks, buildings, and other non-drivable areas. However, instance segmentation is vital for understanding and predicting the behavior of individual vehicles, pedestrians, and cyclists. Knowing that there are multiple vehicles around the robot is one thing; knowing *which* vehicle is braking, *which* pedestrian is about to cross, and *which* cyclist is swerving requires instance-level awareness. This detailed perception enables safer path planning and more responsive control for autonomous vehicles.

Furthermore, for robots operating in human environments, such as domestic service robots or assistive robots for the elderly, understanding individual objects and their semantic context is paramount for natural and safe interaction. If a robot is assisting a person in the kitchen, it needs to identify not just "a cup" (semantic) but *which* cup the person is reaching for (instance). This allows the robot to offer appropriate assistance, such as moving other objects out of the way or handing over the correct item. The ability to distinguish between a specific glass of water and other cups on a table, for instance, contributes significantly to the robot's utility and user experience.

The development and application of semantic and instance segmentation techniques are heavily reliant on the availability of large, annotated datasets. Creating these datasets is a labor-intensive process, often involving human annotators painstakingly labeling each pixel or delineating precise masks for every object in thousands of images. Techniques like active learning and semi-supervised learning are being explored to reduce this annotation burden. Active learning, for example, allows the model to query for labels on the most informative unlabeled data points, thereby optimizing the annotation effort.

Moreover, the robustness of these segmentation models in real-world robotic applications is a significant area of research. Robots often operate under challenging conditions that differ from the controlled environments of data centers where models are trained. These challenges include:

Varying Lighting Conditions: Shadows, glare, and low-light environments can significantly alter the appearance of objects and surfaces, impacting segmentation accuracy.
Occlusions: Objects being partially or fully hidden by other objects are a common problem. Instance segmentation, particularly methods that can infer object presence from partial views, is crucial here.

Viewpoint Changes: Objects appearing from different angles can be challenging for models trained on specific viewpoints.
Scale Variations: Objects appearing at different distances from the sensor require models that can handle multi-scale information effectively.
Texture and Material Properties: Reflective, transparent, or featureless surfaces can be difficult to segment accurately.
Dynamic Environments: Moving objects and rapidly changing scenes require segmentation models that can operate at high speeds and adapt to new instances.

To address these challenges, researchers employ several strategies. Data augmentation, as mentioned in the context of object detection, plays a vital role in making segmentation models more robust. This includes applying random transformations like brightness, contrast, and saturation adjustments, geometric transformations (rotations, scaling, shearing), and even more advanced techniques like simulating fog or rain.

The integration of multi-modal sensor data is another powerful approach. For instance, combining data from RGB cameras with depth sensors (like LiDAR or stereo cameras) provides richer information about the scene. Depth information can help disambiguate objects based on their distance, even if their visual appearance is similar, and can significantly improve segmentation in low-light conditions where color information might be unreliable. Fusing semantic and depth information can lead to more robust and accurate segmentation, especially for complex 3D scenes.

The ongoing evolution of deep learning architectures and training methodologies continues to push the boundaries of semantic and instance segmentation. Researchers are exploring lightweight architectures for deployment on resource-constrained robotic platforms, developing more efficient attention mechanisms to focus on relevant parts of the scene, and improving techniques for few-shot or zero-shot segmentation, enabling robots to segment novel object categories with minimal or no prior training data. The ultimate goal is to equip robots with a perception system that can understand the world at a level of detail and nuance comparable to human vision, unlocking their potential for more autonomous, intelligent, and versatile operation across a vast range of applications.

Beyond the pixel-level understanding provided by semantic and instance segmentation in 2D images, robots often require a more profound comprehension of their surroundings, one that captures the true three-dimensional geometry of objects and environments. This is where **3D Perception and Point Cloud Processing** become indispensable. Unlike images that represent a projection of the 3D world onto a 2D plane, 3D sensors directly capture depth information, providing a more faithful representation of spatial relationships.

The primary source of this 3D data in robotics typically comes from two main types of sensors: **LiDAR (Light Detection and Ranging)** and **depth cameras** (such as stereo cameras or structured light sensors). LiDAR sensors emit laser pulses and measure the time it takes for these pulses to return after reflecting off surfaces. This allows them to generate a dense collection of 3D points, each with associated coordinates (x, y, z) and often intensity information, forming what is known as a **point cloud**. Depth cameras, on the other hand, typically work by triangulating points based on disparity in stereo images or by projecting a known pattern and analyzing its distortion. While often providing a lower resolution than LiDAR, depth cameras can be more cost-effective and are sometimes preferred for indoor applications.

A raw point cloud, while rich in information, can be overwhelming. It often contains millions of points, and the data can be noisy, redundant, or incomplete. Therefore, a series of preprocessing steps are crucial before meaningful analysis can take place. These steps aim to clean, structure, and simplify the data, making it more amenable to subsequent processing and machine learning algorithms.

Filtering is a fundamental preprocessing step. Noise, often introduced by sensor limitations or environmental factors (e.g., reflections from dust particles), can manifest as outlier points. Statistical outlier removal techniques are commonly employed, where for each point, the algorithm calculates the average distance to its nearest neighbors. If this average distance exceeds a certain threshold, the point is deemed an outlier and removed. Another common filtering technique is **voxel filtering** or **uniform downsampling**. This involves dividing the 3D space into a grid of small cubes (voxels) and then representing all points within a single voxel by a single representative point (e.g., the centroid of the points within that voxel). This dramatically reduces the number of points while preserving the overall structure of the point cloud, making subsequent processing much faster. Adaptive downsampling methods also exist, which reduce density more in sparse regions and less in dense regions, preserving more detail where it matters.

Once the point cloud is cleaned and downsampled, the next challenge is often to understand how different point clouds relate to each other in space. This is the domain of
point cloud registration. In robotics, a robot equipped with a 3D sensor will move through an environment, acquiring a series of point clouds over time. To build a complete map of the environment or to track the robot's own motion, these individual scans must be aligned. This process, known as **simultaneous localization and mapping (SLAM)**, relies heavily on accurate registration. A common algorithm for pairwise registration is the **Iterative Closest Point (ICP)** algorithm. ICP works by iteratively finding correspondences between points in two overlapping point clouds and then computing the rigid

transformation (rotation and translation) that minimizes the distance between these corresponding points. While effective, ICP can be sensitive to the initial guess and may converge to a local minimum if the point clouds are not well-aligned initially. More robust methods like **normal distributions transform (NDT)** or methods that leverage distinctive geometric features (e.g., detecting planes or corners) are often used to obtain a better initial alignment before running ICP.

After preparing the point cloud, the focus shifts to extracting meaningful geometric information. This involves identifying structures and features within the data. One of the most fundamental tasks is **plane detection**. Many robotic environments, such as floors, walls, and tabletops, are composed of large planar surfaces. Algorithms like **Random Sample Consensus (RANSAC)** are widely used for this purpose. RANSAC iteratively selects a random subset of points, fits a plane model to this subset, and then counts how many other points in the cloud lie close to this plane (inliers). This process is repeated many times, and the plane with the most inliers is considered the dominant plane. Detecting planes is crucial for tasks like ground segmentation (identifying drivable surfaces for autonomous vehicles) or for removing supporting structures to isolate objects of interest.

Beyond planes, robots need to understand the curvature and local geometry of surfaces. This is achieved by estimating **normal vectors** for each point. The normal vector indicates the orientation of the surface at that point. It is typically computed by fitting a local plane to a neighborhood of points around the target point and using the plane's normal as the estimate. These normals are invaluable for shading, surface reconstruction, and feature extraction. For example, regions with high curvature (e.g., edges, corners) can be identified by analyzing the variation of normal vectors in a local neighborhood.

Once normals are computed, more complex surface representations can be built.
Surface reconstruction aims to create a continuous surface from the discrete set of points. Algorithms like **Poisson surface reconstruction** or **ball pivoting** can generate a mesh (a collection of vertices, edges, and faces) that approximates the underlying surface. This is useful for creating detailed 3D models of objects or environments, which can then be used for tasks like collision detection, path planning, or even for 3D printing or augmented reality applications.

The geometric features extracted from point clouds are not just for visualization; they are critical inputs for higher-level reasoning and decision-making. For instance, identifying geometric primitives like cylinders, spheres, or boxes can directly inform object recognition. Understanding the relative positions and orientations of these primitives can help a robot infer the pose of an object.

While traditional geometric algorithms are powerful, the complexity and variability of real-world 3D scenes often demand more sophisticated approaches. This is where **deep learning architectures specifically designed for point cloud data** have revolutionized 3D perception. Unlike 2D image processing, which readily benefits from convolutional neural networks (CNNs) due to their inherent grid-like structure, raw point clouds are unordered sets of points, making direct application of CNNs challenging.

One of the pioneering architectures for processing point clouds directly was **PointNet**. PointNet operates by processing each point independently through a shared Multi-Layer Perceptron (MLP) to learn point-wise features. To achieve permutation invariance (meaning the network's output is independent of the order of points in the input cloud), it uses a symmetric function, typically max pooling, to aggregate these point features into a global feature vector. This global feature vector can then be used for tasks like point cloud classification (e.g., classifying the entire scene) or part segmentation (assigning each point to a specific part of an object). PointNet also introduced **PointNet++**, which addresses the limitation of PointNet by capturing local geometric structures. It does this by applying PointNet recursively on hierarchical partitions of the point cloud, allowing it to learn features at different scales.

Building on the success of PointNet, other architectures emerged that leveraged different ways to represent and process point cloud data.

Voxel-based methods, such as **VoxelNet**, discretize the 3D space into a grid of voxels. Features are then extracted within each voxel, and a 3D CNN can be applied to these voxel features. This approach bridges the gap between point-wise processing and the highly effective 3D convolutional operations. However, voxelization can lead to loss of detail if the voxel resolution is too coarse and can be computationally expensive if the resolution is very fine.

More recent architectures, such as **PointRCNN**, **SECOND (Sparsely Embedded Convolutional Detection)**, and **PV-RCNN (Point-Voxel Feature Set Abstraction)**, aim to combine the strengths of different approaches. For example, PointRCNN uses a two-stage process: it first generates 3D bounding box proposals directly from the point cloud and then refines these proposals using PointNet-like architectures. SECOND uses sparse convolutions to efficiently process voxelized features, reducing computational cost. PV-RCNN integrates point-based and voxel-based features, leveraging the strengths of both for tasks like 3D object detection.

The applications of these 3D perception techniques in robotics are vast and continually expanding. In **autonomous driving**, LiDAR-based point cloud processing is fundamental for detecting vehicles, pedestrians, cyclists, and obstacles. Semantic segmentation of point clouds can identify drivable road surfaces, distinguish between different types of infrastructure, and even infer the intent of other road users based on their 3D motion.

In
robot manipulation, 3D perception allows robots to grasp objects with precision. By generating a detailed 3D model of an object, a robot can determine its exact pose, identify stable grasp points, and plan a collision-free trajectory for its manipulator. This is crucial for tasks ranging from picking and placing components in manufacturing to handling delicate items in logistics or even assisting with household chores.

For robots operating in
unstructured or unknown environments, such as inspection robots in mines or search-and-rescue robots in disaster zones, 3D perception is vital for navigation and exploration. Building accurate 3D maps of the environment allows the robot to understand its surroundings, avoid obstacles, and plan efficient paths to reach its objectives. Techniques like occupancy grid mapping or 3D SLAM, heavily reliant on point cloud processing, are central to these capabilities.

Asset management and inventory tracking in warehouses can be significantly enhanced by 3D perception. Robots equipped with 3D sensors can quickly scan shelves and identify individual items, their quantities, and their locations, automating tasks that were previously manual and error-prone. Similarly, in construction and infrastructure monitoring, drones equipped with LiDAR can create precise 3D models of buildings or bridges, aiding in progress tracking, defect identification, and structural analysis.

Furthermore, the integration of 3D perception with other sensing modalities, such as RGB cameras, leads to even more robust and informative robotic systems. Fusing 2D semantic segmentation with 3D point cloud data can provide a richer understanding of the scene, where semantic labels from images are projected onto the 3D geometry. This can improve the accuracy of 3D object detection and segmentation, especially for objects that are visually similar but geometrically distinct. For instance, color information from an RGB camera can help distinguish between different types of foliage in a point cloud, or it can be used to add texture to a reconstructed 3D surface.

The challenges in 3D perception are as significant as its potential.
Occlusions remain a persistent problem; parts of objects or environments may be hidden from the sensor's view. While multiple sensor viewpoints or sensor fusion can mitigate this, it's an ongoing area of research. **Reflective and transparent surfaces** are notoriously difficult to capture accurately with LiDAR and some depth sensors, leading to sparse or erroneous data. The sheer volume of data generated by high-resolution 3D sensors also poses a computational challenge, requiring efficient algorithms and powerful hardware for real-time processing, especially on embedded robotic platforms.

The development of more compact, affordable, and higher-resolution 3D sensors, coupled with advancements in deep learning and efficient algorithms for point cloud processing, continues to drive progress in this field. As robots become more prevalent in diverse environments, their ability to perceive and understand the world in its full three-dimensional complexity will be paramount to their success and their integration into our daily lives. The ongoing research in point cloud filtering, registration, feature extraction, and deep learning-based 3D analysis is laying the groundwork for robots that can interact with the physical world with unprecedented awareness and intelligence.

Visual Odometry (VO) represents a significant leap in a robot's ability to understand its own movement within an environment, acting as a crucial bridge between perception and localization. While previous discussions focused on how robots interpret their surroundings through 3D point clouds and semantic understanding, VO directly addresses the challenge of estimating the robot's pose (position and orientation) by analyzing sequences of images. Essentially, it allows a robot to "see" its way through the world, inferring its trajectory by observing how features in the environment shift from one camera frame to the next. This capability is foundational for many autonomous systems, especially those relying on cameras as their primary sensing modality.

At its core, Visual Odometry is about tracking visual features. When a robot moves, the scene captured by its camera undergoes a transformation. Points, edges, or more complex visual patterns that were in one location in a previous frame will appear in a different location in the current frame. By detecting and tracking these distinctive visual elements across consecutive frames, VO algorithms can deduce the camera's motion. This process typically involves several key stages: feature detection, feature matching, and finally, pose estimation.

Feature detection is the initial step where salient points of interest are identified within an image. These features need to be robust, meaning they should be reliably detectable even under varying lighting conditions, small rotations, or slight scale changes. Classic feature detectors like SIFT (Scale-Invariant Feature Transform), SURF (Speeded Up Robust Features), ORB (Oriented FAST and Rotated BRIEF), and FAST (Features from Accelerated Segment Test) are commonly employed. These algorithms look for corners, blobs, or other distinctive patterns that are unlikely to be ambiguous. For instance, a sharp corner of a table or a unique texture on a wall are excellent candidates for visual features. The goal is to extract a set of keypoints, each associated with a descriptor that captures its local appearance, allowing for reliable recognition in subsequent frames.

Once features have been detected in two consecutive frames, the next crucial step is
feature matching. This involves finding corresponding features between the two images. If a feature point in frame `k` is observed again in frame `k+1`, and its

appearance is sufficiently similar, it's considered a match. This matching can be performed using the descriptors computed during feature detection. For example, a descriptor vector for a feature in frame `k` can be compared against the descriptor vectors of all features in frame `k+1`. Algorithms like k-Nearest Neighbors (k-NN) or Brute-Force matching are often used. To improve accuracy and reduce false matches, various techniques are applied, such as ratio tests (e.g., Lowe's ratio test for SIFT), which discard matches where the nearest neighbor is too similar to the second nearest neighbor, indicating ambiguity. Geometric consistency checks, like ensuring that matched features adhere to a plausible epipolar geometry, are also vital for filtering out erroneous correspondences.

With a set of reliable feature correspondences between two frames, the system can then proceed to **pose estimation**. This is the process of calculating the rigid transformation (a combination of rotation and translation) that maps the 3D points observed in the first frame to their observed positions in the second frame. This is where the geometric principles of computer vision come into play. For a monocular (single camera) VO system, this often involves solving the Perspective-n-Point (PnP) problem or, more commonly in VO, recovering the essential or fundamental matrix to estimate the relative camera pose.

The **Essential Matrix (E)** relates corresponding points in two calibrated camera views. It encodes the relative rotation and translation between the two camera poses. Given a set of matched points `(x1, y1)` in the first image and `(x2, y2)` in the second image, and knowing the camera's intrinsic parameters (focal length, principal point), the Essential Matrix can be estimated. Common algorithms like the 8-point algorithm or 5-point algorithm are used for this purpose. Once the Essential Matrix is recovered, it can be decomposed into rotation `R` and translation `t` vectors that represent the relative camera motion. However, for monocular VO, the scale of the translation is inherently ambiguous; the system can only recover the direction of motion, not its magnitude. This is a significant limitation that requires external information or further processing to overcome.

For stereo VO systems (using two calibrated cameras with a known baseline), the situation is different. With a known baseline, the absolute scale of the motion can be recovered directly, making stereo VO generally more robust and accurate than monocular VO. The process involves triangulating the 3D positions of matched features in the stereo pair and then comparing these 3D positions across frames to estimate the relative transformation.

The
Perspective-n-Point (PnP) problem is another fundamental task in pose estimation, though it's more directly applied when you know the 3D coordinates of points and their 2D projections. In VO, we often don't directly know the 3D coordinates of our features.

However, once a partial 3D structure is built (e.g., by triangulating a few features), PnP can be used to estimate the pose of subsequent frames by matching the 2D projections of these known 3D points in the new frame. Algorithms like the P3P (Perspective-3-Point) or EPnP (Efficient PnP) are efficient solvers for this problem.

The output of a VO system is a sequence of estimated poses, which can be chained together to form a trajectory. This trajectory represents the robot's path through the environment as perceived by its camera. While VO provides continuous motion estimates, it's important to acknowledge its limitations. VO systems are susceptible to
drift. Each pose estimation is based on the previous one, and any small error in a single frame's estimation accumulates over time. This means that even with accurate measurements, the estimated trajectory will gradually diverge from the true path. This drift is a significant problem for long-term navigation and mapping.

To combat drift, VO is often integrated into a broader **Simultaneous Localization and Mapping (SLAM)** framework. In SLAM, VO serves as the "front-end," providing a consistent, albeit potentially drifting, estimate of the robot's motion. The "back-end" of a SLAM system then refines this estimate. This refinement typically involves two key mechanisms: **loop closure detection** and **map optimization**.

Loop closure detection is the process by which a robot recognizes that it has returned to a previously visited location. When a loop is detected, it signifies a globally consistent path. Imagine a robot navigating a room. If it moves in a circle and returns to its starting point, VO might report a small displacement, but the environment's appearance would be very similar to when it started. Loop closure detection algorithms compare the current scene's appearance with a database of previously seen scenes. If a significant match is found, it means the robot has closed a loop. This detection provides a strong constraint: the robot's current pose should be consistent with its pose at the time the loop was closed.

Once a loop is detected and confirmed, the SLAM system can use this information to **optimize the entire trajectory and map**. This is often achieved using techniques like **graph optimization**, where the poses and the map points are represented as nodes in a graph, and the constraints (derived from VO measurements and loop closures) are represented as edges. Minimizing the error across this graph, using methods like Bundle Adjustment, simultaneously corrects the drift in the trajectory and refines the 3D map of the environment. This process allows SLAM systems to achieve globally consistent maps and highly accurate localization, overcoming the inherent drift of VO alone.

The choice between monocular, stereo, or even multi-camera VO systems depends heavily on the application requirements, the available hardware, and the complexity of the operating environment. Monocular VO is attractive due to its low hardware cost (a single camera is inexpensive) and its ability to work in environments with significant texture. However, it requires careful handling of scale ambiguity and can be less robust to fast motions or lack of texture. Stereo VO, while requiring two calibrated cameras, provides direct scale recovery and generally offers better accuracy and robustness, making it a popular choice for many robotic platforms.

Furthermore, VO systems can be broadly categorized into
sparse and **dense** approaches. Sparse VO tracks a limited number of salient features (like corners), which is computationally efficient and robust to dynamic environments (as outliers are easier to discard). Dense VO, on the other hand, attempts to track every pixel or a large number of pixels in the image. This can provide a more accurate estimate of the motion and can be used to reconstruct a denser representation of the environment. However, dense VO is computationally much more demanding and is more sensitive to lighting changes and dynamic objects.

The development of VO and its integration into SLAM systems have revolutionized robot navigation. It allows robots to explore unknown territories, build maps, and localize themselves with increasing autonomy and accuracy. From self-driving cars that perceive their surroundings through cameras to drones that map complex terrain, visual odometry forms a critical component of their perception and navigation pipelines. The continuous advancements in feature descriptors, matching algorithms, pose estimation techniques, and loop closure strategies, coupled with the increasing computational power available, are pushing the boundaries of what's possible with camera-based localization.

The ability to perceive and interpret the actions and presence of humans is no longer a secondary concern but a paramount necessity for the advancement of robotics, especially in domains where robots operate alongside people. As robots move beyond structured industrial environments and into the dynamic, unpredictable world of human society, their capacity to understand human presence, intentions, and activities becomes as critical as their ability to navigate and manipulate objects. This subsection delves into the sophisticated techniques employed by modern robots to achieve this vital form of perception, often termed Human and Activity Recognition (HAR).

At the foundational level, a robot must first be aware that a human is present. This is the domain of **human detection**. Early approaches relied on a variety of feature-based methods. For instance, Haar-like features, popularized by the Viola-Jones object detection framework, were highly effective for face detection and could be extended to body detection. These methods involve training classifiers on a large dataset of images, learning to identify patterns of light and dark regions that are characteristic of human forms. Another common technique involved using histograms of oriented gradients (HOG) descriptors, which capture the shape of objects by looking at the distribution of local intensity gradients. When combined with a Support Vector Machine (SVM) classifier, HOG features proved highly successful for pedestrian detection. These methods, while computationally intensive, offered a robust way to identify the presence and approximate location of humans within an image.

More contemporary methods have largely shifted towards deep learning, specifically Convolutional Neural Networks (CNNs). Architectures such as YOLO (You Only Look Once), Faster R-CNN, and SSD (Single Shot MultiBox Detector) have revolutionized object detection, including human detection. These networks are trained end-to-end on massive datasets (like COCO or ImageNet) and can achieve remarkable speed and accuracy. They work by directly predicting bounding boxes around detected objects and classifying them. For example, a YOLO network processes an entire image in a single pass, dividing it into a grid and predicting bounding boxes and class probabilities for each grid cell. This allows for real-time human detection, a critical requirement for many interactive robotic applications. The output of these detectors is typically a set of bounding boxes, each indicating a region in the image that likely contains a human. This information is invaluable for subsequent steps, such as tracking and activity recognition.

Once a human is detected, the next challenge is to

track their movement over time. This is essential for understanding their trajectory, predicting their future position, and analyzing their actions. Tracking algorithms build upon detection, using the information from previous frames to predict the location of the human in the current frame. A common approach is to use a Kalman filter or its more sophisticated variants, like the Extended Kalman Filter (EKF) or Unscented Kalman Filter (UKF). These filters maintain a probabilistic estimate of the human's state (e.g., position, velocity) and update this estimate as new observations (detections) become available.

However, the dynamic nature of human movement and potential occlusions (where a human might be temporarily hidden by an obstacle) pose significant challenges for traditional filtering methods. More advanced tracking techniques often involve sophisticated data association strategies. For example, the Multiple Hypothesis Tracking (MHT) algorithm maintains multiple possible interpretations of the scene over time, aiming to resolve ambiguities. Another powerful paradigm is the
SORT (Simple Online and Realtime Tracking) algorithm and its successor, **DeepSORT**. SORT uses Kalman filters for motion prediction and the Hungarian algorithm for data association, efficiently linking detections to existing tracks. DeepSORT enhances this by incorporating appearance information using a deep neural network, allowing it to re-identify individuals even after they have been occluded for a period. The ability to maintain consistent identity across frames is

crucial for understanding individual actions within a group.

Beyond just tracking their position, robots need to understand the
pose of a human. Human pose estimation refers to the task of inferring the location of an individual's body joints (e.g., shoulders, elbows, wrists, hips, knees) in an image or a 3D space. This provides a much richer representation of human posture and movement than a simple bounding box. Early methods involved detecting individual body parts (like heads, hands, torsos) and then assembling them into a coherent skeleton. More recently, deep learning has again led to significant breakthroughs.

2D Pose Estimation models, such as OpenPose and HRNet, can predict the 2D coordinates of a human's keypoints directly from an RGB image. These networks typically employ a bottom-up or top-down approach. Bottom-up methods detect all keypoints in the image first and then group them into skeletons. Top-down methods first detect the person and then predict keypoints within the person's bounding box. These methods are highly effective for tasks where 2D information is sufficient, such as analyzing gestures or identifying common postures.

For robots equipped with 3D sensors like depth cameras or LiDAR,
3D Pose Estimation becomes possible. This is particularly valuable as it provides a more accurate and

robust representation of a human's configuration in space, irrespective of their distance from the camera or their orientation in the image plane. 3D pose estimation can be achieved either by directly regressing 3D joint coordinates from RGB-D data or by lifting 2D pose estimates into 3D. Techniques often involve specialized neural network architectures designed to process point clouds or depth images. Some methods first estimate the 2D pose and then use a separate network to infer the depth of each joint, or they might directly predict the 3D coordinates of keypoints. The availability of 3D pose information allows for a more nuanced understanding of human embodiment and movement. For instance, a robot can differentiate between a person reaching out to grasp an object and simply stretching their arm.

The ultimate goal of human perception for robotics is often
activity recognition – understanding what a human is doing. This goes beyond mere detection or pose estimation to classifying a sequence of actions or a specific behavior. This typically involves analyzing temporal patterns in the detected human presence, their trajectory, and their pose over time.

One of the most powerful classes of models for sequence analysis are
Recurrent Neural Networks (RNNs), particularly Long Short-Term Memory (LSTM) and Gated Recurrent Unit (GRU) networks. These networks are designed to process sequential data by maintaining an internal memory of past events. When applied to human activity

recognition, an RNN can take a sequence of detected human poses (or other features like optical flow) as input and output a probability distribution over a set of predefined actions (e.g., walking, sitting, eating, waving). For example, a sequence of joint angles and their temporal derivatives over several seconds could be fed into an LSTM, which might then classify the activity as "drinking coffee."

Another significant advancement in temporal modeling comes from
Temporal Convolutional Networks (TCNs). TCNs leverage causal convolutions to process sequences, offering advantages in parallelism and gradient flow compared to traditional RNNs. They can capture long-range temporal dependencies effectively, making them suitable for recognizing complex activities that unfold over extended periods.

The input to these temporal models can vary. It might be a sequence of bounding box coordinates, velocity vectors, or, more commonly and powerfully, the sequences of 2D or 3D human poses. Extracting meaningful features from pose sequences is crucial. This can involve analyzing the relative positions and velocities of body joints, the overall body orientation, or the dynamics of specific limbs. For instance, recognizing "opening a door" might involve identifying a sequence where the hand reaches for a handle, then applies a rotational motion, followed by a translation of the door.

Beyond recognizing predefined actions, there is growing interest in recognizing human **intentions**. This is a more challenging but critical aspect for safe human-robot collaboration. Intention recognition aims to predict what a human *will* do next, based on their current actions, trajectory, and context. For example, if a human is approaching a table where a robot is working, and their hand trajectory suggests they are reaching for a tool, a robot should infer the intention to collaborate or take the tool and adjust its behavior accordingly (e.g., pause, offer the tool, or move out of the way).

Intentions can be inferred by modeling the human's goals and predicting their optimal path or actions to achieve those goals. This often involves combining current sensory data with a model of human behavior and preferences. Techniques from reinforcement learning and probabilistic graphical models can be employed here, where the robot tries to infer the human's underlying reward function or desired outcome.

The fusion of different sensor modalities (vision, depth, LiDAR, sometimes even audio) significantly enhances the robustness and accuracy of human and activity recognition. For example, a depth camera can provide precise distance information, making pose estimation more accurate and helping to resolve ambiguities that might arise in 2D images. LiDAR can offer reliable human detection and tracking in low-light conditions where cameras might struggle. Combining these streams of information through sensor fusion techniques can lead to a more comprehensive and reliable understanding of the human presence and their actions.

The applications of robust human and activity recognition in robotics are vast and growing:

Collaborative Robotics (Cobots): Robots in shared workspaces need to understand human movements to avoid collisions, anticipate their actions, and work in tandem. Recognizing when a human is reaching for a shared tool, or when they are about to enter the robot's workspace, is crucial for safety and efficiency.

Service Robots: Robots in homes, hospitals, or public spaces need to interact naturally with people. Recognizing if someone needs assistance (e.g., an elderly person falling, a person looking confused) allows the robot to offer help. Understanding if a person is entering a room or approaching the robot can trigger appropriate greetings or behaviors.

Autonomous Vehicles: While not always considered "robots" in the traditional sense, autonomous vehicles rely heavily on recognizing pedestrians, cyclists, and other road users, and predicting their behavior (e.g., crossing the road, turning). This is a direct application of human detection, tracking, and intent prediction.

Security and Surveillance: Robots deployed for monitoring can use activity recognition to detect unusual or suspicious behavior, such as loitering or unauthorized access.

Human-Robot Interaction (HRI) Research: Understanding how humans interact with robots, and how robots can adapt their behavior based on human actions and expressions, is a key area of HRI research. HAR provides the perceptual foundation for this.

The development of sophisticated human and activity recognition systems is an ongoing process. Challenges remain, including recognizing subtle actions, understanding complex multi-person interactions, dealing with occlusions and cluttered environments, and generalizing models to unseen activities and individuals. However, the progress driven by deep learning and advanced sensor technologies has brought us closer to robots that can not only coexist with humans but truly understand and collaborate with them in a safe and intuitive manner. The ability to recognize human presence, track their movements, infer their posture, and ultimately understand their actions is a cornerstone for the future of intelligent autonomous systems operating in human-centric environments.

Chapter 6: State Estimation and Localization

In the intricate dance between a robot and its environment, understanding the robot's own condition is paramount. This is the domain of **state estimation**, a fundamental problem in robotics that seeks to ascertain the robot's current "state" with the highest possible confidence. But what constitutes this "state"? At its core, the state encapsulates all the essential information needed to describe the robot's condition at a given moment. This typically includes its pose – its position and orientation in space – but can extend to its velocity, acceleration, joint angles (for articulated robots), and even internal parameters like battery level or sensor calibration status. Essentially, the state vector is a mathematical representation, a collection of numbers, that captures the robot's instantaneous reality.

The challenge, however, lies in the fact that this reality is never perfectly known. Robots operate in a world characterized by inherent uncertainties. Sensor readings, the primary source of information about the environment and the robot's interaction with it, are invariably noisy. A camera might misinterpret pixel values, a LiDAR might return a slightly inaccurate distance, or an IMU might drift over time. These inaccuracies mean that each individual measurement is an imperfect glimpse of the truth. Furthermore, our understanding of the robot's own dynamics – how it moves and how its actuators affect its motion – is also an approximation. The equations of motion are simplifications of complex physical phenomena, and factors like friction, unmodeled forces, or slippage can introduce deviations. This means that even if we perfectly knew the robot's state at one moment, predicting its state at the next moment using a motion model will also be subject to errors.

Therefore, the goal of state estimation is not to achieve perfect knowledge, which is often unattainable, but to produce the
most likely estimate of the robot's state, given the available, albeit imperfect, information. This estimate is a fusion of predictions derived from the robot's motion model and corrections provided by incoming sensor measurements. It's a continuous process of refinement, where the robot constantly updates its belief about its own whereabouts and condition. This probabilistic approach acknowledges the inherent uncertainties and quantifies them, allowing the robot to make decisions based on a well-understood level of confidence.

The importance of accurate state estimation cannot be overstated. It forms the bedrock upon which many other critical robotic capabilities are built. Consider navigation: to chart a course from point A to point B, a robot must first know where it is (its current state). Without a reliable estimate of its position and orientation, any path planning or trajectory following becomes meaningless. Similarly, in robot control, actuators are commanded to achieve desired movements. The control system needs to know the robot's current velocity and position to ensure that these commands are executed accurately. If the robot's state estimate is wrong, the control system will be operating on faulty information, leading to incorrect actions and potentially unstable behavior. Even for tasks like manipulation, knowing the robot's precise end-effector pose relative to an object is crucial for successful grasping.

The process of state estimation is inherently a sequence of operations that unfold over time. At each discrete time step, the robot typically performs two main operations: prediction and update. The **prediction** step uses the robot's motion model to forecast its state at the next time step, based on its current estimated state and any control inputs that were applied. This prediction naturally incorporates the expected uncertainty, often resulting in a broader probability distribution of the possible states. Following the prediction, the **update** step incorporates new sensor measurements. The robot compares the sensor readings it expects to receive, given its predicted state, with the actual readings. The difference between these two informs a correction to the predicted state, refining the estimate and reducing the uncertainty. This cycle of prediction and update is repeated continuously, allowing the robot to maintain an up-to-date and accurate estimate of its state as it moves and interacts with its environment.

Several mathematical frameworks have been developed to formalize and implement state estimation. These frameworks provide the tools to manage the probabilistic nature of the problem, balancing the information from motion models and sensor measurements. While the detailed mechanisms of these frameworks will be explored further, it's important to acknowledge their role here. Broadly, these approaches fall into categories that either assume a linear system with Gaussian noise (like the classic Kalman Filter) or handle non-linearities in various ways (such as the Extended Kalman Filter, Unscented Kalman Filter, Particle Filters, and others). Each framework offers a different trade-off in terms of computational complexity, accuracy, and robustness to different types of noise and system dynamics. The choice of framework often depends on the specific characteristics of the robot, its sensors, and the environment in which it operates.

The output of any state estimation process is not a single, definitive value for each state variable, but rather a probability distribution. This distribution represents the robot's "belief" about its state. For instance, instead of saying "the robot is at x=5.2 meters," a state estimator might say, "the robot's x-position is estimated to be 5.2 meters, with a standard deviation of 0.1 meters." This quantification of uncertainty is vital. It allows higher-level systems, such as planners, to understand the reliability of the robot's perceived location. A planner might choose a more conservative path if the uncertainty in the robot's position is high, or it might confidently execute a precise maneuver if the uncertainty is low.

In essence, state estimation is the robot's internal compass and speedometer, constantly calibrated by its senses and its understanding of physics. It is the process by which a robot builds a coherent and reliable picture of itself within the world, a prerequisite for any intelligent action it aims to perform. This ongoing refinement of its internal representation of its own condition is a continuous dialogue between its models of the world and the noisy echoes it receives back from reality, a dialogue that is essential for its very survival and functionality. The fidelity of this internal representation directly impacts the robot's ability to navigate complex spaces, interact safely with its surroundings, and achieve its operational objectives. Without robust state estimation, a robot would be akin to a ship sailing without a rudder or a compass, adrift and incapable of purposeful movement. This foundational capability is what enables robots to move beyond pre-programmed paths in controlled environments and venture into the dynamic, unpredictable real world.

The Kalman Filter stands as a cornerstone in the field of state estimation, providing an elegant and computationally efficient framework for inferring the unobserved state of a dynamic system from a series of noisy measurements. Its power lies in its recursive nature, meaning it processes measurements sequentially and updates its state estimate without needing to store the entire history of past observations. This makes it exceptionally well-suited for real-time applications, where computational resources can be constrained and data arrives continuously.

At its heart, the Kalman Filter operates under the

assumption that the system dynamics are linear and that the noise corrupting both the system's evolution and the measurements is Gaussian. This is a significant simplification of reality, as most real-world systems exhibit non-linearities, and their noise characteristics may not always conform to a perfect Gaussian distribution. Nevertheless, for systems that closely approximate these assumptions, the Kalman Filter provides the optimal minimum mean-square error estimate.

The Kalman Filter algorithm proceeds in a two-step cycle: prediction and update. The **prediction** step projects the current state estimate and its associated uncertainty forward in time, using a mathematical model of the system's dynamics. If we denote the state vector at time step $k-1$ as $\mathbf{x}{k-1}$ *and the control input at time step k as $\mathbf{u}*k$, the predicted state $\hat{\mathbf{x}}_k^-$ at time k can be expressed as:

$$ \hat{\mathbf{x}}
k^- = Fk \hat{\mathbf{x}}{k-1} + Bk \mathbf{u}_k $$

Here, $F
k$ is the state transition matrix, which describes how the state evolves from time $k-1$ to k in the absence of control inputs and noise. $B*k$ is the control input matrix, mapping the control vector $\mathbf{u}*k$ to the state space. Crucially, the prediction step also accounts for the uncertainty in the state. This uncertainty

*is typically represented by a covariance matrix, $\mathbf{P}$. The predicted covariance matrix $\mathbf{P}*k^-$ is updated according to:

$$ \mathbf{P}
*k^- = F*k \mathbf{P}*{k-1} F*k^T + Q_k $$

where Q_k is the process noise covariance matrix, which quantifies the uncertainty in the system dynamics model. The superscript 'T' denotes the transpose of a matrix. This equation shows that the uncertainty in the predicted state generally increases compared to the previous state's uncertainty, reflecting the accumulation of errors due to unmodeled dynamics and noise.

Following the prediction, the
update step incorporates a new measurement to refine the predicted state. Let's say we have a measurement vector $\mathbf{z}_k$ obtained at time k. This measurement is related to the state through a measurement model:

$$ \mathbf{z}
*k = H*k \mathbf{x}*k + \mathbf{v}*k $$

Here, $H
*k$ is the measurement matrix, which maps the true state $\mathbf{x}*k$ to the measurement space.
$\mathbf{v}*k$ represents the measurement noise, which is assumed to be zero-mean Gaussian with covariance $R*k$. The Kalman Filter uses the innovation (or

measurement residual), which is the difference between the actual measurement and the expected measurement based on the predicted state:

$$ \mathbf{y}
k = \mathbf{z}k - Hk \hat{\mathbf{x}}k^- $$

The Kalman Gain, denoted by K_k, is then computed. This gain determines how much weight the filter gives to the innovation when updating the state estimate. A higher Kalman Gain means the filter trusts the measurement more, while a lower gain means it relies more on the prediction. The Kalman Gain is calculated as:

$$ K
k = \mathbf{P}k^- Hk^T (Hk \mathbf{P}k^- Hk^T + R_k)^{-1} $$

The state estimate is then updated by incorporating the innovation, scaled by the Kalman Gain:

$$ \hat{\mathbf{x}}
k = \hat{\mathbf{x}}k^- + Kk \mathbf{y}k $$

Finally, the covariance matrix is updated to reflect the reduced uncertainty after incorporating the measurement:

$$ \mathbf{P}

$$k = (I - K_k H_k) \mathbf{P}_k^-$$

where I is the identity matrix. The process then repeats for the next time step.

The beauty of this recursive formulation is that it provides an optimal estimate at each step without requiring past measurements or states. It elegantly balances the information from the system's motion model (prediction) with the information from the sensor measurements (update). The resulting state estimate is the one that minimizes the mean squared error, given the linearity and Gaussian noise assumptions.

However, the strict assumptions of linearity and Gaussian noise limit the direct applicability of the standard Kalman Filter to many practical robotic systems. Robotic systems, by their very nature, often involve non-linear transformations. For example, the relationship between wheel odometry and robot pose involves trigonometric functions (sine and cosine for rotation), which are inherently non-linear. Similarly, sensor models, like range and bearing measurements from a LiDAR, are often non-linear functions of the robot's position and the target's position.

When faced with non-linearities, the standard Kalman Filter can produce significantly suboptimal or even divergent results. This is because the linear propagation of the mean and covariance through non-linear functions is not accurate. The assumption that the probability distribution of the state remains Gaussian after a non-linear transformation is generally false.

To address these limitations, several variants of the Kalman Filter have been developed. Two of the most prominent and widely used are the Extended Kalman Filter (EKF) and the Unscented Kalman Filter (UKF). These filters offer different approaches to handle non-linear system dynamics and measurement models, extending the reach of Kalman filtering to a broader range of real-world problems.

The
Extended Kalman Filter (EKF) is perhaps the most intuitive extension of the standard Kalman Filter to non-linear systems. The core idea behind the EKF is to linearize the non-linear system and measurement models around the current state estimate. This is typically achieved using a first-order Taylor series expansion.

Let's consider a non-linear system described by:

$$ \mathbf{x}_k = f(\mathbf{x}_{k-1}, \mathbf{u}_k) + \mathbf{w}_{k-1} $$

$$ \mathbf{z}_k = h(\mathbf{x}_k) + \mathbf{v}_k $$

Here, $f(\cdot)$ and $h(\cdot)$ represent the non-linear state transition and measurement functions, respectively. $\mathbf{w}
{k-1}$ and $\mathbf{v}k$ are still assumed to be zero-mean Gaussian noise with covariances $Q{k-1}$ and Rk.

In the prediction step of the EKF, the non-linear state transition function f is linearized around the previous state estimate $\hat{\mathbf{x}}
{k-1}$. This is done by computing the Jacobian matrices of f with respect to $\mathbf{x}$ and $\mathbf{u}$ evaluated at $\hat{\mathbf{x}}{k-1}$ and $\mathbf{u}k$. Let $Fk = \frac{\partial f}{\partial \mathbf{x}}|{\hat{\mathbf{x}}{k-1}, \mathbf{u}k}$ and $Bk = \frac{\partial f}{\partial \mathbf{u}}|{\hat{\mathbf{x}}{k-1}, \mathbf{u}_k}$. The predicted state is then calculated using the linearized dynamics:

$$ \hat{\mathbf{x}}
k^- = f(\hat{\mathbf{x}}{k-1}, \mathbf{u}_k) $$

Note that the non-linear function f is used to predict the state mean, not its linearized version. However, the covariance propagation relies on the linearized dynamics. The predicted covariance is updated using the Jacobian F_k:

$$ \mathbf{P}
k^- = Fk \mathbf{P}{k-1} Fk^T + Q_{k-1} $$

In the update step, the non-linear measurement function h is linearized around the predicted state $\hat{\mathbf{x}}
k^-$. The Jacobian of h with respect to $\mathbf{x}$, evaluated at $\hat{\mathbf{x}}k^-$, is denoted by $Hk = \frac{\partial h}{\partial \mathbf{x}}|{\hat{\mathbf{x}}_k^-}$. The measurement residual is computed using the non-linear measurement function:

$$ \mathbf{y}
k = \mathbf{z}k - h(\hat{\mathbf{x}}_k^-) $$

The Kalman Gain is then computed using the linearized measurement model's Jacobian H_k:

$$ K
k = \mathbf{P}k^- Hk^T (Hk \mathbf{P}k^- Hk^T + R_k)^{-1} $$

The state estimate is updated using the residual and Kalman Gain, again with the non-linear measurement function used to calculate the residual:

$$ \hat{\mathbf{x}}
k = \hat{\mathbf{x}}k^- + Kk \mathbf{y}k $$

The covariance is updated as in the standard Kalman Filter:

$$\mathbf{P}_k = (I - K_k H_k) \mathbf{P}_k^-$$

While the EKF is widely used and conceptually straightforward, it has a significant drawback: the accuracy of the linearization depends heavily on how well the first-order Taylor approximation represents the non-linear functions. If the non-linearities are severe or the state uncertainty is large, the linearization can be a poor approximation, leading to filter divergence or biased estimates. The computation of Jacobians can also be complex and error-prone, especially for intricate systems.

This leads us to the **Unscented Kalman Filter (UKF)**, which offers an alternative and often superior approach to handling non-linearities. Instead of linearizing the functions, the UKF uses a deterministic sampling technique called the unscented transform to capture the mean and covariance of the state distribution. The unscented transform is based on the observation that it is often easier to approximate a probability distribution than to linearize a non-linear function.

The UKF works by selecting a set of carefully chosen "sigma points" that capture the mean and covariance of the current state estimate. These sigma points are then propagated through the true non-linear system and measurement functions. The transformed sigma points are then used to reconstruct the mean and covariance of the predicted or updated state. This approach can capture the posterior mean and covariance more accurately than linearization, especially for highly non-linear systems, and it does not require the computation of Jacobians.

For a state vector $\mathbf{x}$ with mean $\mathbf{m}$ and covariance $\mathbf{P}$, a set of $2n+1$ sigma points $\{\xi
i\}{i=0}^{2n}$ are generated (where n is the dimension of the state vector). These sigma points are calculated as:

$$ \xi_0 = \mathbf{m} $$

$$ \xii = \mathbf{m} + (\sqrt{\frac{n}{ \lambda}}
\Sigmai)_{\text{col}} \quad \text{for } i = 1, \dots, 2n
$$

Here, $\Sigma
i$ are the columns of the matrix square root of $\mathbf{P}$ (e.g., Cholesky decomposition), and λ is a scaling parameter that can be tuned to adjust the spread of the sigma points. The corresponding weights for reconstructing the mean and covariance are $W0^m = \frac{\lambda}{n+\lambda}$ for the first sigma point and $Wi^m = \frac{1}{2(n+\lambda)}$ for the remaining $2n$ sigma points. Similarly, for covariance reconstruction, $W0^c = \frac{\lambda}{n+\lambda} + (1-\alpha^2+\beta) $ and $W_i^c = \frac{1}{2(n+\lambda)}$ for $i=1, \dots, 2n$. The parameters α and β are also

tuning parameters. Typically, $\alpha \in [0.001, 1]$, β is chosen based on the prior knowledge of the distribution (e.g., $\beta=2$ for Gaussian), and λ is set to $3-n$.

In the
prediction step of the UKF:
1. Generate $2n+1$ sigma points from the current state estimate $\hat{\mathbf{x}}{k-1}$ and its covariance $\mathbf{P}{k-1}$.
2. Propagate each sigma point $\xii^{k-1}$ through the non-linear state transition function, including the control input: $\xii^{,k} = f(\xii^{k-1}, \mathbf{u}k) + \mathbf{w}_{k-1}$.
3. Reconstruct the predicted mean $\hat{\mathbf{x}}k^-$ and covariance $\mathbf{P}k^-$ from the transformed sigma points and their associated weights.

$$ \hat{\mathbf{x}}
k^- = \sum{i=0}^{2n} Wi^m \xii^{,k} $$
$$ \mathbf{P}k^- = \sum{i=0}^{2n} Wi^c (\xii^{,k} - \hat{\mathbf{x}}k^-)(\xii^{,k} - \hat{\mathbf{x}}k^-)^T + Q{k-1} $$

In the
update step of the UKF:
1. Generate $2n+1$ sigma points from the predicted state $\hat{\mathbf{x}}k^-$ and its covariance $\mathbf{P}k^-$.
2. Propagate each sigma point $\xii^{-,k}$ through the non-linear measurement function: $\zetai^k = h(\xii^{-,k}) + \mathbf{v}k$.

3. Reconstruct the predicted measurement mean $\hat{\mathbf{z}}k$ and measurement covariance $\mathbf{P}{zz,k}$ from the transformed sigma points.

$$ \hat{\mathbf{z}}
k = \sum{i=0}^{2n} Wi^m \zetai^k $$
$$ \mathbf{P}{zz,k} = \sum{i=0}^{2n} Wi^c (\zetai^k - \hat{\mathbf{z}}k)(\zetai^k - \hat{\mathbf{z}}k)^T + Rk $$

4. Compute the cross-covariance matrix between the state and the measurement:

$$ \mathbf{P}
{xz,k} = \sum{i=0}^{2n} Wi^c (\xii^{-,k} - \hat{\mathbf{x}}k^-)(\zetai^k - \hat{\mathbf{z}}_k)^T $$

5. Calculate the Kalman Gain:

$$ K
k = \mathbf{P}{xz,k} \mathbf{P}_{zz,k}^{-1} $$

6. Update the state estimate and covariance:

$$ \hat{\mathbf{x}}
k = \hat{\mathbf{x}}k^- + Kk (\mathbf{z}k - \hat{\mathbf{z}}_k) $$
$$ \mathbf{P}k = \mathbf{P}k^- - Kk \mathbf{P}{zz,k} K_k^T $$

The UKF typically offers better accuracy than the EKF for non-linear systems, particularly when the non-linearities are strong or the state uncertainty is large. It also avoids the need for analytical Jacobian calculations, simplifying implementation for complex models. However, the UKF involves propagating $2n+1$ sigma points, which can increase the computational cost compared to the EKF, especially for high-dimensional state spaces.

Beyond the EKF and UKF, other variants and related algorithms exist, such as the Cubature Kalman Filter (CKF), which is similar in spirit to the UKF but uses cubature rules for integration, and particle filters, which can handle arbitrary non-linearities and non-Gaussian noise by representing the probability distribution as a set of weighted samples. However, the EKF and UKF remain the most widely adopted extensions of the Kalman filter due to their balance of performance and computational tractability.

When implementing these filters in a robotic context, several practical considerations arise. The choice of state vector is crucial; it should encompass all relevant information for the task at hand, such as position, orientation, velocity, and potentially sensor biases. The accuracy of the system and measurement models is paramount. Even with sophisticated filters, a flawed model will lead to poor estimation. Careful tuning of the process noise covariance (Q) and measurement noise covariance (R) matrices is often necessary. These matrices represent the designer's belief about the uncertainty in the models and sensors. Incorrectly set noise covariances can lead to a filter that is too sluggish (high noise settings) or too sensitive to measurement noise (low noise settings).

For systems with significant non-linearities, especially in orientation representation, using Euler angles can lead to issues like gimbal lock and trigonometric singularities. Modern robotics applications often prefer using quaternions or rotation matrices for representing orientation, which are generally more robust and avoid these problems, though they can also introduce their own complexities in the state dynamics and measurement models.

The computational burden of these filters can also be a limiting factor, especially for high-dimensional state spaces or resource-constrained embedded systems. Efficient implementation, potentially using optimized linear algebra libraries, is important. In some cases, approximations or simplifications to the filter equations might be employed to reduce computational load.

Furthermore, the stability and convergence of these filters are not always guaranteed, especially for the EKF in the presence of strong non-linearities. Robustness checks and monitoring of the filter's performance, such as observing the magnitude of the innovation or the evolution of the covariance matrix, can help detect potential issues.

In summary, the Kalman Filter and its variants, the EKF and UKF, are indispensable tools for state estimation in robotics. While the standard KF is optimal for linear systems with Gaussian noise, the EKF and UKF extend its applicability to non-linear scenarios by employing linearization and deterministic sampling, respectively. The choice between EKF and UKF often hinges on the degree of non-linearity, computational constraints, and desired accuracy, with the UKF generally offering superior performance for highly non-linear systems. Understanding the underlying principles, the assumptions, and the practical considerations of these filters is essential for building reliable and capable robotic systems.

The realm of state estimation, as we have explored, has traditionally been dominated by the Kalman Filter and its derivatives like the Extended Kalman Filter (EKF) and the Unscented Kalman Filter (UKF). These methods, while remarkably effective under specific assumptions—linearity and Gaussian noise for the standard Kalman Filter, or approximate linearity for the EKF and improved handling of Gaussian distributions for the UKF—encounter significant limitations when faced with systems that deviate substantially from these idealized conditions. Many real-world robotic systems exhibit dynamics that are highly non-linear, and their associated noise distributions can be far from Gaussian, often exhibiting multi-modal characteristics, heavy tails, or even being completely unknown. In such scenarios, the approximations made by the EKF and UKF can lead to degraded performance, filter divergence, or simply an inability to represent the true state distribution accurately.

It is precisely in these challenging, yet ubiquitous, situations that
Particle Filters, also known as Sequential Monte Carlo (SMC) methods, emerge as a profoundly powerful and flexible alternative. Unlike the parametric approach of Kalman filters, which rely on estimating the mean and covariance of a Gaussian distribution, particle filters adopt a non-parametric representation of the probability distribution. They approximate the posterior probability distribution of the system's state using a set of random samples, often referred to as *particles*. Each particle represents a hypothesis about the current state of the system, and it is associated with a weight that signifies its likelihood or importance. By employing a sufficient number of particles, a particle filter can, in principle, approximate any probability distribution, regardless of its complexity, non-linearity, or non-Gaussian nature. This inherent flexibility makes particle filters exceptionally well-suited for a wide array of challenging state estimation problems in robotics, including simultaneous localization and mapping (SLAM), target tracking, and particularly, robot localization in environments with significant uncertainty.

The fundamental concept behind a particle filter is to

represent the probability distribution of the state vector $\mathbf{x}
k$ at time step k, denoted by $p(\mathbf{x}k | \mathbf{z}{1:k})$, where $\mathbf{z}{1:k}$ represents the sequence of measurements up to time k, as a set of N weighted particles $\{\mathbf{x}k^{(i)}, wk^{(i)}\}{i=1}^N$. Here, $\mathbf{x}k^{(i)}$ is the i-th particle representing a possible state at time k, and $wk^{(i)}$ is its corresponding importance weight. These weights are normalized such that $\sum{i=1}^N w_k^{(i)} = 1$. The collection of these weighted particles forms an empirical approximation of the posterior probability distribution.

The particle filter algorithm operates recursively, much like the Kalman filter, involving a prediction step and an update step. However, the mechanics of these steps are fundamentally different, relying on sampling and re-weighting rather than analytical propagation of mean and covariance.

Let's consider a general non-linear, non-Gaussian state-space model:

$$ \mathbf{x}k = f(\mathbf{x}{k-1}, \mathbf{u}k, \mathbf{v}{k-1}) $$
$$ \mathbf{z}k = h(\mathbf{x}k, \mathbf{e}_k) $$
where f is the non-linear state transition function, $\mathbf{u}k$ is the control input, $\mathbf{v}{k-1}$ is the process noise, h is the non-linear measurement function, and $\mathbf{e}_k$ is the measurement noise. Crucially, the noise terms $\mathbf{v}$ and

$\mathbf{e}$ are not restricted to being Gaussian and can follow any arbitrary distribution.

The Prediction Step (Sampling)

The goal of the prediction step is to propagate the particle set from the previous time step, $\{ \mathbf{x}{k-1}^{(i)}, w{k-1}^{(i)} \}{i=1}^N$, to represent the *prior* distribution at the current time step, $p(\mathbf{x}k | \mathbf{z}{1:k-1})$. This is achieved by drawing new states for each particle from the state transition model. For each particle i from the previous time step, we generate a new state sample $\tilde{\mathbf{x}}k^{(i)}$ by sampling from the process model:
$$ \tilde{\mathbf{x}}k^{(i)} \sim p(\mathbf{x}k | \mathbf{x}{k-1}^{(i)}, \mathbf{u}k) $$
This sampling can be done in various ways depending on the specific form of f and the distribution of $\mathbf{v}{k-1}$. If f is deterministic given $\mathbf{x}{k-1}$ and $\mathbf{u}k$, and $\mathbf{v}{k-1}$ has a known distribution, then we can directly sample $\mathbf{v}{k-1}$ and compute $\tilde{\mathbf{x}}k^{(i)} = f(\mathbf{x}{k-1}^{(i)}, \mathbf{u}k) + \mathbf{v}{k-1}$. If the process noise is not explicitly modeled or is complex, specialized sampling techniques might be employed. The key is that each $\tilde{\mathbf{x}}k^{(i)}$ represents a potential state at time k consistent with the system's dynamics and the previous state estimate.

The weights from the previous time step, $w {k-1}^{(i)}$, are typically carried forward to these predicted particles initially, or they are adjusted in the next step. For now, we consider the set of predicted particles $\{\tilde{\mathbf{x}}k^{(i)}\}_{i=1}^N$.

The Update Step (Weighting)

The update step incorporates the new measurement $\mathbf{z}
k$ to refine the predicted particle set and approximate the posterior distribution $p(\mathbf{x}k | \mathbf{z}{1:k})$. This is achieved by assigning a new weight to each predicted particle $\tilde{\mathbf{x}}k^{(i)}$ based on how well it explains the current measurement $\mathbf{z}k$. The weight of a particle is made proportional to the likelihood of observing the measurement $\mathbf{z}k$ given that the system is in state $\tilde{\mathbf{x}}_k^{(i)}$. This likelihood is determined by the measurement model:
$$ wk^{(i)} \propto w{k-1}^{(i)} \cdot p(\mathbf{z}k / \tilde{\mathbf{x}}k^{(i)}) $$
Here, $p(\mathbf{z}k / \tilde{\mathbf{x}}k^{(i)})$ is the measurement likelihood function, derived from the measurement model h and the distribution of the measurement noise $\mathbf{e}k$. For example, if $\mathbf{e}k$ is Gaussian with covariance Rk, then $p(\mathbf{z}k | \tilde{\mathbf{x}}k^{(i)}) = \mathcal{N}(\mathbf{z}k; h(\tilde{\mathbf{x}}k^{(i)}),

Rk)$. The proportionality constant is determined by normalizing the weights such that they sum to one:
$$ wk^{(i)} = \frac{w{k-1}^{(i)} \cdot p(\mathbf{z}k | \tilde{\mathbf{x}}k^{(i)})}{\sum{j=1}^N w{k-1}^{(j)} \cdot p(\mathbf{z}k | \tilde{\mathbf{x}}k^{(j)})} $$
After this re-weighting, the set of weighted particles $\{ \tilde{\mathbf{x}}k^{(i)}, wk^{(i)} \}{i=1}^N$ *forms an approximation of the posterior distribution* $p(\mathbf{x}k | \mathbf{z}_{1:k})$.

The state estimate at time k can then be derived from this weighted set of particles. A common way to obtain a point estimate is to compute the expected value of the state, which is approximated by a weighted sum of the particles:

$$ \hat{\mathbf{x}}k = \sum{i=1}^N wk^{(i)} \tilde{\mathbf{x}}k^{(i)} $$
Similarly, estimates for covariance and other statistics can be computed.

The Resampling Step

A critical issue that arises with particle filters is the degeneracy problem. Over successive time steps, the weights of most particles tend to become very small, while a few particles may acquire very large weights. This leads to a situation where the effective number of particles diminishes, and the particle set can no longer adequately represent the posterior distribution. This phenomenon is exacerbated in non-linear systems where particle drift can occur.

To combat this degeneracy, a **resampling** step is introduced, typically performed after the update step, or periodically. The core idea of resampling is to eliminate particles with low weights and to duplicate particles with high weights. This is done by drawing a new set of N particles $\{ \mathbf{x}*k^{(i)} \}*{i=1}^N$ from the current weighted set $\{ \tilde{\mathbf{x}}*k^{(i)}, w*k^{(i)} \}*{i=1}^N$ in a way that the probability of drawing particle $\mathbf{x}*k^{(i)}$ is proportional to its weight $w_k^{(i)}$. Essentially, it regenerates the particle set such that all particles have equal weights ($1/N$).

Several resampling strategies exist, each with its own computational characteristics and statistical properties. A common and effective method is **multinomial resampling**. In this approach, N new particles are drawn independently from the current set, with replacement, where the probability of selecting particle i is given by its weight $w_k^{(i)}$. Particles with higher weights are more likely to be selected, and thus duplicated, while particles with low weights are likely to be discarded.

Another efficient method is **systematic resampling**. It draws a single random number u from a uniform distribution between 0 and $1/N$. Then, the new particles are generated by sequentially selecting particles with cumulative weights u, $u + 1/N$, $u + 2/N$, ..., $u + (N-1)/N$. This method reduces the variance of the sampling process compared to multinomial resampling.

After resampling, all particles $\{\mathbf{x} *k^{(i)}\}*{i=1}^N$ are assigned equal weights, $w_k^{(i)} = 1/N$. The state estimate can then be computed using these resampled particles, often leading to a more stable and representative approximation of the posterior distribution. The choice of resampling strategy and its frequency of application can significantly impact the performance and computational cost of the particle filter.

Particle Filters for Robot Localization: The Monte Carlo Localization (MCL) Approach

One of the most prominent applications of particle filters in robotics is
Monte Carlo Localization (MCL). In this context, the particle filter is used to estimate the robot's pose (position and orientation) within a known map of the environment, given a sequence of sensor measurements.

Let the state vector $\mathbf{x}
*k$ represent the robot's pose at time k, often defined as (x, y, θ) in a 2D Cartesian coordinate system. The particles $\{\mathbf{x}*k^{(i)}\}_{i=1}^N$ represent different hypotheses about the robot's current pose.

The MCL algorithm proceeds as follows:

1.
Initialization: At time $k=0$, a set of N particles $\{ \mathbf{x}0^{(i)} \}{i=1}^N$ is generated to represent the initial belief about the robot's pose. This initialization can be:
Uniform: Particles are spread randomly across the entire map, representing complete uncertainty about the initial pose.
Gaussian: If there is some prior knowledge about the initial pose (e.g., from a sensor or a rough guess), particles can be sampled from a Gaussian distribution centered around this prior pose.
Specific: If the robot's initial pose is known with high certainty, all particles can be initialized to this pose.
All initial particles are assigned equal weights $w_0^{(i)} = 1/N$.

2.
Prediction (Motion Update): When the robot executes a motion command $\mathbf{u}k$ *(e.g., velocity commands, wheel odometry readings), this motion is applied to each particle. The motion model f describes how the robot's pose changes based on the motion command and introduces some uncertainty due to imperfect actuation and odometry noise. For each particle $\mathbf{x}*{k-1}^{(i)}$, a new pose $\tilde{\mathbf{x}}_k^{(i)}$ is sampled:
$$ \tilde{\mathbf{x}}k^{(i)} \sim p(\mathbf{x}k | \mathbf{x}{k-1}^{(i)}, \mathbf{u}k) $$

This distribution typically incorporates noise that models the discrepancies between the commanded motion and the actual motion. For example, in a 2D robot, a motion command might be a linear and angular velocity, and the resulting pose update for a particle might be perturbed by Gaussian noise on the translational and rotational components.

3.

Update (Measurement Update): After the motion update, a new sensor measurement $\mathbf{z}_k$ is obtained (e.g., range and bearing readings from a LiDAR, distances from sonar sensors, or visual features). The likelihood of this measurement is computed for each predicted particle $\tilde{\mathbf{x}}_k^{(i)}$ using the measurement model h. The measurement model defines the probability of observing $\mathbf{z}_k$ given the robot is at pose $\tilde{\mathbf{x}}_k^{(i)}$ and considering the robot's sensor capabilities and the map. For a LiDAR sensor, for instance, the measurement model would involve simulating the LiDAR scan from the particle's pose in the map and comparing it with the actual scan $\mathbf{z}_k$.

The weight of each particle is updated based on this measurement likelihood:

$$ w_k^{(i)} \propto w_{k-1}^{(i)} \cdot p(\mathbf{z}_k | \tilde{\mathbf{x}}_k^{(i)}) $$

Then, the weights are normalized:

$$ w_k^{(i)} = \frac{w_k^{(i)}}{\sum_{j=1}^N w_k^{(j)}} $$

Particles whose poses are consistent with the sensor measurements will receive higher weights, while those that are not will receive lower weights.

4.
Resampling: To prevent particle degeneracy, a resampling step is performed. Particles are resampled based on their updated weights $wk^{(i)}$. *This ensures that the particles with high weights are retained and potentially duplicated, while particles with low weights are discarded. After resampling, all particles are assigned equal weights* $wk^{(i)} = 1/N$.

5.
State Estimation: The robot's estimated pose at time k can be computed as the weighted average of the resampled particles:
$$ \hat{\mathbf{x}}k = \sum{i=1}^N wk^{(i)} \mathbf{x}k^{(i)} $$
Since all weights are equal after resampling, this simplifies to the average of the resampled particle poses. The uncertainty in the pose estimate can be approximated by the variance of the particle positions and orientations.

The process repeats for the next time step, with the resampled particles $\{ \mathbf{x} k^{(i)} \}{i=1}^N$ becoming the basis for the prediction step at time $k+1$.

Advantages of Particle Filters for Localization

The power of MCL lies in its ability to handle complex scenarios that are intractable for Kalman filters:

Non-linearities: The motion and measurement models can be arbitrarily non-linear. For example, robot motion involving complex turning maneuvers or range measurements from sensors that have non-linear relationships with distance are handled naturally.

Non-Gaussian Noise: The noise distributions associated with motion and sensing can be non-Gaussian. This is common in real-world robotics; for instance, odometry errors might not always be perfectly Gaussian, and sensor readings can be affected by outliers or multi-modal noise.

Multi-modal Distributions: MCL can represent and track multi-modal distributions, which can arise in situations where the robot might be in one of several distinct possible locations (e.g., in a corridor with multiple possible junctions, or when localization has been lost and regained). Kalman filters, being limited to unimodal Gaussian representations, struggle to cope with such scenarios.

Arbitrary Maps: MCL works effectively with various types of maps, including occupancy grid maps, feature-based maps, or even raw sensor data. The measurement model's implementation is crucial for adapting to different map representations.

Challenges and Considerations for Particle Filters

Despite their significant advantages, particle filters also present challenges:

Computational Cost: The primary drawback of particle filters is their computational complexity. To achieve a good approximation of the posterior distribution, a large number of particles (N) is often required, especially in high-dimensional state spaces. This can lead to high computational demands, which might be prohibitive for real-time applications on resource-constrained robots. The prediction, weighting, and resampling steps all need to be performed for each of the N particles.
The Curse of Dimensionality: As the dimension of the state space increases (e.g., for 3D robots or when estimating additional state variables like velocities or biases), the number of particles required to maintain a good representation of the distribution grows exponentially. This phenomenon, known as the "curse of dimensionality," can severely limit the applicability of standard particle filters to high-dimensional problems.
Particle Depletion and Smearing: While resampling combats degeneracy, it can also lead to "particle depletion" or "smearing." If the resampling step is too aggressive or occurs too frequently, the diversity of the particle set can be reduced, potentially leading to a loss of important information about the state distribution. Conversely, if resampling is not done often enough, degeneracy can still become an issue.

Tuning and Choice of Parameters: The performance of a particle filter can be sensitive to the choice of parameters, such as the number of particles (N), the specific sampling strategies for prediction and resampling, and the noise models used in the motion and measurement updates. Careful tuning is often required to achieve optimal performance.
Systematic Errors: While particle filters can handle arbitrary noise distributions, they are still susceptible to systematic errors in the motion or measurement models. If the models themselves are fundamentally flawed, even a large number of particles may not be able to correct for these systematic biases.

Enhancements and Variations

To address some of these challenges, several enhancements and variations of the basic particle filter have been developed:

Importance Sampling Strategies: Different importance sampling proposals can be used to improve the efficiency of particle filters. Instead of using the standard proposal distribution $p(\mathbf{x}_k | \mathbf{x}_{k-1}^{(i)}, \mathbf{u}_k)$, one might use a proposal that is closer to the posterior distribution, such as one derived from a Kalman filter's prediction step, which can help reduce variance.

Adaptive Particle Filters: These filters adjust the number of particles dynamically based on the complexity of the state distribution or the convergence of the filter.
Hierarchical Particle Filters: For high-dimensional problems, hierarchical approaches can break down the state space into smaller, more manageable subspaces.
Rao-Blackwellized Particle Filters: These filters exploit conditional independence in the state space. If a subset of the state variables can be tracked optimally using a Kalman filter (e.g., linear Gaussian dynamics for certain parts of the state), a Rao-Blackwellized particle filter can use a combination of Kalman filtering and particle filtering to reduce the dimensionality of the particle approximation. For example, in SLAM, the robot's pose can be estimated using particle filtering, while the map landmarks are estimated using Kalman filters.
Optimized Resampling Strategies: Research has focused on developing more efficient and statistically sound resampling techniques, such as stratified resampling or optimal resampling, to minimize resampling noise.

In summary, particle filters, particularly through the application of Monte Carlo Localization, represent a significant advancement in state estimation for robotic systems operating in uncertain and non-linear environments. Their ability to represent arbitrary probability distributions makes them a versatile tool where Kalman filter-based approaches falter. While computational cost and the curse of dimensionality remain important considerations, ongoing research and algorithmic improvements continue to enhance their practical utility, making them indispensable for complex robotic perception and navigation tasks. The flexibility to model non-linear dynamics and non-Gaussian noise without requiring analytical linearization makes them a cornerstone in modern robotics for robust state estimation.

The inherent limitations of relying solely on dead reckoning, as provided by odometry, become readily apparent over extended periods or in the presence of accumulating errors. While odometry offers a continuous, albeit noisy, estimate of relative motion, its drift—the tendency for the estimated position to diverge from the true position—is an unavoidable consequence of sensor inaccuracies and unmodeled dynamics. To counteract this drift and achieve global, absolute localization, mobile robots must integrate external sources of information that provide absolute or globally consistent cues about their position. Among the most effective and widely adopted of these external cues are **landmarks**.

Landmarks are distinct, identifiable features in the environment that can be reliably detected and recognized by the robot's sensors. They serve as spatial anchors, providing direct observations of the robot's position relative to a known, fixed point in the environment. The concept is intuitive: if a robot knows where it is in relation to a recognizable object, and it knows where that object is in a global map, it can significantly refine its own pose estimate. This integration of odometry with landmark-based measurements forms the bedrock of many robust mobile robot localization systems.

The Synergy of Odometry and Landmarks for Localization

The combination of odometry and landmark observations creates a powerful complementary system for state estimation. Odometry provides a high-frequency, continuous estimate of the robot's motion, effectively predicting its next state based on its current belief. This is crucial for maintaining a smooth and responsive pose estimate between landmark observations. However, odometry alone cannot resolve the accumulated drift. Landmark observations, on the other hand, occur less frequently and are typically less precise than odometry over very short intervals, but they provide absolute positional information that can correct the odometric drift.

The typical workflow involves using odometry to predict the robot's pose at a given time step. Then, when a sensor detects a landmark, this measurement is used to update the predicted pose, effectively "pulling" the robot's estimated position towards a location consistent with both the observed landmark and the known map. This process can be framed within various state estimation frameworks. While the previous discussion highlighted the flexibility of particle filters (Monte Carlo Localization) for non-linear and non-

Gaussian scenarios, the integration of odometry and landmarks is also a fundamental building block for Kalman filter-based localization, particularly the Extended Kalman Filter (EKF), and even forms the basis for more advanced techniques.

Let's consider the state vector $\mathbf{x}
k$ which, at time k, encapsulates the robot's pose. For a 2D robot, this is typically $\mathbf{x}k = [xk, yk, \thetak]^T$, representing its position (xk, yk) and orientation $\thetak$ in a global coordinate frame. The state estimation process can be broken down into two primary phases: prediction and update, driven by odometry and landmark measurements, respectively.

Prediction using Odometry

At each time step k, the robot receives odometry information, usually in the form of wheel encoder readings or velocity commands. This information, denoted as $\mathbf{u}
k$, represents the motion executed by the robot since the last time step. The odometry system provides a measurement of this motion, but it is subject to errors (e.g., wheel slippage, inaccuracies in encoder counts). A motion model, f, describes how the robot's pose is expected to change based on this motion command $\mathbf{u}k$, and it incorporates process noise $\mathbf{v}_{k-1}$ to account for the uncertainties.

The prediction step uses the odometry to project the robot's state from time $k-1$ to time k:

$$ \mathbf{x}
k^- = f(\mathbf{x}{k-1}, \mathbf{u}k) + \mathbf{v}{k-1}
$$

Here, $\mathbf{x}
k^-$ denotes the predicted state at time k before incorporating any new sensor measurements. The process noise $\mathbf{v}{k-1}$ is typically modeled as a Gaussian distribution with a covariance matrix $\mathbf{Q}_k$. In an EKF, the state transition function f is linearized using its Jacobian matrix to propagate the mean and covariance of the state.

If $\mathbf{x}
{k-1}$ is represented by a mean $\hat{\mathbf{x}}{k-1}$ and covariance $\mathbf{P}_{k-1}$, the predicted mean and covariance at time k are:

$$ \hat{\mathbf{x}}
k^- = f(\hat{\mathbf{x}}{k-1}, \mathbf{u}_k) $$
$$ \mathbf{P}k^- = \mathbf{F}k \mathbf{P}{k-1}
\mathbf{F}k^T + \mathbf{Q}_k $$

where $\mathbf{F}
k = \left. \frac{\partial f}{\partial \mathbf{x}}
\right|{\hat{\mathbf{x}}{k-1}, \mathbf{u}k}$ is the Jacobian of the motion model with respect to the state.

Update using Landmark Measurements

The true power of localization lies in integrating external measurements. When the robot's sensors detect a landmark, this provides a crucial piece of information for correcting the odometry-induced drift. A landmark measurement $\mathbf{z}_k$ typically consists of the observed range and bearing to the landmark, or perhaps its pixel coordinates in an image. However, there are several critical sub-problems that must be addressed before this measurement can be used for updating the robot's pose:

1. **Landmark Detection:** The robot's sensors (e.g., LiDAR, cameras, sonar) must first reliably detect features that can be classified as landmarks. This involves signal processing and feature extraction. The quality and robustness of this detection phase are paramount.

2. **Data Association:** This is arguably the most challenging aspect. When a landmark is detected, the system needs to determine *which* known landmark in its map it corresponds to. If the robot has M known landmarks in its map, and it observes a feature, it must decide if this feature is a previously seen landmark j (and if so, which one), or if it is a new, previously unmapped landmark. This matching process can be

probabilistic, considering the likelihood of observing the feature given the robot's current estimated pose and the map. Common data association techniques include nearest neighbor, probabilistic data association filters, and the Joint Probabilistic Data Association (JPDA) method. Incorrect data association can lead to severe localization errors.

3.
Measurement Model: Once a measurement $\mathbf{z}k$ has been reliably associated with a known landmark j, a measurement model h is used to relate the robot's pose $\mathbf{x}k$ to the observed landmark. The measurement model describes the expected observation $\hat{\mathbf{z}}k$ given the robot's pose and the known landmark's position in the map. For example, if the landmark j is at a known map coordinate $(x{Lj}, y{Lj})$, and the robot is at $(xk, yk, \thetak)$, the expected range r and bearing ϕ to the landmark would be:

$$ r = \sqrt{(x {Lj} - xk)^2 + (y{Lj} - yk)^2} $$
$$ \phi = \operatorname{atan2}(y{Lj} - yk, x{Lj} - xk) - \theta_k $$

The actual measurement $\mathbf{z}k = [r{obs}, \phi{obs}]^T$ will differ from this expected measurement due to sensor noise $\mathbf{e}k$. This noise is typically modeled as a Gaussian distribution with covariance $\mathbf{R}_k$.

The update step then uses this landmark measurement

to refine the predicted state $\mathbf{x}_k^-$. In an EKF, this involves calculating the Kalman gain and updating the state estimate and its covariance:

$$ \mathbf{K}_k = \mathbf{P}_k^- \mathbf{H}_k^T (\mathbf{H}_k \mathbf{P}_k^- \mathbf{H}_k^T + \mathbf{R}_k)^{-1} $$

$$ \hat{\mathbf{x}}_k = \hat{\mathbf{x}}_k^- + \mathbf{K}_k (\mathbf{z}_k - h(\hat{\mathbf{x}}_k^-)) $$

$$ \mathbf{P}_k = (\mathbf{I} - \mathbf{K}_k \mathbf{H}_k) \mathbf{P}_k^- $$

Here, $\mathbf{H}_k = \left. \frac{\partial h}{\partial \mathbf{x}} \right|_{\hat{\mathbf{x}}_k^-}$ is the Jacobian of the measurement model with respect to the state. The term $(\mathbf{z}_k - h(\hat{\mathbf{x}}_k^-))$ represents the measurement residual, i.e., the difference between the observed landmark measurement and the expected measurement based on the predicted robot pose.

Types of Landmarks and Sensors

The effectiveness of landmark-based localization is heavily dependent on the types of landmarks and the sensors used to detect them.

Artificial Landmarks: These are features explicitly placed in the environment to aid localization. Examples include:
Beacons: Radio frequency (RF) beacons emitting signals that can be triangulated or used for distance estimation.
QR Codes or AprilTags: Visual markers that can be detected by cameras, providing not only position but also potentially unique identifiers. These are excellent for data association as they are inherently distinct.
Infrared (IR) Emitters/Detectors: Simple active markers.
Pre-defined Structures: Pillars, specific colored panels, or distinctive architectural elements designed for easy recognition.

Artificial landmarks offer the advantage of being highly reliable and distinct, simplifying data association significantly. However, they require pre-installation and maintenance, which may not be feasible in all environments.

Natural Landmarks: These are naturally occurring features in the environment that are sufficiently distinctive to be detected and recognized. Examples include:
Corners and Edges: Detected by LiDAR or cameras.
Distinctive Textures: Areas of wall with unique patterns or colors, detected by cameras.
Geometric Shapes: Doorways, windows, furniture, or unique structural elements.

Visual Features (e.g., SIFT, SURF points): Keypoints extracted from camera images.

Natural landmarks are advantageous as they do not require pre-installation. However, their reliability can be variable. Environmental changes (lighting, occlusion, temporary objects) can affect their detectability and appearance, making data association more challenging.

Sensors for Landmark Detection:
LiDAR (Light Detection and Ranging): Provides precise range and bearing measurements, excellent for detecting geometric features like walls, corners, and doorways. LiDAR scans can be compared against a pre-existing map (e.g., a point cloud or occupancy grid) to identify landmarks.
Cameras (Monocular, Stereo, RGB-D): Cameras are versatile. Monocular cameras can detect visual features and their appearance (color, texture). Stereo cameras can infer depth, aiding in the detection of 3D geometric features. RGB-D cameras (like those on Microsoft Kinect or Intel RealSense) provide both color and depth information, facilitating rich feature extraction.
Sonar (Ultrasonic Sensors): Less precise than LiDAR but cost-effective for detecting larger objects and surfaces. They can be used to detect wall features.
Infrared (IR) Sensors: Can detect IR emitters or significant temperature differences.
Radio Frequency (RF) Receivers: Used for detecting beacons.

The choice of sensor dictates the types of landmarks that can be effectively utilized. For instance, a camera might be used to detect QR codes or distinctive visual textures, while a LiDAR might be used to detect geometric features in a warehouse.

The Challenge of Data Association

Data association is a critical bottleneck in landmark-based localization. If a robot incorrectly associates an observed feature with a known landmark, its pose estimate will be corrupted. Consider a scenario where a robot observes a feature that appears similar to landmark j, but it is actually landmark k. If the system mistakenly identifies it as landmark j, the update step will try to correct the pose as if the robot were near landmark j, leading to a large error. Over time, these incorrect associations can cause the robot's estimated pose to diverge significantly from its true pose, potentially leading to localization failure.

Several strategies exist to mitigate data association errors:

Probabilistic Methods: Instead of making hard assignments, probabilistic methods maintain a set of

hypotheses about which landmark the measurement corresponds to. The probability of each hypothesis is updated based on new measurements.
Feature Descriptors: For visual landmarks, robust feature descriptors (e.g., SIFT, ORB) can be used to match observed features to features in a database or map, providing a measure of similarity.
Spatial Consistency: The consistency of landmark measurements over time can be used. If a landmark consistently appears at a certain relative bearing and range from the robot, it increases confidence in its association.
Landmark Richness: Environments with many distinct landmarks simplify data association. When multiple landmarks are visible simultaneously, their relative positions can be used to disambiguate associations. For example, if a robot sees two landmarks, \$A\$ and \$B\$, it knows its relative pose to both. If it only has a rough idea of its location, it can narrow down the possibilities for \$A\$ and \$B\$ to a few candidate locations. The configuration of \$A\$ and \$B\$ relative to each other can then uniquely identify their true locations and, by extension, the robot's pose. This is sometimes referred to as "landmark geometry" or "map ambiguity resolution."

Example: Integrating Odometry and Landmark Measurements in an EKF

Let's illustrate with a simplified example of an EKF-based localization system using both odometry and a

single type of landmark (e.g., a detectable beacon at a known fixed position $(x
*L, y*L)$).

The state vector is $\mathbf{x} = [x, y, \theta]^T$.

The process model is based on odometry $\mathbf{u}_k = [\Delta r, \Delta \phi]^T$ (change in forward distance and change in turn angle).
The measurement model for a beacon is $\mathbf{z}*k = [r*{obs}, \phi*{obs}]^T$, where $r*{obs}$ is the observed distance and ϕ_{obs} is the observed bearing to the beacon.

Initialization:
Assume an initial estimate $\hat{\mathbf{x}}*0 = [\hat{x}*0, \hat{y}*0, \hat{\theta}*0]^T$ and its covariance $\mathbf{P}_0$.

At each time step k:

1.
Prediction (Odometry):
Receive odometry $\mathbf{u}*k = [\Delta r*k, \Delta \phi_k]$.
The motion model is:
$$ \hat{x}*k^- = \hat{x}*{k-1} + \Delta r*k \cos(\hat{\theta}*{k-1}) $$
$$ \hat{y}*k^- = \hat{y}*{k-1} + \Delta r*k \sin(\hat{\theta}*{k-1}) $$

$$ \hat{\theta}k^- = \hat{\theta}{k-1} + \Delta \phi_k $$

The process noise covariance $\mathbf{Q}k$ models uncertainties in Δrk and $\Delta \phik$. The Jacobian $\mathbf{F}k$ is computed and used to update the covariance $\mathbf{P}_k^-$.

2.

Landmark Detection and Data Association:

The robot's sensors detect a landmark. Suppose it is confidently associated with the known beacon at (xL, yL).

3.

Measurement Update:

The actual measured range and bearing to the beacon are $\mathbf{z}k = [r{obs,k}, \phi_{obs,k}]^T$.
The expected measurement model $h(\mathbf{x}k^-)$ based on the predicted pose $\hat{\mathbf{x}}k^- = [\hat{x}k^-, \hat{y}k^-, \hat{\theta}_k^-]^T$ is:

$$ r{exp} = \sqrt{(xL - \hat{x}k^-)^2 + (yL - \hat{y}_k^-)^2} $$

$$ \phi{exp} = \operatorname{atan2}(yL - \hat{y}k^-, xL - \hat{x}k^-) - \hat{\theta}k^- $$

The measurement residual is $\mathbf{y}k = \mathbf{z}k - h(\hat{\mathbf{x}}k^-)$. Note that the angular residual $\phi{obs,k} - \phi_{exp}$ needs to be handled carefully to account for angle wrapping (e.g., between $-\pi$ and π).
The Jacobian of the measurement model $\mathbf{H}k$ is computed with respect to $\hat{\mathbf{x}}k^-$.
The measurement noise covariance $\mathbf{R}_k$ models uncertainties in the range and bearing sensors.

The Kalman gain $\mathbf{K}_k$ is calculated.
The state estimate is updated: $\hat{\mathbf{x}}_k = \hat{\mathbf{x}}_k^- + \mathbf{K}_k \mathbf{y}_k$.
The covariance is updated: $\mathbf{P}_k = (\mathbf{I} - \mathbf{K}_k \mathbf{H}_k) \mathbf{P}_k^-$.

This process repeats, with odometry continuously predicting the pose and landmark observations periodically correcting the estimate, thus mitigating drift and achieving a more accurate and globally consistent localization.

Challenges and Advanced Considerations

While the integration of odometry and landmarks is powerful, several challenges and advanced considerations arise:

Sensor Fusion: In practice, robots often use multiple types of sensors and landmarks. Effectively fusing information from LiDAR, cameras, and odometry requires sophisticated sensor fusion techniques. This might involve using a single state estimator that can process different types of measurements or employing a multi-stage approach where information is passed between different modules.
Dynamic Environments: Environments are rarely static. Moving objects, changing lighting conditions, or

temporary obstructions can affect landmark detection and data association. Robust systems often incorporate mechanisms to detect and handle dynamic elements, or to operate with partial information when landmarks are occluded.

Landmark Map Uncertainty: The accuracy of the landmark map itself plays a crucial role. If the map is inaccurate, even perfect sensor measurements and estimation algorithms will lead to incorrect localization. Techniques for simultaneously building and localizing (SLAM) address this by estimating both the robot's pose and the map of landmarks concurrently.

Computational Efficiency: Especially in environments with many landmarks, the computational cost of data association and the measurement update can become significant, particularly if a probabilistic approach is used. Efficient algorithms and data structures are necessary to keep the computation manageable for real-time operation.

Initialization: Similar to particle filters, Kalman filter-based localization often requires a good initial guess of the robot's pose. If the initial estimate is too far from the true pose, the filter may diverge. Techniques like "global localization" or "kidnapped robot problem" solvers are used to re-localize the robot when it has lost track of its pose.

Outliers: Sensor measurements can occasionally contain significant outliers due to sensor malfunctions or unusual environmental conditions. Robust estimation techniques, such as using M-estimators or a mixture of Gaussians for measurement noise, can help the filter be less sensitive to such outliers.

In conclusion, combining odometry with external landmark observations represents a fundamental and highly effective strategy for achieving robust robot localization. Odometry provides the necessary high-frequency motion updates, while landmarks serve as absolute reference points that correct accumulated drift. The success of this approach hinges on reliable landmark detection, accurate data association, and a well-tuned state estimation framework, such as the Kalman Filter. This synergistic integration forms the backbone of many navigation systems, enabling mobile robots to operate reliably and accurately in diverse and often challenging environments.

The previous discussion has established the critical role of landmarks in mobile robot localization. By integrating odometry with landmark observations, a robot can significantly improve its pose estimation accuracy and overcome the limitations of drift inherent in dead reckoning. However, this approach implicitly assumes that a map of the environment, containing the locations of these landmarks, is already available. In many real-world scenarios, this is a luxury the robot cannot afford. The robot might be deployed into an entirely unknown environment, where no prior map exists. In such situations, the robot faces a dual challenge: it must not only determine its own position but also construct a consistent representation of the environment simultaneously. This is the essence of the problem addressed by Simultaneous Localization and Mapping, commonly abbreviated as SLAM.

At its core, SLAM is a bootstrapping problem. Imagine a robot placed in a completely empty, unknown room. It has no idea where it is, and it has no map of the room. Its initial pose estimate is essentially random, and its map is empty. To move and explore, it needs to make some decisions about its initial pose and then start building a map. As it moves, its pose estimate will drift, just like with odometry. However, as it perceives landmarks (either natural or artificial), it can use these observations to refine its pose estimate relative to those landmarks. The critical insight of SLAM is that these landmarks, as they are observed and their relative positions determined, also provide the building blocks for the map. Conversely, as the map of landmarks becomes more consistent, it provides a more stable and accurate reference frame for localizing the robot. This creates a tightly coupled, interdependent relationship between mapping and localization.

This interplay is often described as a "chicken-and-egg" problem. Accurate localization requires a reliable map to provide reference points. Conversely, constructing a consistent and accurate map requires the robot to know its own position with sufficient precision to correctly place observed features into the map. If the robot's pose estimate is inaccurate, then features observed at different times might be incorrectly placed relative to each other, leading to a distorted and erroneous map. Similarly, if the map is inaccurate, the robot's attempts to localize against it will be flawed. SLAM algorithms are designed to break this cycle by iteratively refining both the robot's pose estimate and the map representation concurrently.

The fundamental components of any SLAM system are sensing, mapping, and data association, working in concert. The
sensing component is responsible for acquiring raw data from the robot's environment using its onboard sensors. These sensors could be LiDAR, cameras, sonar, radar, or a combination thereof. The type of sensor heavily influences the kind of map that can be built and the features that can be reliably detected. For instance, a LiDAR sensor might be used to scan for geometric features like walls, corners, and doorways, leading to a geometric map, often represented as a point cloud or an occupancy grid. A camera, on the other hand, might detect visual features such as distinctive textures, colors, or specific objects, leading to a feature-based map or a semantic map.

The
mapping component takes the processed sensor data and integrates it into a representation of the environment. This representation can take many forms. As mentioned, it could be a geometric map, which describes the physical structure of the environment, such as an occupancy grid where each cell indicates the probability of being occupied by an obstacle. Alternatively, it could be a feature-based map, where the map consists of a collection of landmarks, each with a unique identifier and a precisely estimated position and uncertainty. Semantic maps go a step further by not

only representing the geometry but also the meaning or function of different parts of the environment, identifying objects like chairs, tables, or doors. The choice of map representation is dictated by the intended application and the capabilities of the sensing system.

Data association, as discussed in the context of localization, is equally, if not more, critical in SLAM. When a robot perceives a feature or landmark in its environment, it must correctly associate this observation with an existing entry in its map, or decide if it is a new landmark that needs to be added to the map. This is a non-trivial task, especially in large or complex environments where similar features might appear multiple times. Incorrect data association can lead to the merging of distinct landmarks or the creation of duplicate entries for the same landmark, corrupting the map and leading to severe localization errors. Robust data association strategies are therefore a cornerstone of successful SLAM systems, often employing probabilistic methods to manage uncertainty and ambiguity.

The process of SLAM can be broadly categorized into two main families of approaches: filter-based methods and optimization-based methods.

Filter-based SLAM aims to maintain a probabilistic estimate of the robot's state (its pose) and the map's state (the locations of landmarks) over time. This is analogous to how Kalman filters or particle filters are used for pure localization. In SLAM, the state vector is augmented to include not only the robot's pose but also the parameters describing the map. For example, if the map consists of N landmarks, and each landmark is represented by its 2D position (x_i, y_i), the state vector might look like:

$$ \mathbf{X}_k = [\mathbf{x}_{robot,k}^T, \mathbf{m}_1^T, \mathbf{m}_2^T, \dots, \mathbf{m}_N^T]^T $$

where $\mathbf{x}_{robot,k}$ is the robot's pose at time k, and $\mathbf{m}_i = [x_i, y_i]^T$ is the map-based representation of landmark i.

The SLAM process unfolds in a predictive-update cycle. At each time step k:

1.
Prediction: The robot moves based on its odometry or other motion commands. The state (robot pose and map) is propagated forward using a motion model, incorporating process noise. If the robot's pose estimate drifts, the positions of all landmarks in the map will also be shifted accordingly.

2.
Update: The robot observes new sensor data. For each observed feature, data association is performed to determine if it corresponds to an existing landmark in the map or if it is a new landmark.

If it's a known landmark: The observation is used to update the robot's pose and the position of that landmark simultaneously. This is where the interdependence of localization and mapping is most evident. The correction to the robot's pose based on the landmark measurement also refines the map, and a refined map (due to corrected landmark positions) then leads to a more accurate pose estimate.
If it's a new landmark: The robot uses its current best estimate of its pose to add this new landmark to the map. This new landmark is then also associated with an uncertainty.

The Extended Kalman Filter (EKF) was one of the earliest and most widely used filters for SLAM, particularly for feature-based maps. The EKF-SLAM approach maintains a mean and covariance for the entire state vector (robot pose + all landmark positions). However, as the number of landmarks grows, the state vector and its covariance matrix become very large, leading to significant computational complexity ($O(N^2)$ for landmark updates with N landmarks). This quadratic complexity is a major drawback.

To overcome the limitations of the EKF, other filter-based approaches have been developed. Particle Filters (also known as Monte Carlo Localization in its localization-only form, and FastSLAM when applied to SLAM) offer a way to handle non-linearities and non-Gaussian noise more effectively. In FastSLAM, a particle represents a hypothesis about the robot's trajectory and the map. Each particle maintains its own map. For a particle representing a robot's trajectory, the map is often represented as a set of features. The key idea is that for each particle, the landmark observations are conditioned on the particle's assumed trajectory. This allows for a more flexible and often more robust representation of uncertainty, especially in complex environments. However, particle filters can be computationally expensive due to the large number of particles required to represent the probability distribution accurately.

More recently,
optimization-based SLAM approaches have gained significant traction, particularly for constructing highly accurate and globally consistent maps. Instead of maintaining a probabilistic estimate that is updated incrementally at each time step, optimization-based methods frame the entire SLAM problem as a large-scale optimization task. The goal is to find the robot's trajectory and the map that best explain the observed sensor measurements, minimizing a cost function that penalizes inconsistencies.

This involves constructing a graph where nodes represent robot poses at different time steps and landmark positions, and edges represent constraints derived from sensor measurements (odometry constraints between consecutive poses, and landmark constraints between a robot pose and a landmark). The objective is to find a configuration of nodes that minimizes the total error across all constraints. This is often formulated as a non-linear least-squares problem.

A typical optimization-based SLAM formulation might seek to minimize a function like:

$$ \min_{\mathbf{X}} \sum_{i} \| \mathbf{z}_i - h(\mathbf{x}_{robot}, \mathbf{m}) \|_{\mathbf{\Sigma}_i}^2 $$

where $\mathbf{X}$ is the set of all robot poses and landmark positions, $\mathbf{z}_i$ are the sensor measurements, h is the measurement model, and $\mathbf{\Sigma}_i$ is the measurement uncertainty.

Popular optimization techniques used in SLAM include Levenberg-Marquardt and Gauss-Newton algorithms. These methods often involve iterative refinement of the robot's trajectory and map. One of the significant advantages of optimization-based methods is their ability to perform "global" optimization, often by leveraging techniques like pose graph optimization. This means they can correct for accumulated drift over long trajectories by considering all measurements simultaneously, leading to more consistent and accurate maps than traditional filter-based approaches, especially when dealing with loops (returning to a previously visited area).

However, optimization-based methods can also be computationally demanding. For very large-scale environments, the resulting optimization problem can be enormous, requiring efficient sparse matrix techniques and sometimes distributed computing to solve in a practical timeframe. Furthermore, like filter-based methods, they are sensitive to data association errors. If a wrong association is made, it introduces a significant error into the optimization problem, potentially corrupting the entire map.

Regardless of the specific algorithmic approach (filter-based or optimization-based), the core challenge of SLAM remains the need to handle uncertainty in both the robot's pose and the map. The reliability of SLAM systems is directly tied to the accuracy and robustness of their sensing modalities, the sophistication of their data association mechanisms, and the efficiency and correctness of their state estimation and map building algorithms. As we delve deeper into the algorithms, we will explore these facets in more detail, understanding how robots can effectively navigate and build knowledge of their surroundings from scratch.

Chapter 7: Simultaneous Localization and Mapping (SLAM)

The Simultaneous Localization and Mapping (SLAM) problem can be formally defined as the task of constructing or updating a map of an unknown environment while simultaneously estimating the robot's pose within that map, using only onboard sensors and the data derived from them. This mutual dependency necessitates a concurrent solution; without a map, localization is impossible beyond short-term dead reckoning, and without accurate localization, map construction will be flawed. The fundamental goal is to estimate the robot's trajectory, $\mathbf{X}{1:T} = \{\mathbf{x}1, \mathbf{x}2, \dots, \mathbf{x}T\}$, and the map, $\mathcal{M}$, from a sequence of sensor measurements, $\mathbf{Z}{1:T}$, and motion commands, $\mathbf{U}{1:T}$. Here, $\mathbf{x}t$ represents the robot's pose (position and orientation) at time step t, and $\mathbf{u}{t}$ is the motion command applied between time $t-1$ and t.

Mathematically, SLAM can be framed as inferring the joint posterior probability distribution of the robot's trajectory and the map given the sensor and motion data:

$$ P(\mathbf{X}{1:T}, \mathcal{M} | \mathbf{Z}{1:T}, \mathbf{U}_{1:T}) $$

This probability distribution encapsulates all the knowledge about the robot's state and the environment. In practice, directly computing this full posterior is often intractable due to the high dimensionality and complex dependencies involved. Therefore, various approximation techniques and algorithmic frameworks are employed, broadly categorized as filter-based and optimization-based methods, as previously introduced.

Filter-based SLAM approaches, such as Extended Kalman Filter (EKF) SLAM and Particle Filter SLAM (e.g., FastSLAM), maintain an evolving estimate of the robot's state and map over time. At each time step t, the process involves predicting the next state based on the motion command $\mathbf{u}t$ and the previous state estimate, and then updating this prediction using the new sensor measurements $\mathbf{z}t$. The prediction step typically involves propagating the robot's pose using a motion model, which accounts for the odometry and its associated uncertainty. If the robot's pose estimate is $\hat{\mathbf{x}}{t-1}$ and its covariance is $\mathbf{P}{t-1}$, the predicted pose $\hat{\mathbf{x}}t^-$ and its covariance $\mathbf{P}t^-$ are computed using the motion model f:
$$ \hat{\mathbf{x}}t^- = f(\hat{\mathbf{x}}{t-1}, \mathbf{u}_t) $$
$$ \mathbf{P}t^- = \mathbf{J}f \mathbf{P}{t-1} \mathbf{J}f^T + \mathbf{Q} $$
where $\mathbf{J}_f$ is the Jacobian of the motion model with respect to the state, and $\mathbf{Q}$ is the covariance of the process noise.

The update step utilizes sensor measurements $\mathbf{z}
t$ to refine the predicted state. This is where the map plays a crucial role. If the sensor detects features that can

be associated with existing landmarks in the map, these measurements provide information to correct both the robot's pose and the map's landmark positions. For a feature-based map where $\mathcal{M} = \{\mathbf{m}1, \dots, \mathbf{m}N\}$ and $\mathbf{m}i$ is the position of the i-th landmark, the state vector includes the robot's pose and all landmark positions. When a measurement $\mathbf{z}t$ corresponding to landmark j is observed, the Kalman filter update equations are used to refine the state vector $\mathbf{X}t$ and its covariance $\mathbf{P}t$. The measurement model h relates the robot's pose $\mathbf{x}t$ and the landmark position $\mathbf{m}j$ to the expected measurement: $\hat{\mathbf{z}}t = h(\mathbf{x}t, \mathbf{m}j)$. The update involves calculating the Kalman gain and then updating the state and covariance.

A significant challenge in filter-based SLAM is the computational complexity, particularly for EKF-SLAM, which grows quadratically with the number of landmarks N. The state vector size is $6+3N$ (for 3D poses and 3D landmarks) or $3+2N$ (for 2D poses and 2D landmarks), and the covariance matrix is $(6+3N) \times (6+3N)$ or $(3+2N) \times (3+2N)$. Each landmark observation requires updating the state and covariance, and the computational cost for updating the covariance matrix is $O(N^2)$ and for updating the state is $O(N)$. This becomes prohibitive for environments with many landmarks.

Particle filters, such as FastSLAM, offer an alternative by

representing the posterior distribution using a set of weighted samples (particles). Each particle represents a hypothesis about the robot's entire trajectory and the map. For a particle s, its trajectory is $\{\mathbf{x}_{1}^{(s)}, \dots, \mathbf{x}_{T}^{(s)}\}$ and its map is $\mathcal{M}^{(s)}$. The key innovation in FastSLAM is that the map is often represented as a collection of conditional particle filters, one for each landmark. This decomposition significantly reduces computational complexity, making it more scalable than EKF-SLAM. However, particle filters require a sufficient number of particles to accurately represent the posterior distribution, which can still lead to high computational demands, especially in high-dimensional state spaces.

Optimization-based SLAM methods reframe the problem as a global optimization task. Instead of incrementally updating a state estimate, these methods aim to find the robot's trajectory and map that best explain all the sensor and motion data collected over a period. This is typically formulated as a non-linear least-squares problem, minimizing an objective function that penalizes discrepancies between sensor measurements and predictions derived from the estimated states and map.

The formulation often involves constructing a factor graph, where nodes represent robot poses and landmark positions, and edges (factors) represent constraints derived from measurements. Odometry provides constraints between consecutive robot poses, while landmark observations provide constraints between a robot pose and a landmark. The objective is to find the configuration of nodes (poses and landmark positions) that minimizes the sum of squared errors of these constraints, weighted by their uncertainties:

$$ \min{\mathbf{X}{1:T}, \mathcal{M}} \sum{t=1}^{T} \| \mathbf{z}{odometry, t} - h{odom}(\mathbf{x}{t-1}, \mathbf{x}t) \|^2 + \sum{k} \| \mathbf{z}{landmark, k} - h{lmk}(\mathbf{x}{t{k}}, \mathbf{m}{ik}) \|^2 $$
Here, $h{odom}$ and $h{lmk}$ are the odometry and landmark measurement models, respectively. The term $\mathbf{z}{landmark, k}$ is the k-th landmark observation, which associates a specific robot pose $\mathbf{x}{tk}$ with a landmark $\mathbf{m}{i_k}$. This formulation allows for global correction of accumulated drift.

Techniques like pose graph optimization are a prominent example. In this approach, only the robot's poses are considered as nodes in the graph, and constraints are derived from odometry and loop closures. The map itself is often implicitly represented or constructed separately after the trajectory is optimized. This is particularly effective for large-scale environments where building a detailed map during the initial exploration is challenging. However, constructing a highly accurate and dense map usually requires considering the map's structure within the optimization.

The optimization problem is typically solved using iterative methods such as Gauss-Newton or Levenberg-

Marquardt. These methods iteratively refine the robot's trajectory and map estimates until convergence. A key advantage is their ability to achieve globally consistent maps, as they consider all available information simultaneously. This is particularly important for addressing the challenge of loop closure.

However, optimization-based methods can also be computationally demanding, especially for very large environments. The resulting optimization problem can be sparse but massive, requiring efficient solvers and potentially distributed computing. Moreover, like filter-based methods, they are highly susceptible to errors in data association. A single incorrect data association can introduce a large error into the objective function, leading to significant distortions in the optimized trajectory and map.

Key Challenges in SLAM:

Several fundamental challenges plague all SLAM algorithms, irrespective of their underlying mathematical framework.

1.
Data Association: This is arguably the most critical and difficult challenge. When a robot observes a feature (e.g., a corner, a distinctive visual point, or a landmark), it must determine if this observation corresponds to a

previously mapped feature or if it is a new, unmapped feature. In large or perceptually aliased environments, where similar features appear multiple times, incorrect data association can lead to severe consequences. If a new observation is mistakenly associated with an existing map feature, it can bias the landmark's estimated position. Conversely, if a new observation of an existing landmark is treated as a new feature, the map will become redundant and inconsistent, with multiple entries for the same physical entity. This can lead to significant localization errors and a corrupted map. Probabilistic methods, such as using the Mahalanobis distance between predicted and observed feature locations, and employing techniques like the Nearest Neighbor (NN) rule or Joint Probabilistic Data Association (JPDA), are used to mitigate this, but ambiguity remains a persistent issue, especially in complex scenarios.

2.

Loop Closure: This refers to the problem of recognizing that the robot has returned to a previously visited location. When a robot completes a loop in its trajectory, it has the opportunity to correct for the accumulated drift in its pose estimate. Without loop closure, the errors in the robot's pose estimate and the map's landmark positions grow over time, leading to a map that is internally inconsistent and a robot that is poorly localized. Detecting a loop closure reliably is challenging. It requires comparing the current sensory input with historical sensory data or map features to find a match. Once a loop closure is detected, it provides a strong constraint that can be used to significantly improve the accuracy of the robot's trajectory and the global consistency of the map. This is often achieved by re-evaluating a portion of the robot's past trajectory and map using the new constraint, a process central to optimization-based SLAM. The effectiveness of loop closure detection is heavily dependent on the distinctiveness of the environment and the robustness of the feature extraction and matching algorithms.

3.
Computational Complexity and Scalability: As mentioned, SLAM algorithms need to process a continuous stream of sensor data and update both the robot's pose and the map. For environments with many landmarks or long trajectories, the state space can become enormous, leading to prohibitive computational costs. EKF-SLAM's quadratic complexity with the number of landmarks is a prime example. While particle filters offer better theoretical scalability, they still require a large number of particles for accurate representation. Optimization-based methods can become computationally intensive as the size of the optimization problem grows, requiring efficient sparse solvers and careful management of graph structures. The need for real-time performance in many applications necessitates algorithms that are not only accurate but also computationally efficient and scalable to large environments.

4.
Sensor Noise and Model Uncertainties: All real-world sensors are subject to noise, which introduces uncertainty into the measurements. Similarly, the

motion models used to predict the robot's movement are imperfect, and the relationship between sensor readings and map features (measurement models) is also subject to approximations and inaccuracies. These uncertainties propagate through the SLAM system, affecting both the robot's pose estimate and the map's accuracy. For example, noisy odometry readings will cause the robot's predicted position to drift, and noisy landmark measurements will lead to inaccurately placed landmarks in the map. SLAM algorithms must explicitly model and manage these uncertainties, typically using probabilistic frameworks (like Gaussian distributions in Kalman filters or particle distributions in particle filters), to produce reliable estimates. The accuracy of the SLAM system is directly dependent on the quality of the sensor data and the fidelity of the underlying motion and measurement models.

5.
Landmark Representation and Map Quality: The choice of how to represent the map and its constituent landmarks significantly impacts the performance and capabilities of a SLAM system. Feature-based maps, relying on distinct, identifiable landmarks (e.g., corners, visual features), are often used in EKF-SLAM and some optimization approaches. However, in visually or geometrically uniform environments, finding and tracking such features can be difficult. Occupancy grids, representing the environment as a grid of cells with occupancy probabilities, are common in sensor-based SLAM (e.g., using LiDAR). They provide a dense representation of the environment but can be memory-intensive and less suited for accurate localization based

on sparse features. More advanced approaches involve semantic maps that label map elements with their meaning (e.g., "door," "chair"), which can aid in higher-level reasoning and navigation but add complexity to the SLAM process. The quality of the map – its accuracy, resolution, and consistency – is a direct outcome of the SLAM algorithm's performance and the sensor data used.

These challenges highlight the inherent difficulty of the SLAM problem. It is a problem that requires not only sophisticated mathematical techniques but also a deep understanding of sensor characteristics, environmental properties, and computational limitations. The development of robust and efficient SLAM algorithms is an ongoing area of research, driving progress in autonomous navigation, robotics, and augmented reality.

Feature-based Simultaneous Localization and Mapping (SLAM) is a prominent paradigm in robotics that tackles the SLAM problem by representing the environment as a collection of distinct, salient features. These features, often referred to as landmarks, can be points of interest in an image, corners in a scene, or specific geometric primitives. The core idea is to detect these landmarks in the sensor readings, establish correspondences between observations of the same landmark across different time steps or viewpoints, and then leverage these correspondences to simultaneously estimate the robot's trajectory and the three-dimensional (3D) positions of these landmarks. This approach is particularly well-suited for scenarios where the environment contains rich visual or geometric cues that can be reliably extracted and tracked.

The process of feature-based SLAM can be broadly decomposed into several key stages. First, **feature detection** involves identifying distinctive points or regions within the sensor data that are likely to be invariant to changes in viewpoint or illumination to some extent. For vision-based systems, popular feature detectors include SIFT (Scale-Invariant Feature Transform), SURF (Speeded Up Robust Features), ORB (Oriented FAST and Rotated BRIEF), and FAST (Features from Accelerated Segment Test). These algorithms aim to find keypoints that possess unique local characteristics. For instance, a corner in an image is a good candidate for a feature because its local neighborhood has strong gradients in multiple directions, making it less likely to be confused with other parts of the image. The output of a feature detector is a set of 2D or 3D coordinates of these keypoints, often accompanied by a descriptor that encodes the local appearance around the keypoint. This descriptor is crucial for matching features across different sensor readings.

Following feature detection, **feature description** is performed. The descriptor is a compact representation of the local image patch surrounding the detected keypoint. For successful matching, these descriptors must be robust to variations in scale, rotation, and illumination. SIFT, for example, computes a histogram of gradient orientations in a local neighborhood, creating a descriptor that is invariant to these transformations. SURF uses Haar wavelet responses. ORB combines the FAST keypoint detector with the BRIEF descriptor and adds orientation invariance. The quality and distinctiveness of the feature descriptor directly impact the accuracy of subsequent data association.

The most critical and often most challenging stage is **feature matching**, also known as data association. Given a set of features detected in the current sensor reading and a set of features already present in the map (or detected in previous readings), the goal is to determine which current features correspond to which existing map features. This is achieved by comparing the descriptors of the detected features with the descriptors of the map features. A common approach is to use a distance metric (e.g., Euclidean distance, Hamming distance) to quantify the similarity between descriptors. Features whose descriptors are sufficiently close are considered potential matches. To reduce ambiguity and improve reliability, several techniques are employed. The nearest neighbor approach, where each feature is matched to the closest descriptor in the map, is a common starting point. However, this can lead to incorrect matches, especially in environments with repetitive patterns. More robust methods include ratio tests (e.g., Lowe's ratio test for SIFT), which discard matches where the distance to the nearest neighbor is too close to the distance to the second nearest neighbor, indicating a lack of distinctiveness. Geometric constraints, such as epipolar geometry for stereo vision or fundamental matrix constraints for monocular vision, can also be used to filter out incorrect matches based on the expected geometric relationship between corresponding points.

Once features are detected, described, and matched, the SLAM system uses this information to update its state. The state typically includes the robot's pose (position and orientation) and the 3D locations of the landmarks. **Extended Kalman Filter (EKF) SLAM** was one of the earliest and most influential feature-based SLAM algorithms. In EKF-SLAM, the state vector comprises the robot's pose and the 3D coordinates of all landmarks observed so far. When the robot moves, its pose estimate is predicted using a motion model (e.g., odometry), and the uncertainty in this prediction is propagated. When a sensor measurement provides a correspondence between a robot feature and a map landmark, the Kalman filter update step is applied. This update refines both the robot's pose estimate and the position estimate of the associated landmark, simultaneously reducing their respective uncertainties. The measurement model in EKF-SLAM relates the robot's pose and the landmark's 3D position to the expected sensor measurement (e.g., bearing and range to the landmark).

The EKF-SLAM algorithm iterates through these prediction and update steps. However, it suffers from significant drawbacks. The state vector grows linearly with the number of landmarks, N. For a 2D robot and 3D landmarks, the state vector size is $3 + 3N$, and the covariance matrix is $(3+3N) \times (3+3N)$. The computational complexity for updating the covariance matrix is $O(N^2)$ per sensor observation, which quickly becomes prohibitive for environments with a large number of landmarks. Furthermore, EKF-SLAM is prone to divergence if linearization errors become too large or if data association errors are made. Despite its limitations, EKF-SLAM laid the groundwork for many subsequent feature-based SLAM approaches by demonstrating the feasibility of solving the SLAM problem using a probabilistic filtering framework.

To overcome the scalability issues of EKF-SLAM, **Particle Filter SLAM**, notably **FastSLAM**, emerged as a powerful alternative. FastSLAM represents the posterior probability distribution of the robot's trajectory and map using a set of weighted particles. Each particle represents a complete hypothesis about the robot's past trajectory and the configuration of the map. A key innovation in FastSLAM is the efficient representation of the map. Instead of maintaining a single, monolithic map estimate, FastSLAM decomposes the map into a collection of independent landmark estimates, typically represented by individual Kalman filters (or particle filters for more complex landmark models). Each particle in the FastSLAM algorithm then maintains its own set of these landmark filters.

When the robot moves, each particle's trajectory is updated based on the motion model. When a new sensor measurement of a landmark is observed, the update step is applied to the corresponding landmark filter within

each particle. The weight of a particle is updated based on the likelihood of observing the new landmark measurement given the particle's current hypothesis. Resampling of particles is performed periodically to discard low-weight particles and focus computational resources on more promising hypotheses. This approach offers significant computational advantages. While the number of particles M still affects accuracy, the complexity per landmark observation becomes $O(M)$, and the overall complexity is often $O(M \cdot L)$, where L is the number of observed landmarks, which is far more scalable than the $O(N^2)$ of EKF-SLAM. FastSLAM has been instrumental in enabling feature-based SLAM in larger and more complex environments.

A more recent and highly effective paradigm for feature-based SLAM is
Graph-Based SLAM (also known as Pose Graph Optimization). Instead of maintaining a consistent probabilistic estimate over time using filters, graph-based SLAM formulates the problem as a global optimization task. The environment and the robot's trajectory are represented as a graph. The nodes in the graph typically represent the robot's poses at different time steps, and the edges represent constraints between these poses. These constraints are derived from sensor measurements, primarily odometry and loop closures.

In a typical graph-based SLAM system,
odometry measurements provide relative motion constraints between consecutive poses. For example, if the robot moves from pose $\mathbf{x}_t$ to pose $\mathbf{x}_{t+1}$, the odometry sensor provides an

estimate of the transformation $\Delta \mathbf{x}_t$ such that $\mathbf{x}_{t+1} \approx \mathbf{x}_t \oplus \Delta \mathbf{x}_t$, where $\oplus$ denotes the composition of poses. This constraint is represented as an edge in the pose graph, with an associated uncertainty reflecting the sensor noise.

The true power of graph-based SLAM comes from **loop closures**. When the robot revisits a previously mapped location, it can establish a constraint between the current pose and an earlier pose in its trajectory. This loop closure constraint is crucial for correcting the accumulated drift in the robot's pose estimates. Detecting loop closures typically involves matching the current sensor observations (e.g., camera images, LiDAR scans) with historical observations or map features. Once a loop closure is detected, a new edge is added to the pose graph, connecting the current pose node to the previously visited pose node.

The problem then becomes finding the set of robot poses that best satisfy all the constraints in the graph. This is typically formulated as a non-linear least-squares optimization problem:

$$ \min_{\mathbf{X}_{1:T}} \sum_{i=1}^{T-1} \| \mathbf{z}_{odom, i} - (\mathbf{x}_i^{-1} \oplus \mathbf{x}_{i+1}) \|^2 + \sum_{j \in \text{loops}} \| \mathbf{z}_{loop, j} - (\mathbf{x}_{k_j}^{-1} \oplus \mathbf{x}_{l_j}) \|^2 $$

Here, $\mathbf{x}
i$ represents the pose at time i, $\mathbf{z}{odom, i}$ is the odometry measurement between $\mathbf{x}i$ and $\mathbf{x}{i+1}$, and $\mathbf{z}{loop, j}$ is the loop closure constraint between poses $\mathbf{x}{kj}$ and $\mathbf{x}{l_j}$. The term $\mathbf{x}^{-1}$ denotes the inverse of a pose. The objective is to minimize the sum of squared errors of these constraints, weighted by their uncertainties.

This non-linear least-squares problem is usually solved using iterative optimization techniques such as the Gauss-Newton method or the Levenberg-Marquardt algorithm. These algorithms start with an initial estimate of the robot's trajectory (often obtained from odometry alone) and iteratively refine the poses to reduce the overall error in the graph.

A significant advantage of graph-based SLAM is its ability to achieve global consistency. By optimizing the entire trajectory simultaneously, it can effectively correct for accumulated drift and produce a map that is globally accurate. This is particularly effective for large-scale environments and for applications where a globally consistent map is essential. Furthermore, the graph structure can be very sparse, allowing for efficient optimization using specialized solvers.

When using feature-based SLAM, the representation of the map itself can vary. In EKF-SLAM and some particle filter implementations, the map is explicitly represented by the 3D positions of the detected landmarks. These landmarks are typically point features. However, for robustness and to handle environments with fewer distinct point features, more complex landmark representations can be used. For example, line segments or planar surfaces can also serve as landmarks. Extracting and representing these geometric primitives requires more sophisticated feature extraction algorithms.

In graph-based SLAM, the focus is primarily on optimizing the robot's poses. The map can be constructed in a few ways. One approach is to perform pose graph optimization first, obtaining a globally consistent robot trajectory, and then use this trajectory to build a map by projecting sensor data or detected features into the optimized coordinate frame. Alternatively, landmark positions can also be included as nodes in the optimization graph, making the problem a joint optimization of poses and landmark locations. This increases the size of the optimization problem but can lead to a more accurate and consistent map directly.

The choice between filter-based (like EKF-SLAM and FastSLAM) and optimization-based (like graph-based SLAM) approaches for feature-based SLAM depends on the application requirements. Filter-based methods are inherently online and can provide a continuous estimate of the robot's state and map as data arrives, making them suitable for real-time navigation.

However, they may struggle with accumulating errors over long trajectories and can have computational limitations. Optimization-based methods, on the other hand, excel at producing globally consistent maps and trajectories by considering all data at once. They are often used for offline map building or when a highly accurate, globally consistent map is required. Modern SLAM systems often combine elements of both, using filters for real-time operation and periodic optimization for global consistency.

The success of feature-based SLAM hinges critically on the quality and robustness of the underlying feature detection and matching algorithms. Environments with rich textures and distinctive geometric features are ideal. However, in visually monotonous or highly dynamic environments, feature-based methods can struggle. For instance, in a corridor with uniform walls, it can be difficult to find stable features. In such cases, other SLAM paradigms, such as those based on direct sensor measurements (e.g., LiDAR scans or dense optical flow), might be more appropriate.

Furthermore, the problem of **ambiguity in data association** remains a significant challenge. Even with sophisticated feature descriptors and matching techniques, it is possible to incorrectly associate a detected feature with a wrong landmark in the map. This can lead to significant errors that propagate through the system, potentially causing divergence or producing a corrupted map. Advanced techniques like Multiple Hypothesis Tracking (MHT) or Bayesian filtering approaches try to manage these ambiguities by maintaining multiple hypotheses for each feature association, but these methods can increase computational complexity.

The development of feature-based SLAM has been a cornerstone of progress in robotics and computer vision, enabling applications ranging from autonomous navigation in indoor environments to augmented reality systems and robotic exploration. The continuous refinement of feature descriptors, matching algorithms, and optimization techniques continues to push the boundaries of what is possible, making robots more capable of understanding and navigating the complex world around them.

Direct and indirect SLAM methods represent two fundamental philosophical approaches to solving the Simultaneous Localization and Mapping problem. As we've explored feature-based (indirect) SLAM, it's crucial to understand its counterpart, direct SLAM, and how these paradigms differ in their core methodologies, strengths, and weaknesses. The distinction lies primarily in how they process raw sensor data to extract information for state estimation and map building.

Indirect methods, as discussed, break down the problem into distinct stages: sensing, feature detection, feature description, feature matching (data association), and finally, state estimation and map update. The environment is abstracted into a sparse set of salient landmarks. This abstraction simplifies the subsequent optimization or filtering steps, as it reduces the amount of data to be processed. However, this reliance on intermediate feature extraction introduces its own set of challenges, most notably the sensitivity to the quality and robustness of the feature detector and descriptor, and the inherent difficulty in solving the data association problem, especially in environments lacking distinct features or exhibiting significant appearance changes.

In contrast,
direct SLAM methods bypass the explicit feature extraction step altogether. Instead, they operate directly on the raw sensor measurements – be it pixels from a camera, intensity values, or points from a LiDAR or depth sensor. The core idea is to directly relate consecutive sensor frames or multiple frames to estimate the robot's motion and build a map. This is typically achieved by minimizing an error metric that quantifies the difference between observed sensor data and what is predicted based on a current estimate of the robot's pose and the map.

The primary error metric in direct methods often relates to
photometric consistency for vision-based systems or **geometric consistency** for range-based sensors. For monocular or stereo vision, photometric consistency implies that the brightness (or color) of a point in the scene should remain consistent across different viewpoints, assuming uniform illumination. For a pixel

(or a set of pixels) observed at time t with intensity $It(u, v)$ at image coordinates (u, v) and projected to a 3D point $\mathbf{P}$ with depth dt, and then re-projected into a subsequent frame at time $t+1$ at image coordinates (u', v'), the photometric error would be:

$$ e
{photo} = It(u, v) - I_{t+1}(u', v') $$

The goal in direct visual SLAM is to find the robot's motion (e.g., a transformation $\mathbf{T}_{t, t+1}$ that maps points from frame t to frame $t+1$) that minimizes the sum of these photometric errors over a set of consistent pixels. This optimization can be performed locally between two frames (e.g., for motion estimation) or globally across multiple frames to refine both pose and map information.

For range sensors like LiDAR or depth cameras (e.g., RGB-D cameras), the error metric is often geometric. If a point cloud from frame t is transformed by the estimated motion $\mathbf{T}
{t, t+1}$ to align with a point cloud from frame $t+1$, the geometric error is the distance between corresponding points. This can be formulated as minimizing the sum of squared distances between points in the transformed point cloud Pt and points in the target point cloud P_{t+1}.

$$ e

{geo} = \sum{\mathbf{p}i \in Pt} \| \mathbf{p}i - \text{NN}(\mathbf{T}{t, t+1} \mathbf{p}i, P{t+1}) \|^2
$$

where $\text{NN}(\mathbf{q}, P)$ denotes the nearest neighbor of point $\mathbf{q}$ in point set P. This formulation is the basis of algorithms like Iterative Closest Point (ICP) and its variants, which are fundamental to direct methods using range data.

The direct approach offers several inherent advantages. Firstly, it can utilize the entire sensor's information, rather than relying on a sparse subset of features. This makes direct methods potentially more robust in environments with sparse features or repetitive patterns where feature-based methods might fail. For instance, in a corridor with uniform walls, a direct method might still find enough structure to estimate motion based on subtle gradients or depth variations, whereas a feature-based method might find no reliable features. Secondly, by avoiding the feature detection and matching stages, direct methods can potentially simplify the overall pipeline and reduce computational overhead, especially if efficient optimization techniques are employed. Furthermore, direct methods naturally lend themselves to dense or semi-dense mapping, producing richer environmental representations than sparse feature maps.

However, direct methods are not without their own set of challenges. The primary difficulty lies in their sensitivity to photometric or geometric inconsistencies that are not related to motion. For example, illumination changes in the environment can significantly corrupt photometric error minimization in visual SLAM. Similarly, sensor noise or outliers in depth readings can degrade geometric error minimization. These sensitivities necessitate robust error modeling, outlier rejection techniques, and often, the use of denser and more stable information across multiple frames.

The computational demands of direct methods can also be substantial. Minimizing photometric or geometric error typically involves solving a large-scale optimization problem, often non-linear. The complexity scales with the number of pixels or points being used, which can be orders of magnitude larger than the number of features in an indirect approach. This has historically made real-time direct SLAM more challenging, especially on embedded systems with limited computational power.

Let's delve into specific instances of direct SLAM, categorizing them by sensor modality.

Direct Monocular SLAM:

Monocular SLAM is inherently challenging because a single camera lacks direct depth information. Indirect methods solve this by triangulating sparse features to recover their 3D positions. Direct monocular SLAM, however, must infer depth from the raw pixel data and motion. This is typically achieved by exploiting the photometric consistency constraint over time.

A pioneering work in this area is **Direct Sparse Odometry (DSO)**. While the name suggests "sparse," DSO operates directly on pixels, but it optimizes only for a sparse subset of pixels that exhibit strong photometric gradients. This selection aims to find pixels that are more likely to provide reliable photometric information and are less prone to aliasing or quantization effects. DSO formulates the problem as a non-linear least-squares optimization that simultaneously estimates the camera poses and the 3D structure (depth) of the observed scene points for these selected pixels. The photometric error is minimized, and the uncertainty of the photometric measurements is carefully modeled. DSO is known for its robustness to illumination changes by explicitly modeling photometric variations.

Another significant development is **Semi-Direct Visual Odometry (SVO)**. SVO is a hybrid approach that uses features for efficient motion tracking but operates directly on pixels for depth estimation and map refinement. In SVO, features are detected to track motion between frames. However, instead of triangulating these features to build a map, SVO refines the depth of these feature points by minimizing the photometric error in a local patch around the feature in subsequent frames. This allows it to leverage the robustness of feature tracking while also benefiting from direct photometric optimization for depth estimation. SVO's key innovation is its ability to perform efficient local mapping and odometry by tracking features and refining their depth, and then using a separate bundle adjustment process for global map consistency.

Direct Stereo SLAM:

Stereo cameras provide inherent depth information through triangulation of corresponding pixels in the left and right images. Direct stereo SLAM methods can leverage this to directly estimate motion and build dense or semi-dense maps.

An influential example is **LSD-SLAM (Large-Scale Direct Monocular SLAM)**, which, despite its name, also has a stereo variant. In the stereo version, LSD-SLAM builds a semi-dense map by directly extracting depth information from stereo pairs. It then tracks camera motion by minimizing photometric errors across multiple stereo frames, similar to the monocular approach but with the advantage of direct depth cues. The "semi-dense" aspect means it focuses on regions of the image with significant intensity gradients, rather than just sparse features or the entire image. This allows it to build a more detailed map than sparse methods while remaining computationally tractable.

The core idea here is to establish a correspondence between pixels in the left and right images (stereo matching) to get an initial depth estimate. Then, using this depth, these 3D points are projected into subsequent frames. The photometric consistency between the observed pixels and the projected pixels is then used to estimate the camera motion. This process is iterative and aims to refine both the camera poses and the depth estimates of the scene points to achieve a consistent map.

Direct RGB-D SLAM:

RGB-D cameras, such as Microsoft Kinect or Intel RealSense, provide both color (RGB) and depth information for each pixel, simplifying the direct SLAM problem by providing direct 3D point clouds. This makes direct methods particularly well-suited for RGB-D data.

One of the most prominent direct RGB-D SLAM systems is
KinectFusion. KinectFusion is a real-time system that reconstructs dense 3D models of environments using a depth camera. It operates by maintaining an explicit volumetric representation of the scene (e.g., a truncated signed distance function or TSDF). The system tracks camera motion using a volumetric ICP approach, which aligns the current depth frame with the existing

volumetric map. By minimizing the geometric error between the observed depth points and the rendered depth from the TSDF volume, KinectFusion achieves robust and accurate camera pose estimation. The input depth frames are then integrated into the TSDF volume to continuously update and refine the 3D map.

KinectFusion is designed for relatively small to medium-sized environments and operates on a fixed-size voxel grid. While it excels at producing dense and visually appealing reconstructions, it suffers from accumulated drift over long trajectories and does not explicitly handle loop closures. This means that if the robot revisits a previously mapped area, KinectFusion will typically create a duplicate map rather than merging it, due to the accumulated pose errors.

To address the limitations of systems like KinectFusion, more advanced direct RGB-D SLAM systems have emerged. For example,

BundleFusion builds upon the volumetric approach but incorporates a global optimization framework, similar to graph-based SLAM. It collects a series of camera poses and their associated depth frames, and then performs a global optimization (global bundle adjustment) over all poses and the scene geometry. This allows for the correction of accumulated drift and the creation of globally consistent dense maps, especially when loop closures are detected. This hybrid approach combines the efficiency of direct methods for local tracking with the accuracy of global optimization.

Comparison of Direct and Indirect Methods:

The choice between direct and indirect SLAM methods often depends on the specific application, the sensor modalities available, and the environmental characteristics.

Robustness:

Indirect (Feature-Based): Generally more robust to illumination changes in visual SLAM due to relying on local appearance descriptors. However, they struggle in environments with few distinct features (e.g., white walls, repetitive patterns) or where features are highly dynamic (e.g., moving objects). Data association ambiguity is a major failure point.

Direct: More robust in feature-poor environments as they utilize all available data. However, they are more sensitive to illumination changes (photometric error) and sensor noise (geometric error). Robust outlier rejection and careful error modeling are crucial.

Map Representation:

Indirect: Typically produces sparse maps of landmark locations (points, lines, planes). While these are efficient for localization, they are less descriptive for other applications requiring detailed environmental understanding.

Direct: Naturally lends itself to dense or semi-dense map representations (e.g., point clouds, voxels, mesh). This provides a richer and more detailed environmental model, suitable for applications like augmented reality, robot manipulation, or detailed scene understanding.

Computational Complexity:
Indirect: The complexity is largely dependent on the number of features and the complexity of the matching and optimization algorithms. EKF-SLAM scales quadratically with the number of landmarks, while FastSLAM and Graph-based SLAM offer better scalability. Feature detection and description add an initial computational cost.
Direct: Can be computationally intensive due to the optimization of a large number of variables (poses and possibly scene structure) over many pixels or points. However, efficient optimization techniques and careful pixel selection (e.g., DSO, SVO) can make them real-time capable. Systems like KinectFusion are real-time by design but limited in scope, while BundleFusion offers global consistency at a higher computational cost.

Sensitivity to Initialization:
Indirect: Often requires an initial estimate of motion (e.g., from odometry) to bootstrap feature matching and pose estimation. A poor initial guess can lead to incorrect data associations.
Direct: Can sometimes be initialized more easily from scratch, especially with stereo or RGB-D data, as the depth information or stereo disparity provides an initial

geometric scale. However, a highly inaccurate initial pose can still lead to poor optimization results.

Handling of Dynamic Environments:
Both direct and indirect methods face challenges with dynamic environments. Features detected on moving objects can lead to incorrect map updates. Direct methods can be confused by moving objects that change the photometric or geometric properties of the scene. Specialized techniques for dynamic object detection and removal are often required for both paradigms.

In practice, many modern SLAM systems adopt hybrid approaches that combine the strengths of both direct and indirect methods. For instance, some systems might use feature detection for initial pose estimation and loop closure detection, while using direct methods for dense mapping and local pose refinement between keyframes. This "best of both worlds" approach aims to achieve robust, accurate, and dense SLAM capabilities across a wider range of environments and conditions. The ongoing research in SLAM continues to explore innovative ways to fuse sensor data, optimize state estimation, and build comprehensive environmental models, pushing the boundaries of what autonomous systems can perceive and navigate.

The quest for robust and reliable Simultaneous Localization and Mapping (SLAM) systems hinges on addressing a fundamental limitation: the accumulation of errors over time. As a robot navigates an environment, even small inaccuracies in pose estimation at each step compound, leading to significant drift in both its estimated trajectory and the generated map. This drift is particularly problematic for applications requiring long-term operation, consistent environmental understanding, or revisits to previously explored areas. To counter this, the concept of **loop closure detection and correction** becomes paramount. A loop closure occurs when a robot revisits a location it has previously mapped. Successfully identifying these revisited locations and leveraging that information to correct accumulated errors is a cornerstone of building globally consistent and accurate maps.

At its heart, loop closure detection is a recognition problem: can the robot determine, with high confidence, that its current sensory input corresponds to a previously observed state? This requires the system to store a history of its perceived environment and to efficiently query this history. The challenge lies in the fact that a revisited location may be observed under different conditions—varying illumination, different viewpoints, or even partial occlusion. Therefore, the methods employed must be robust to these variations.

One of the most successful paradigms for loop closure detection, particularly in visual SLAM, relies on the idea of creating descriptive signatures or "place descriptors" for different locations. These descriptors are compact, high-dimensional representations that capture the essential visual characteristics of a scene, aiming to be invariant to minor changes in viewpoint, scale, and illumination. A popular and powerful technique for generating such descriptors is the **Bag-of-Words (BoW) model**. Inspired by natural language processing, where documents are represented as a collection of words, the BoW model for visual scenes treats distinct visual features (like SIFT, SURF, or ORB keypoints) as "visual words."

The process of building a visual Bag-of-Words model begins with extracting a large number of local features from a diverse set of images captured from the environment. These features are then clustered using algorithms like k-means, where each cluster center represents a "visual word." The collection of all cluster centers forms the visual vocabulary. Once this vocabulary is established, any new image can be represented by counting the occurrences of each visual word within it. This results in a histogram, where each bin corresponds to a visual word, and the value in the bin represents the frequency of that word in the image. This histogram serves as the place descriptor.

When a robot encounters a new scene, it extracts local features, quantizes them according to the visual vocabulary to obtain a histogram, and then compares this histogram against a database of previously stored histograms. The comparison is typically done using a similarity metric, such as the chi-squared distance or cosine similarity. If a stored histogram is found to be sufficiently similar to the current one, a potential loop closure is flagged. The robustness of the BoW approach stems from its inherent invariance to the spatial arrangement of features; it only cares about the presence and frequency of visual words, making it resilient to changes in viewpoint and minor deformations.

Beyond the Bag-of-Words model, other place recognition techniques exist. For instance, **visual place recognition** can also be achieved by directly comparing image patches or by using global image descriptors. However, BoW remains a cornerstone due to its computational efficiency and the relative robustness it offers. For range-based sensors like LiDAR, similar principles apply, although the "features" are derived from geometric properties of the point cloud (e.g., scan lines, curvature, normals) rather than visual appearance. These geometric features can be clustered to form a vocabulary, and scan descriptors can be built as histograms of these geometric words, enabling loop closure detection in 3D space.

Once a potential loop closure is detected—that is, a strong match is found between the current observation and a past observation—the system enters the crucial phase of **loop closure correction**. A simple detection isn't enough; the system must definitively confirm that it is indeed the same location and then use this information to rectify the accumulated errors in the robot's pose graph.

The pose graph is a probabilistic representation of the

robot's trajectory and its spatial relationships. It consists of nodes, each representing the robot's pose (position and orientation) at a specific point in time, and edges, which represent the relative motion constraints between consecutive poses. These constraints are derived from the robot's odometry (e.g., wheel encoders, IMU integration) or visual/inertial odometry estimates. The problem is that these relative constraints are not perfect; each one carries an associated uncertainty. As these uncertainties accumulate along a path, the global poses can drift significantly from their true locations.

A loop closure introduces a new, powerful constraint into the pose graph. If the robot is at pose i and revisits a previously mapped location corresponding to pose j, this forms a new relative pose constraint between i and j. Crucially, this constraint is not about the motion
between i and j, but rather about the fact that they represent the *same physical location*. This "closure" provides a global correction signal. The goal of loop closure correction is to adjust all the poses in the graph such that all the relative constraints (both the original odometry-based ones and the new loop closure constraint) are satisfied as closely as possible, while minimizing the overall error and respecting the uncertainties of each constraint.

This problem is formulated as a
non-linear optimization problem. Given a set of nodes (poses) and edges (constraints), we seek to find

the optimal set of poses that minimizes an objective function representing the total error in the graph. The error is typically defined as a sum of squared errors, where each error term corresponds to an edge in the graph. For an odometry edge between pose i and pose $i+1$ with relative transformation $\Delta \mathbf{T}{i, i+1}$, the error is related to the difference between the estimated relative transformation and the transformation implied by the optimized poses: $\|\mathbf{T}i^{-1} \mathbf{T}{i+1} - \Delta \mathbf{T}{i, i+1}\|^2$. Similarly, for a loop closure edge between pose i and pose j with a detected relative transformation $\Delta \mathbf{T}{i, j}^{\text{loop}}$, the error is: $\|\mathbf{T}i^{-1} \mathbf{T}j - \Delta \mathbf{T}{i, j}^{\text{loop}}\|^2$.

The objective function to be minimized is thus:

$$ \min {\mathbf{T}1, \dots, \mathbf{T}N} \sum{i=1}^{N-1} \| \mathbf{T}i^{ 1} \mathbf{T}{i+1} - \Delta \mathbf{T}{i, i+1} \|^2{\Sigma{i, i+1}} + \sum{(i, j) \in \mathcal{L}} \| \mathbf{T}i^{-1} \mathbf{T}j - \Delta \mathbf{T}{i, j}^{\text{loop}} \|^2{\Sigma_{i, j}^{\text{loop}}} $$

where N is the number of poses, $\mathbf{T} k$ is the pose of the robot at time k, $\Delta \mathbf{T}{i, i+1}$ is the relative motion constraint from odometry, $\Delta \mathbf{T}_{i, j}^{\text{loop}}$ is the relative motion constraint from loop closure, $\mathcal{L}$ is the set of detected loop closures, and Σ represents the information matrices (inverses of the covariance matrices) that weigh each constraint according to its uncertainty.

Solving this non-linear least-squares problem is computationally demanding, especially for large graphs. Common optimization techniques employed include **Levenberg-Marquardt (LM)** or **Gauss-Newton** methods. These are iterative algorithms that approximate the solution by linearizing the objective function and solving a system of linear equations at each step.

The Levenberg-Marquardt algorithm is particularly popular for pose graph optimization. It provides a robust way to balance between the Gauss-Newton method (which is fast but can diverge if the initial guess is poor) and the gradient descent method (which is slow but guaranteed to converge). LM interpolates between these two by introducing a damping parameter. At each iteration, it updates the poses by solving a linear system derived from the Jacobian of the error function. The damping parameter is adjusted dynamically: if an iteration successfully reduces the error, the damping is decreased; if it fails, the damping is increased to move towards a more stable gradient descent step.

The process of pose graph optimization can be broken down into several steps:

1. **Graph Construction:** As the robot moves, nodes (poses) and odometry edges are added to the pose graph.
2. **Loop Closure Detection:** Periodically, or when a potential loop closure is detected via place recognition, a candidate loop closure edge is identified.
3. **Loop Closure Verification:** To avoid adding erroneous constraints, loop closures are often verified. This might involve re-observing the candidate location and performing a local pose optimization around it to ensure consistency. In visual SLAM, this could involve matching features between the current frame and the frames around the detected loop closure candidate.
4. **Graph Optimization:** Once a loop closure is confirmed and added as an edge, the entire pose graph is optimized. This is typically performed periodically, or when a sufficient number of loop closures have been detected, or when the accumulated drift becomes too large. The optimization adjusts all poses to minimize the overall error.
5. **Map Update:** After the poses are optimized, the map itself needs to be updated to reflect the corrected trajectories. For feature-based maps, this means re-triangulating features using the refined poses. For direct methods that build dense maps, the integration process needs to be re-run or adjusted based on the corrected poses.

The scale of the optimization problem can be a bottleneck. For very long trajectories, the pose graph can become enormous, making full graph optimization computationally prohibitive in real-time. Techniques such as

incremental optimization or **sub-map optimization** are employed to manage this complexity. Incremental optimization involves only re-optimizing a part of the graph around the newly added constraints, rather than the entire graph. Sub-map optimization might divide the graph into smaller, interconnected components, optimizing each component and then resolving the connections.

Another critical aspect of loop closure correction is ensuring that the corrected map and trajectory are not only consistent but also accurate. The quality of the loop closure constraint itself is paramount. If the loop closure detection is faulty, or if the estimated relative transformation of the loop closure is inaccurate, the optimization process could potentially corrupt the entire map. Therefore, robust methods for verifying loop closures and estimating their associated uncertainties are essential. For visual loop closures, this might involve re-projecting landmarks from the old view into the new view (and vice-versa) and re-estimating the relative pose using a small bundle adjustment around the loop closure to refine its pose and uncertainty.

Furthermore, the choice of optimization solver and its parameters can significantly impact the performance and convergence speed. Advanced solvers often incorporate techniques for handling sparsity in the Hessian matrix (which arises from the structure of the pose graph), block matrix operations, and parallel processing to accelerate the optimization.

In summary, loop closure detection and correction form a vital feedback mechanism for SLAM systems. By enabling robots to recognize previously visited locations and to correct their internal state (poses and map) based on this recognition, these techniques transform a drift-prone system into one capable of generating globally consistent and accurate representations of the environment. The interplay between robust place recognition, accurate constraint estimation, and efficient non-linear optimization forms the backbone of this crucial capability, enabling long-term autonomy and reliable navigation.

The advent and rapid evolution of deep learning have begun to profoundly reshape the landscape of Simultaneous Localization and Mapping (SLAM). While traditional SLAM algorithms have achieved remarkable success, they often rely on handcrafted features and explicit geometric models that can be brittle in the face of real-world complexities such as extreme illumination changes, dynamic environments, and sensor noise. Deep learning offers a powerful paradigm for learning robust representations and discriminative models directly from data, leading to significant improvements across various components of SLAM pipelines. This subsection delves into the recent advances in deep learning for SLAM, exploring how neural networks are enhancing feature detection and matching, improving place recognition for loop closure, enabling end-to-end SLAM solutions, and facilitating the creation of richer, semantically aware maps.

One of the most impactful applications of deep learning in SLAM has been in the domain of **feature detection and description**. Traditional methods, such as SIFT (Scale-Invariant Feature Transform) and SURF (Speeded Up Robust Features), while effective, are computationally intensive and can struggle with robustness in challenging conditions. Deep learning-based feature detectors and descriptors, trained on vast datasets of images, have demonstrated superior performance. Networks like SuperPoint, trained for feature detection and description jointly, learn to identify salient keypoints and simultaneously extract rich, discriminative descriptors. These learned features often exhibit greater invariance to viewpoint, scale, and illumination changes compared to their handcrafted counterparts. For instance, SuperPoint uses an "interest point" detector that outputs a heatmap of feature locations and a "descriptor" network that generates a descriptor for each detected point. This allows for more reliable matching between consecutive frames, which is crucial for accurate visual odometry. Another notable architecture is D2-Net, which learns to detect keypoints and then extracts descriptors for these keypoints from a larger receptive field, enabling more context-aware descriptions. The benefits extend to the matching process itself. Learned matching networks can directly predict correspondences between features in two images, potentially bypassing the need for expensive nearest-neighbor searches with traditional descriptors. Techniques like Learned Denoising-based Feature Detector (LF-Net) also show promise in improving feature robustness by incorporating denoising within the feature extraction pipeline. The ability to extract more distinctive and invariant features directly from raw sensor data significantly enhances the reliability of pose estimation, especially in environments where classical methods might fail. This improvement in the foundational element of feature extraction cascades into better performance for subsequent SLAM tasks, from odometry estimation to loop closure detection. The datasets used to train these deep feature extractors are critical; they often comprise millions of image pairs with ground truth correspondences or known camera motion, enabling the networks to learn generalized representations that perform well across a wide range of visual conditions.

Beyond local features, deep learning has revolutionized **place recognition** for loop closure. The Bag-of-Words (BoW) model, while a significant advancement, relies on hand-engineered visual words and can still be sensitive to changes in appearance. Deep learning offers more sophisticated ways to represent entire scenes or places in a way that is highly discriminative and robust to variations. Deep Convolutional Neural Networks (CNNs) can be trained to extract global image descriptors, often referred to as "learned global features" or "image embeddings," that capture the semantic content and overall appearance of a scene. Networks like NetVLAD, which builds upon the VLAD (Vector of Locally Aggregated Descriptors) descriptor but uses a learned metric and network architecture, have shown state-of-the-art performance in place recognition tasks. NetVLAD, for instance, is trained using large-scale geolocated image datasets, learning to produce descriptors that are invariant to viewpoint and illumination changes. Another approach involves using the features extracted from intermediate layers of pre-trained CNNs (e.g., ResNet, VGG) as image descriptors. These features, when properly aggregated or processed, can provide a dense and discriminative representation of the visual scene. For example, GeM (Generalized-Mean pooling) is a pooling operation that can be applied to CNN feature maps to generate a global image descriptor. These learned global descriptors can then be compared using standard distance metrics to identify previously visited locations, enabling more accurate and robust loop closures. Furthermore, deep learning can be employed to learn a similarity function directly, rather than relying on generic distance metrics. Siamese networks or triplet networks can be trained to output embeddings such that embeddings of the same place are close in the feature space, while embeddings of different places are far apart. This learned metric space can be more effective than Euclidean or cosine similarity for place recognition. The robustness gained from these deep place recognition methods is crucial for achieving globally consistent maps, as it allows SLAM systems to correctly identify revisited locations even under challenging environmental conditions that would typically confuse traditional methods. This is particularly important in long-term SLAM scenarios where the environment might undergo significant changes (e.g., seasonal variations, construction) over time.

The concept of

end-to-end SLAM represents a more ambitious frontier where deep learning aims to replace traditional modular SLAM pipelines with a single, unified neural network. Instead of separate modules for feature extraction, matching, pose estimation, and mapping, end-to-end systems aim to learn the entire process of inferring camera poses and/or generating maps directly from raw sensor inputs (e.g., camera images, LiDAR scans). One prominent approach in this area is **deep visual odometry (DVO)**. Networks like Droid-SLAM are designed to take sequences of images and output the relative camera poses between them. These networks often leverage recurrent neural network (RNN) architectures, such as LSTMs or GRUs, or more recently, transformer architectures, to process the temporal information in image sequences. They can also incorporate feature extraction modules (often CNN-based) and optical flow estimation. Some DVO systems are trained purely supervisedly, using ground truth pose information from datasets like KITTI or EuRoC. Others are trained in a self-supervised manner, where the network learns to minimize photometric reprojection errors between frames. For example, a network might predict depth maps and camera poses, and then use these predictions to warp one frame to align with another, minimizing the pixel-wise intensity difference. This approach bypasses the need for explicit feature points and geometric constraints, learning a more direct mapping from visual input to motion. While end-to-end DVO can achieve impressive performance, especially in terms of speed and robustness to certain types of visual degradation, it presents unique challenges. The learned models can be less interpretable than traditional methods, making it difficult to diagnose failures. Furthermore, accumulating errors can still be an issue, and achieving global consistency often requires integration with loop closure mechanisms, which can be challenging to implement within a purely end-to-end framework. Some recent work attempts to integrate loop closure directly into the end-to-end learning process or proposes architectures that can learn to perform global optimization implicitly.

Another exciting application of deep learning is in the generation of
semantic maps. Traditional SLAM systems typically build geometric maps (e.g., point clouds, occupancy grids, sparse feature maps) that describe the physical structure of the environment. However, these maps often lack rich semantic information that would be highly beneficial for robot navigation and interaction. Deep learning excels at semantic understanding tasks, such as object detection, semantic segmentation, and instance segmentation. By integrating these capabilities into SLAM, robots can build maps that not only know "where" things are but also "what" they are. For example, a robot can use a semantic segmentation network to classify each pixel in an image (e.g., as "road," "building," "vegetation," "car"). This semantic information can then be fused with the geometric map. Semantic SLAM systems can leverage deep learning for various purposes:
1. **Object Detection and Instance Segmentation:** Networks like YOLO (You Only Look Once) or Mask R-CNN can detect and segment individual objects (e.g.,

chairs, tables, doors, people) within the environment. This allows the map to explicitly represent and locate these objects.

2. **Semantic Segmentation:** Pixel-wise semantic segmentation networks (e.g., DeepLab, U-Net) can provide a dense semantic labeling of the scene. This can be used to generate semantic occupancy grids, where each grid cell is labeled with the dominant semantic class (e.g., a cell might be labeled as "sidewalk" or "grass").

3. **Semantic Scene Understanding:** Beyond individual objects, deep learning can infer higher-level semantic understanding of regions or scenes, such as identifying rooms (kitchen, bedroom) or functional areas (parking lot, street).

The integration of semantic information into SLAM offers several advantages. It can aid in **re-localization**, as semantic landmarks (e.g., "the red sofa," "the entrance door") can be more distinctive and robust than purely geometric features. It can also facilitate **semantic navigation**, allowing robots to plan paths based on object types or functional areas (e.g., "navigate to the kitchen," "avoid the pedestrian zone"). Furthermore, semantic maps can be used for **object manipulation** and interaction, enabling robots to locate and interact with specific objects based on their semantic identity. The process often involves back-projecting semantic labels from image pixels into the 3D map. For example, after a semantic segmentation network classifies pixels in an image, these labels can be associated with 3D points in a point cloud or voxels in an occupancy grid, provided the camera pose and depth information are available. Challenges in semantic SLAM include the computational cost of running deep semantic networks in real-time, the accurate fusion of uncertain semantic labels with geometric data, and the handling of dynamic objects whose semantic labels might change or disappear. Nevertheless, the prospect of robots operating in environments endowed with rich semantic understanding is a significant driver of research in this area.

Beyond these core areas, deep learning is also being explored for other aspects of SLAM, such as **sensor fusion**. By training networks to combine information from multiple sensors (e.g., cameras, LiDAR, IMU), robots can achieve more robust and accurate state estimation. For instance, a neural network can learn optimal ways to fuse visual features with LiDAR-based geometric features, or to leverage IMU data to provide prior motion estimates that help regularize visual odometry. Additionally, **learned priors** are being incorporated into SLAM. For example, a network might learn to predict common environmental structures or object configurations, which can then be used to constrain the optimization process or improve the consistency of the generated map.

Despite the remarkable progress, several research challenges remain in deep learning for SLAM. **Generalization** is a key issue; models trained on one dataset or environment may not perform well in significantly different ones. Creating large-scale, diverse datasets that capture the full complexity of real-world scenarios is crucial but labor-intensive. **Computational efficiency** is another significant hurdle. Running complex deep neural networks in real-time on resource-constrained robotic platforms often requires specialized hardware (e.g., GPUs, TPUs) or efficient network architectures (e.g., MobileNets, EfficientNets) and model compression techniques. **Interpretability and robustness** are also areas of active research. Understanding why a deep learning-based SLAM system fails and how to make it more resilient to adversarial perturbations or novel situations remains a challenge. The problem of **accumulated drift** in end-to-end systems still needs robust solutions that go beyond simple loop closure. Finally, the **integration of semantic and geometric reasoning** in a principled and efficient manner is an ongoing area of exploration, aiming to create truly intelligent spatial understanding for robots.

In conclusion, deep learning is not merely an incremental improvement but a transformative force in SLAM. It has empowered systems with more robust feature extraction, more accurate place recognition, and the potential for end-to-end solutions and semantically rich maps. As research continues to address the remaining challenges, we can expect deep learning to play an increasingly central role in enabling robots to navigate and understand complex, real-world environments with unprecedented reliability and intelligence. The synergy between classical geometric principles and data-driven learning approaches promises a future where SLAM systems are more capable, adaptable, and ubiquitous.

Chapter 8: Motion Planning Fundamentals

The ability of an autonomous system to execute a task hinges on its capacity to move purposefully from one state to another. This fundamental capability is broadly categorized as motion planning, a discipline that equips robots with the intelligence to navigate through their operational space from an initial configuration to a desired goal configuration. Crucially, this journey must be undertaken while adhering to a set of constraints, paramount among which is the avoidance of collisions with static or dynamic obstacles present in the environment. Motion planning, therefore, represents the algorithmic heart that dictates the sequence of movements an agent will perform, effectively bridging the gap between sensing the environment and enacting a physical response. It is the silent choreographer of robotic action, ensuring that a robot can not only perceive its surroundings but also act upon that perception in a meaningful and safe manner. Without robust motion planning capabilities, even the most sophisticated perception systems would be rendered inert, incapable of translating environmental understanding into tangible progress towards a task objective.

At the core of understanding motion planning lies the concept of the
configuration space, often abbreviated as C-space. This abstract but critical construct represents the complete set of all possible states, or configurations, that a robot can assume. For a simple mobile robot operating on a 2D plane, its configuration can be described by its x-coordinate, y-coordinate, and its

orientation (yaw angle), thus requiring three parameters. This translates to a 3-dimensional configuration space. However, for more complex robotic manipulators with multiple joints, the dimension of the C-space escalates rapidly. A robotic arm with six degrees of freedom (DoF), for instance, would have a C-space of six dimensions, where each dimension corresponds to the angle of a specific joint. Visualizing and operating within such high-dimensional spaces presents significant computational and algorithmic challenges. The robot's physical body, with its inherent geometry and degrees of freedom, is mapped into this abstract space. Obstacles in the physical world, which are often visualized as geometric shapes, are "grown" or "mapped" into the C-space to form the **C-obstacles**. These C-obstacles represent all configurations that would result in a collision between the robot and an environmental obstacle. The space devoid of C-obstacles is known as the **free configuration space**, denoted as C-free. The fundamental problem of motion planning then becomes finding a continuous path within C-free from a starting configuration to a goal configuration. This path must be smooth, computationally feasible to generate, and ideally, satisfy certain optimality criteria such as minimum path length or minimum execution time. The dimensionality of the C-space directly impacts the complexity and computational cost of motion planning algorithms. As the number of DoF increases, the free space can become highly fragmented and complex, making it exponentially harder to find a valid path. This is often referred to as the "curse of dimensionality" in robotics.

The paramount importance of motion planning is underscored by its indispensable role in enabling robots to perform tasks effectively and safely in environments that are rarely static or predictable. Real-world scenarios are characterized by dynamic changes: objects may move, new obstacles may appear, and the robot's own state might be subject to uncertainty. A robot tasked with delivering packages in a bustling city, for example, must contend with moving vehicles, pedestrians, changing traffic signals, and construction zones. A surgical robot operating within a human body must navigate around delicate organs and avoid inadvertent contact that could cause harm. An industrial robot on an assembly line must precisely maneuver around other machines, human workers, and work-in-progress components. In all these contexts, the ability to generate a safe and efficient trajectory is not merely desirable but a critical requirement for mission success and, in many cases, for ensuring the safety of humans and property. Motion planning algorithms provide the intelligence to make these movements, transforming a perception of the environment into a concrete sequence of actions. They are the decision-makers that determine "where to go next," ensuring that the robot's actions are aligned with its overarching goals while respecting the physical and environmental constraints. The efficiency and robustness of the motion planning system directly translate into the robot's overall utility and reliability. A slow or error-prone motion planner can render a robot impractical for real-time applications, while a planner that frequently fails to find a path can lead to task abandonment or mission failure.

Furthermore, motion planning serves as a critical bridge between the robot's sensory perception and its effector system, i.e., its actuators. Perception systems provide the robot with an understanding of its current state and the state of its environment. This information is then fed into the motion planning module, which uses it to construct or update the representation of the C-space and identify C-obstacles. Based on this representation, the planner computes a valid path. This path is then often refined into a trajectory, which includes time information and velocity profiles, before being sent to the robot's control system. The control system, in turn, translates the trajectory commands into specific actuator commands (e.g., motor torques or joint velocities) that drive the robot's movement. This perception-action loop is fundamental to autonomous operation. Without motion planning, the perceived world would remain an abstract representation, disconnected from the robot's ability to interact with it. The quality of this connection—how quickly and reliably a robot can translate sensory input into executable motion—is a key determinant of its autonomy and capability. An effective motion planner can leverage subtle cues from perception to optimize movement, such as anticipating the path of a moving obstacle or exploiting temporary openings in a cluttered environment. Conversely, limitations in motion planning can constrain how effectively a robot can utilize the information gathered by its sensors, leading to suboptimal or even unsafe behaviors.

The complexity of motion planning problems can vary significantly depending on the robot's kinematics, the dimensionality of its C-space, the nature of the environment (static vs. dynamic, known vs. unknown), and the presence of uncertainties. For instance, planning for a robot arm with many degrees of freedom in a cluttered industrial setting requires far more sophisticated algorithms than planning for a simple differential-drive mobile robot in an open space. The development of motion planning algorithms has been a cornerstone of robotics research for decades, leading to a rich landscape of approaches. These approaches can broadly be categorized into several families, each with its strengths and weaknesses. Some methods rely on building explicit geometric models of the C-space and then searching for a path within that space. Others operate directly in the robot's configuration space without explicitly constructing it, often employing sampling-based techniques. More recent advancements have also explored the integration of machine learning, particularly deep learning, to enhance various aspects of the motion planning process, from learning cost functions to directly predicting trajectories. The choice of algorithm often depends on the specific application, the available computational resources, and the desired performance characteristics.

One of the earliest and most intuitive approaches to motion planning involves **roadmap-based methods**. These techniques aim to create a graph-like structure, or roadmap, that captures the connectivity of the free configuration space. Once this roadmap is constructed, planning a path from a start to a goal configuration reduces to a graph search problem, typically using algorithms like Dijkstra's or A. Popular roadmap-based methods include visibility graphs and probabilistic roadmaps (PRMs). A visibility graph connects pairs of configurations that are "visible" to each other, meaning a straight line segment between them lies entirely within C-free. While conceptually simple, constructing a visibility graph can be computationally expensive, especially in higher dimensions, as it involves checking for intersections with all obstacles. Probabilistic Roadmaps (PRMs), introduced by Kavraki and Latombe, offer a more scalable approach, particularly for high-dimensional C-spaces. PRMs work by randomly sampling configurations from C-free and connecting "nearby" configurations with simple local paths (e.g., straight lines in C-space). The resulting graph of sampled configurations and their connections serves as an approximate representation of the connectivity of C-free. The more samples and connections are generated, the higher the probability that a path exists between any two configurations in C-free, and the more likely it is that PRM will find such a path if one exists. PRMs are particularly effective for multi-query planning, where a single roadmap can be built and then used for many different start-goal pairs. The quality of the roadmap and the efficiency of pathfinding depend on the sampling strategy and the local planner used to connect samples.

Another significant class of motion planning algorithms falls under the umbrella of **sampling-based planners**, with **Rapidly-exploring Random Trees (RRTs)** being a prime example. Unlike roadmap-based methods that pre-compute a global representation of C-free, RRTs build a tree-like structure incrementally by sampling configurations and extending the tree towards these samples. An RRT starts with a root node at the initial configuration. In each iteration, a random configuration is sampled from C-space. The algorithm then finds the node in the existing tree that is closest to this random sample and attempts to extend the tree from that node towards the sample by a certain step size. If the attempted extension (typically a straight line in C-space) does not intersect any obstacles, a new node is added to the tree at the end of the extension. This process is repeated until the tree reaches the goal configuration or a predefined time limit is met. RRTs are inherently biased towards exploring large, open regions of C-free, which makes them effective at finding paths in high-dimensional spaces where traditional grid-based or roadmap-based methods might struggle. Variations of RRTs, such as RRT-Connect, employ two trees growing simultaneously from the start and goal configurations, which often leads to faster convergence. RRTs are well-suited for single-query planning problems and can handle complex robot kinematics and constraints. However, RRTs do not inherently optimize for path length or smoothness; the paths generated can sometimes be jagged and inefficient. Subsequent algorithms, like Informed RRT *and Lazy RRT*, aim to address these limitations by incorporating sophisticated re-planning and pruning strategies to generate asymptotically optimal and smoother paths.

Beyond these traditional methods, the field has seen the emergence of
optimization-based approaches. These methods frame motion planning as an optimization problem, aiming to find a trajectory that minimizes a cost function subject to constraints. The cost function might penalize path length, execution time, energy consumption, or deviations from desired velocities. Constraints typically include collision avoidance, kinematic limits of the robot, and dynamic feasibility. Trajectory optimization techniques, such as sequential quadratic programming (SQP) or gradient-based optimization, can be applied to find a sequence of configurations that satisfy these criteria. These methods can produce highly smooth and optimized trajectories, which are crucial for many robotic applications, especially those involving high speeds or precise movements. However, optimization-based methods can be sensitive to the initial guess of the trajectory and may get stuck in local minima, failing to find a valid path if the initial guess is poor or the C-space is highly constrained. They also typically require an initial path or a good guess of the solution space to converge effectively.

In recent years, the paradigm of

learning-based motion planning has gained significant traction, particularly with the advancements in deep learning. Instead of relying solely on explicit geometric reasoning or handcrafted algorithms, learning-based approaches aim to learn motion planning policies directly from data. This can take several forms. **Learning a cost function or heuristic** for traditional planners: Neural networks can be trained to predict the cost-to-go from any given configuration, providing a more informed heuristic for search-based planners like A or guiding sampling-based methods. **Learning to generate motion primitives**: Robots can learn a set of basic, reusable motion segments that can be composed to form longer, more complex trajectories. **End-to-end motion planning**: This is perhaps the most ambitious area, where a neural network directly maps sensory inputs and a goal command to a sequence of control actions or a full trajectory. These approaches often leverage deep neural networks, such as convolutional neural networks (CNNs) for processing visual input, recurrent neural networks (RNNs) for handling temporal sequences, or reinforcement learning (RL) agents that learn optimal policies through trial and error in simulated or real environments. Deep reinforcement learning has shown promise in enabling robots to learn complex behaviors and adapt to dynamic environments without explicit programming. For instance, a robot might learn to navigate a maze or avoid dynamic obstacles by receiving rewards for reaching the goal and penalties for collisions. While learning-based methods offer the potential for greater adaptability, generalization, and efficiency in complex scenarios, they also come with their own set of challenges, including the need for large amounts of training data, the difficulty of ensuring safety guarantees, and the interpretability of learned policies.

The ultimate goal of motion planning is to enable autonomous systems to operate reliably and efficiently in the real world. This requires not only finding *a* path but finding a path that is not only collision-free but also optimal in some sense—shortest, fastest, smoothest, most energy-efficient, or most robust to uncertainty. The development of sophisticated motion planning algorithms is therefore a continuous endeavor, driven by the ever-increasing complexity of robotic tasks and environments. The ability to plan motion effectively is a fundamental prerequisite for virtually any autonomous robotic application, from industrial automation and logistics to autonomous driving, exploration, and personal robotics. It is the essential ingredient that allows robots to move beyond being mere tools to become capable agents that can navigate and interact with the world around them. Without it, a robot could perceive a goal but have no means of reaching it safely or efficiently. The ongoing research in this domain continues to push the boundaries, striving for planners that are faster, more robust, more adaptable, and capable of handling increasingly complex robotic systems and unpredictable environments.

The fundamental challenge in robotics lies in enabling a robot to move from a starting state to a desired goal state in a safe and efficient manner. This overarching problem, known as motion planning, is the focus of this chapter. At its heart, motion planning is about defining a sequence of movements—a path or trajectory—that guides the robot through its environment while respecting various constraints, most critically, the avoidance of collisions. To truly grasp the intricacies of motion planning, it is essential to distinguish between two distinct but related spaces: the physical *workspace* and the *configuration space* (often abbreviated as C-space).

The
workspace is the tangible, three-dimensional (or sometimes two-dimensional, for planar robots) environment in which the robot physically exists and operates. It's the familiar world of floors, walls, tables, and any other objects that the robot might encounter. When we visualize a robot, we are usually imagining its form and its position within this physical workspace. For a mobile robot, the workspace might be a factory floor or a living room. For a robotic arm, it might be the area above a workbench or a specific region in space that it needs to reach. Obstacles in the workspace are simply the physical objects that occupy this space, posing potential barriers to the robot's movement. The robot's own physical body, with its dimensions and extent, also occupies a portion of this workspace.

In contrast, the
configuration space is a more abstract mathematical construct. It is a multi-dimensional space where each point represents a unique, complete specification of the robot's state. This state encompasses all the parameters needed to precisely define the robot's pose or configuration. For a simple unicycle-like mobile robot

moving on a flat surface, its configuration can be defined by three parameters: its x-coordinate, its y-coordinate, and its orientation (yaw angle, often denoted as θ). Thus, the configuration space for such a robot is a 3-dimensional space, often represented as C = SE(2), where SE(2) signifies the Special Euclidean group in 2 dimensions. Any point (x, y, θ) in this space uniquely describes the robot's position and heading.

For more complex robots, like a robotic manipulator arm with multiple joints, the dimensionality of the configuration space grows directly with the number of degrees of freedom (DoF). A robotic arm with six rotary joints, for instance, would have a 6-dimensional configuration space. Each dimension corresponds to the angle of a specific joint. A point in this 6-dimensional space, e.g., $(\theta
1, \theta2, \theta3, \theta4, \theta5, \theta6)$, defines the exact angular position of each of the arm's joints, and consequently, the precise pose and shape of the entire arm in the workspace. Visualizing and navigating such high-dimensional spaces can be intuitively challenging, as our human perception is typically limited to 3 spatial dimensions.

The critical insight for motion planning is that the robot's entire body, with all its links and joints, is represented by a single point in its configuration space. The "shape" of the robot in the workspace is implicitly encoded in the coordinates of this point within C-space. This abstraction is immensely powerful because it transforms the complex problem of moving a physically embodied object in a 3D world into the problem of moving a point in a multi-dimensional space.

Just as the workspace contains physical obstacles, the configuration space contains *C-obstacles*. A C-obstacle is a region within the configuration space that corresponds to physically impossible or forbidden configurations of the robot. Specifically, a configuration is part of a C-obstacle if, when the robot is placed in that configuration in the workspace, it would result in a collision between the robot and an environmental obstacle, or a self-collision (for robots with multiple links). Therefore, C-obstacles are essentially the "grown" or "mapped" representations of workspace obstacles into the configuration space.

The concept of C-obstacles is fundamental. Imagine a simple 2D point robot (a robot with no physical dimensions) moving in a 2D workspace with a circular obstacle. The C-space for this robot is just the 2D workspace itself. The C-obstacle would be the set of all points inside the circle, and potentially a boundary region around it if the robot is not truly a point. Now consider a small circular robot moving in the same 2D workspace. Its configuration space is still 2D (x, y). However, the C-obstacle will be larger than the actual workspace obstacle. The C-obstacle will be a region that is the union of all robot positions where *any* part of the robot's circular body overlaps with the workspace obstacle. Effectively, the C-obstacle's boundary is the locus of configurations where the robot is just touching the workspace obstacle.

For a robotic arm, the transformation of workspace obstacles into C-obstacles is more complex. If a link of the robotic arm collides with a table in the workspace, this corresponds to a region in the arm's C-space. The boundary of this C-obstacle is defined by all the configurations where the arm is exactly touching the table. The entirety of the C-obstacle comprises all configurations that lead to any collision. These C-obstacles can be incredibly complex shapes, especially in high-dimensional C-spaces. They can be disconnected, have holes, and be arbitrarily convoluted.

The
free configuration space, denoted as $C*{free}$, is the subset of the configuration space where the robot can exist without any collisions. It is the configuration space minus all the C-obstacles: $C*{free} = C \setminus Obstacles$. The primary goal of motion planning is to find a continuous path within $C*{free}$ that connects a given starting configuration ($C*{start}$) to a desired goal configuration ($C*{goal}$). Both $C*{start}$ and $C*{goal}$ must, of course, reside in $C*{free}$. If either the start or goal configuration itself results in a collision, then no valid path can exist.

Performing motion planning directly in the configuration space offers several advantages:

1. **Simplification of Robot Geometry:** By representing the robot as a single point in C-space, its complex geometry and kinematics are implicitly handled. We don't need to constantly track the position and orientation of each individual link; we only need to track the single point representing the robot's configuration.

2. **Unified Representation:** The robot's entire state is represented by a single point. This allows for a consistent way to define states, transitions, and goals, regardless of the robot's complexity.

3. **Collision Checking:** Collision detection, which is inherently a geometric problem in the workspace, is transformed into a point-in-set problem in C-space. We simply need to determine if a given configuration (a point in C-space) lies within a C-obstacle or in C_{frcc}. While computing the exact shape of C-obstacles can be difficult, checking a single configuration for collision is often computationally more tractable than checking the entire robot body against all workspace obstacles for every possible pose.

4. **Path Planning as Point Navigation:** The complex task of moving a physical object becomes the simpler task of navigating a point through a multi-dimensional space.

Algorithms designed for pathfinding in abstract spaces can be applied directly.

The dimension of the configuration space is a crucial factor influencing the complexity of motion planning. A 2D mobile robot has a 3D C-space. A 6-DOF robotic arm has a 6D C-space. A humanoid robot, with dozens of DoFs, could have a C-space of 30, 40, or even more dimensions. As the dimensionality increases, the "volume" of the configuration space grows exponentially. Furthermore, the free space ($C *{free}$) can become highly fragmented and complex, potentially consisting of many disconnected regions. This phenomenon is often referred to as the "curse of dimensionality." In high-dimensional spaces, C-obstacles can partition $C*{free}$ in very intricate ways, making it exponentially harder for algorithms to find a path between two points. Many classic planning algorithms that rely on discretizing space (like grid-based methods) become computationally infeasible in high dimensions due to the explosion in the number of cells required.

The core problem of motion planning can be rephrased as finding a continuous function $p: [0, 1] \to C *{free}$ such that $p(0) = C*{start}$ and $p(1) = C_{goal}$. This function $p(t)$ describes the evolution of the robot's configuration over time. Often, this path is then converted into a *trajectory*, which is a time-parameterized path $p(t)$ that also specifies velocities, accelerations, and possibly other derivatives of the configuration. The trajectory must not only be collision-free but also kinematically and dynamically feasible for the robot's actuators and mechanical limitations.

Understanding the distinction between workspace and configuration space is foundational. It allows us to abstract away the robot's physical form and focus on its state as a point in a space where obstacles are represented as regions. This abstraction is the bedrock upon which all sophisticated motion planning algorithms are built. The transformation of workspace obstacles into C-obstacles is a conceptual step that simplifies the problem from dealing with complex geometric interactions to navigating a space with forbidden regions. The dimensionality of this C-space is a primary driver of algorithmic complexity, dictating the feasibility and efficiency of various planning strategies. Therefore, before delving into specific algorithms, solidifying this understanding of workspace versus configuration space is paramount for anyone seeking to comprehend how robots navigate their world. The goal is always to find a path in the free configuration space, and the methods employed to achieve this are the subject of the subsequent sections.

The process of moving a robot from an initial state to a goal state can be conceptualized as navigating a landscape. In the physical world, this landscape is the workspace, dotted with obstacles. For the robot's planning system, however, this landscape is re-imagined within the configuration space. Every point in this abstract space represents a complete snapshot of the robot's pose—all its joint angles or its spatial coordinates and orientation. This mapping from physical reality to an abstract state space is what allows us to treat the robot, regardless of its physical complexity, as a single entity whose position in this abstract space needs to be managed.

Consider a simple example: a robot arm designed to pick and place objects on a table. The workspace is the area around the table where the arm operates. The table itself, along with any other objects present, are workspace obstacles. If the arm is a 6-DOF manipulator, its configuration space is 6-dimensional. A specific configuration might be $(30^\circ, -45^\circ, 90^\circ, 0^\circ, 0^\circ, 0^\circ)$, meaning the first joint is at 30 degrees, the second at -45 degrees, and so on. In this 6D space, this single set of angles defines the exact position and orientation of every link of the arm. If, with these joint angles, the arm's gripper would collide with the table, then this particular configuration point is mapped to a C-obstacle. The entirety of the C-obstacle is the set of all such configuration points that would lead to a collision.

The crucial advantage of working in C-space is that it consolidates all collision-related constraints into geometric regions within this abstract space. Instead of performing complex geometric collision checks between potentially many robot links and numerous workspace obstacles at every planning step, the system needs to determine if a configuration point falls within a forbidden region (a C-obstacle). This conceptual simplification is what makes motion planning feasible for complex systems. The problem is transformed from "How does this physical arm move without hitting that physical table?" to "How does this point move in a 6D space without entering the region representing the arm hitting the table?"

The dimensionality of the configuration space is a critical factor. For a mobile robot with three degrees of freedom (x, y, θ), the C-space is 3D. For a six-DOF robotic arm, the C-space is 6D. As the number of DoFs increases, the C-space grows in dimensionality. This exponential increase in dimensions leads to significant computational challenges, often referred to as the "curse of dimensionality." In high-dimensional spaces, the free space (C_{free}) can become highly fragmented, making it very difficult to find paths. Imagine trying to find a path through a vast, complex labyrinth where the walls can twist and turn in many more directions than we can intuitively visualize. This fragmentation means that a path might exist, but algorithms may struggle to discover it efficiently.

The ultimate objective of motion planning is to discover a continuous path within C_{free} from a starting configuration to a goal configuration. This path must be collision-free. Furthermore, for practical robotic applications, the path often needs to be smooth, kinematically feasible (meaning the robot's joints can move at the required speeds and accelerations), and sometimes optimized with respect to certain criteria, such as shortest distance, minimum time, or lowest energy consumption. The transformation into C-space provides the abstract arena where such paths can be sought. The challenge then becomes developing algorithms that can efficiently explore this high-dimensional space, identify the regions of free space, and chart a course through them.

The representation of obstacles in C-space, termed C-obstacles, is not merely a conceptual convenience; it is the cornerstone that allows for algorithmic treatment of collision avoidance. The shape and complexity of these C-obstacles are directly influenced by the geometry of the robot and the obstacles in the workspace. For simple robots, like a point robot, the C-obstacles might closely resemble the workspace obstacles. However, for articulated robots, where the robot's own kinematic chain determines its shape and extent in the workspace, the C-obstacles can become vastly more complex. The boundary of a C-obstacle is defined by the configurations where the robot is in contact with a workspace obstacle or itself.

The definition of $C
{free}$ as the configuration space excluding all C-obstacles is the space within which all valid robot movements must occur. The task of the motion planner is to find a continuous path, $\gamma(t)$, where t is a parameter typically representing time or progress along the path, such that $\gamma(t) \in C*{free}$ for all t in the path's domain, and $\gamma(0) = C*{start}$ and

$\gamma(1) = C${goal}$. This means that at no point along the planned path should the robot be in a state that results in a collision.

The conceptual leap from workspace to configuration space is critical. It allows us to abstract the robot's physical form and treat it as a single point. This point moves through a space where regions are designated as "free" or "occupied" (by C-obstacles). The complexity of this multi-dimensional space, particularly the potential fragmentation of $C
{free}$, *is the primary hurdle that motion planning algorithms must overcome. The dimensionality of C-space is directly determined by the degrees of freedom of the robot. A robot with more joints or more translational/rotational freedom will have a higher-dimensional C-space, leading to greater planning challenges. The development of efficient and robust motion planning algorithms relies heavily on how effectively they can explore and exploit the structure of* C{free}$ in these potentially very high-dimensional spaces.

Pathfinding algorithms represent a cornerstone of motion planning, providing systematic methods to discover feasible routes through an environment. When dealing with the complexities of robot navigation, these algorithms often operate within a discretized representation of the space. One of the most intuitive and widely applied approaches is to transform the continuous configuration space (C_{free}) into a discrete grid. This grid is composed of a finite number of cells, each representing a specific configuration or a small range of configurations. The environment is effectively tiled, and the robot's task of finding a path is reduced to finding a sequence of connected cells from a starting cell to a goal cell.

The resolution of this grid is a critical parameter. A finer grid, meaning more cells, offers a more accurate approximation of the continuous C_{free}. This can lead to more precise and potentially shorter paths. However, a finer grid also dramatically increases the number of cells to be explored, leading to a significant rise in computational cost. Conversely, a coarser grid is computationally cheaper but may miss narrow passages or lead to paths that are suboptimal or even infeasible due to the simplification of obstacle shapes and free space. The choice of grid resolution thus involves a fundamental trade-off between accuracy and computational efficiency.

Within this grid framework, several classic graph search algorithms can be employed. These algorithms treat the grid cells as nodes in a graph and define edges between adjacent cells that are traversable. The "cost" associated with traversing an edge typically represents the distance or effort required to move between the centers of two adjacent cells. For instance, moving between cells horizontally or vertically might have a cost of 1, while moving diagonally might have a cost of $\sqrt{2}$ in a 2D grid, assuming Euclidean distance.

Breadth-First Search (BFS) is a fundamental graph traversal algorithm that systematically explores the grid layer by layer. It starts at the initial cell and explores all its immediate neighbors, then explores the neighbors of those neighbors, and so on. BFS guarantees finding the shortest path in terms of the number of edges (or cells traversed) if all edge costs are uniform. It uses a queue data structure to manage the cells to be visited. When a cell is visited, it is added to the queue, and its unvisited neighbors are then enqueued. The algorithm terminates when it reaches the goal cell or when the queue is empty, indicating that the goal is unreachable. While BFS finds the shortest path in terms of cell count, it does not consider varying traversal costs between cells, making it less suitable for scenarios where moving between different cell types or over different terrains has different associated costs.

Dijkstra's Algorithm is a more general and powerful graph search algorithm that finds the shortest path from a single source node to all other nodes in a graph with non-negative edge weights. Unlike BFS, Dijkstra's algorithm explicitly accounts for the cost of traversing each edge. It maintains a set of visited nodes and a priority queue of nodes to visit, ordered by their tentative distance from the source. At each step, it extracts the node with the smallest tentative distance from the priority queue, marks it as visited, and updates the tentative distances of its unvisited neighbors. If the path through the current node to a neighbor is shorter than the neighbor's current tentative distance, the neighbor's distance is updated. Dijkstra's algorithm guarantees finding the shortest path in terms of total cost. In the context of grid-based pathfinding, Dijkstra's algorithm would efficiently find the path with the minimum cumulative cost from the start cell to the goal cell, considering different costs for moving between adjacent cells.

A *Search Algorithm* is an extension of Dijkstra's algorithm that is particularly well-suited for pathfinding problems and is widely used in robotics. A enhances Dijkstra's by incorporating a heuristic function, $h(n)$, which estimates the cost from a given node n to the goal node. The algorithm prioritizes exploring nodes that are not only close to the start node (as in Dijkstra's) but also seem promisingly close to the goal. It does this by evaluating each node n using a cost function $f(n) = g(n) + h(n)$, where $g(n)$ is the actual cost from the start node to node n, and $h(n)$ is the heuristic estimate to the goal. The algorithm uses a priority queue to select the node with the lowest $f(n)$ value to explore next.

For A

to guarantee finding the optimal (shortest) path, the heuristic function $h(n)$ must be admissible. *An admissible heuristic never overestimates the true cost to reach the goal. Common admissible heuristics for grid-based pathfinding include the Manhattan distance (sum of absolute differences in x and y coordinates) and the Euclidean distance (straight-line distance). The Manhattan distance is admissible for grids where only horizontal and vertical movements are allowed, while the*

Euclidean distance is admissible when diagonal movement is also permitted. If the heuristic is also consistent *(meaning that for any node n and its successor n', $h(n) \le cost(n, n') + h(n')$), A* is guaranteed to be optimal and efficient.

The power of A

lies in its ability to significantly prune the search space compared to Dijkstra's algorithm. By guiding the search towards the goal, it avoids exploring large portions of the grid that are unlikely to be part of the optimal path. This makes A considerably more efficient, especially in large and complex grids. However, the computational cost of A is still dependent on the heuristic's quality and the grid's complexity. In scenarios with very cluttered environments or when the goal is far from the start, the number of explored cells can still be substantial.

Let's consider an illustrative example of applying these grid-based methods. Imagine a simple wheeled mobile robot operating on a flat factory floor represented as a 2D grid. The configuration space for this robot, assuming it can move freely in x and y directions, is effectively reduced to the 2D plane. We discretize this plane into a grid of, say, 1-meter by 1-meter cells. Some cells are marked as occupied due to the presence of static obstacles like machines, walls, or other fixed equipment. The robot starts at a cell $(x
{start}, y{start})$ and needs to reach a goal cell $(x*{goal}, y*{goal})$.

For

BFS, we would initialize a queue with the start cell. We'd then iteratively dequeue a cell, mark it as visited, and enqueue all its unvisited, traversable neighbors (up, down, left, right). The first time we dequeue the goal cell, we've found a path with the minimum number of cell transitions. This might be sufficient if all cells have equal "travel cost."

Using
Dijkstra's algorithm, we would assign a cost to moving between adjacent cells. If moving to an adjacent cell (horizontally or vertically) costs 1 unit, and diagonally costs $\sqrt{2}$ units, Dijkstra's would find the path with the lowest total cost. We would maintain a table of minimum costs to reach each cell from the start and update these costs as we explore.

A *search* would refine this further. If we use the Euclidean distance as our heuristic $h(n)$ (estimating the straight-line distance from cell n to the goal), A would prioritize exploring cells that are "on the way" to the goal. For instance, if the goal is to the northeast of the start, A *would favor exploring cells to the northeast, effectively "focusing" its search effort. If the robot encounters a narrow corridor, A* might explore cells within that corridor more thoroughly if they appear to lead towards the goal, whereas BFS might explore cells in a wider, but ultimately less direct, area first.

The practical implementation of these algorithms in robotics involves several considerations. Firstly, the definition of "traversable" cells needs to be precise. A cell might be considered traversable if its center is in $C *{free}$, or if a significant portion of the cell is in $C*{free}$, or if a path exists from the robot's current configuration to the cell's representative configuration without collision. The collision checking for each cell or potential movement between cells is computationally intensive and requires an accurate model of the robot and its environment.

Secondly, the grid-based approach is often applied to the configuration space, not just the workspace. For a 2D mobile robot, the C-space is 3D (x, y, θ). Discretizing this 3D C-space into a grid of voxels (3D cells) is possible but rapidly becomes computationally prohibitive as the dimensionality increases. A 100x100 grid in 2D has 10,000 cells. A 100x100x100 grid in 3D has 1,000,000 voxels. For higher-dimensional C-spaces (e.g., for robotic arms with multiple joints), a direct grid-based discretization becomes infeasible due to the "curse of dimensionality." This is why A and other graph search methods are often combined with sampling-based motion planning techniques for higher-dimensional problems, or they are applied to simplified representations of the C-space.

Another aspect is the discrete nature of grid-based methods. While they guarantee finding a path within the discrete grid, the resulting path is a sequence of cell centers or transitions between cell centers. This often results in a "jagged" or "staircase" path that may not be smooth or kinematically feasible for a real robot.

Therefore, a post-processing step, such as path smoothing (e.g., using B-splines or Bezier curves) or shortcutting (where the algorithm attempts to connect non-adjacent waypoints directly), is often applied to generate a more fluid trajectory. This smoothing process itself needs to ensure that the smoothed path remains collision-free within the original C_{free}.

Furthermore, grid-based methods are most effective in environments where the obstacles are relatively static and well-defined, and where the robot's dynamic constraints are less critical. For highly dynamic environments where obstacles appear and disappear, or for robots with stringent velocity and acceleration limits, other planning paradigms might be more appropriate or require significant extensions to grid-based approaches (e.g., incorporating temporal information or dynamic replanning).

Despite their limitations, grid-based methods, particularly A, remain fundamental in motion planning education and in many practical applications. They provide a clear and understandable framework for tackling pathfinding problems. Their efficiency can be improved through techniques like hierarchical grids (dividing the space into larger cells and then refining the path within relevant sub-regions) or by using specialized data structures for efficient neighbor discovery and priority queue operations. The core concept of discretizing the problem space and applying graph search algorithms has been immensely influential and forms the basis for many more advanced planning techniques. The ability to efficiently search through these discretized spaces, especially with heuristics that guide the search effectively, is a key skill for any roboticist involved in motion planning. The trade-offs remain central: the granularity of the discretization directly impacts the fidelity of the solution and the computational resources required. For robots operating in structured, relatively low-dimensional environments, grid-based methods offer a robust and often optimal solution.

When the configuration space (C-space) becomes high-dimensional, the grid-based approaches, as discussed previously, quickly run into computational limitations. The curse of dimensionality dictates that the number of grid cells grows exponentially with the number of dimensions. For instance, a simple wheeled robot in a 2D workspace might have a C-space that is 3-dimensional (x, y, orientation θ). A discretized grid in this space would require millions of cells. For a robotic arm with six or more degrees of freedom, the C-space becomes six-dimensional or higher, rendering a uniform grid discretization utterly intractable. In such scenarios, a fundamentally different approach is needed: sampling-based motion planning.

Sampling-based motion planning algorithms tackle the high-dimensionality problem by intelligently exploring the C-space. Instead of attempting to discretize the entire space, these methods "sample" configurations from the C-space and build a graph or tree that represents connectivity in the *free* C-space ($C*{free}$). The core idea is that if a path exists, a sufficiently dense sampling of $C*{free}$ will eventually reveal a sequence of configurations that connect the start to the goal. These algorithms are probabilistic in nature, meaning they do not guarantee finding a path if one exists in a finite time, but they offer a probabilistic completeness guarantee. This means that as the number of samples increases, the probability of finding a path, if one exists, approaches one.

Two of the most influential and widely used sampling-based motion planning algorithms are the Probabilistic Roadmap (PRM) and the Rapidly-exploring Random Tree (RRT).

Probabilistic Roadmap (PRM)

The Probabilistic Roadmap (PRM) algorithm is a two-stage process designed to efficiently answer multiple queries in a static environment. The first stage, called "building the roadmap," involves constructing a graph that captures the connectivity of C_{free}. The second stage, "querying the roadmap," uses this pre-computed graph to find a path between a specific start and goal configuration.

Building the Roadmap:

The roadmap is built by randomly sampling configurations from the C-space. For each sampled configuration, a collision detection test is performed. If the configuration is in C_{free} (i.e., collision-free), it is added as a node to the roadmap graph. Once a sufficient number of nodes are sampled and validated, the algorithm attempts to connect "nearby" nodes. This connection process typically involves a "local planner," which tries to find a simple, collision-free path (often a straight line in the C-space) between two nodes. If such a path exists and is collision-free, it is added as an edge to the roadmap graph, with the weight of the edge representing the length or cost of this local path.

The key components and steps in building the PRM roadmap are:

1.
Sampling: Random configurations are generated within the bounds of the C-space. The sampling strategy can be uniform random sampling across the entire C-space, or it can be biased to sample more densely in areas that are likely to be part of $C{free}$, *or even in regions near the start and goal configurations. For high-dimensional spaces, uniform sampling can be inefficient as it often generates many collision-prone configurations. More advanced sampling techniques, such as Gaussian sampling or sampling from simpler representations of* $C{free}$, can improve efficiency.

2.
Collision Checking: Each sampled configuration is checked for collisions with the environment. This is a critical and often computationally expensive step. A configuration is considered valid if it does not intersect with any obstacles in the environment.

3.
Node Addition: Valid, collision-free configurations are added as nodes to the PRM graph.

4.
Local Planning and Edge Creation: For each new node added, the algorithm searches for existing nodes within a certain "neighbor radius." The radius can be a fixed value, or it can be determined dynamically based on the density of nodes. For every pair of nodes considered neighbors, a local planner is invoked. The local planner attempts to generate a simple, often straight-line, path in C-space between the two nodes. If this path is collision-free, an edge is added to the graph connecting these two nodes. The weight of the edge is typically the length of the local path. It's crucial that the local planner can handle the dimensionality of the C-space. For very high-dimensional spaces, simple straight-line interpolation might not be sufficient, and more sophisticated local planners might be required.

The number of samples and the definition of "nearby" are important parameters. A larger number of samples leads to a more detailed and potentially more connected roadmap, increasing the chances of finding a path. A smaller neighbor radius results in a denser graph with more edges, while a larger radius results in a sparser graph. The connectivity of the roadmap is vital; if C_{free} is disconnected into multiple components, PRM will only find paths within the component containing the start and goal nodes.

Querying the Roadmap:

Once the roadmap is built, answering a query (finding a path from a specific start configuration q_{start} to a specific goal configuration q_{goal}) is relatively fast.

1. **Connect Start and Goal:** The start configuration q_{start} and the goal configuration q_{goal} are first tested for collision. If they are collision-free, they are added to the roadmap. Then, using the same local planner as in the building phase, the algorithm attempts to connect q_{start} to its nearest neighbors in the roadmap and q_{goal} to its nearest neighbors. If successful connections are made, q_{start} and q_{goal} become part of the roadmap graph.

2. **Graph Search:** Standard graph search algorithms, such as Dijkstra's algorithm or A, are then applied to the roadmap graph to find a path from the node representing q_{start} to the node representing q_{goal}. If no path exists between these nodes in the

graph, it implies that either no path exists in C_{free} or the roadmap was not sufficiently dense to capture the connectivity.

Advantages of PRM:

Multi-Query Efficiency: After the initial, potentially time-consuming, roadmap construction, answering multiple path queries is very fast. This makes PRM ideal for scenarios where a robot needs to navigate in a static environment and perform many different tasks.
Handles High Dimensions: PRM is well-suited for high-dimensional C-spaces where grid-based methods fail.
Probabilistic Completeness: As the number of samples increases, the probability of finding a path if one exists approaches 1.

Disadvantages of PRM:

Roadmap Construction Cost: Building the roadmap can be computationally expensive, especially in complex environments or high-dimensional C-spaces. The cost is dominated by sampling and collision checking.
Query Efficiency Dependent on Roadmap: The efficiency of answering queries depends on how well the roadmap represents the connectivity of C_{free}. A sparse or disconnected roadmap can lead to failed queries.

Not Ideal for Dynamic Environments: PRM is designed for static environments. If the environment changes, the entire roadmap needs to be rebuilt, which is inefficient.
Roadmap May Not Be Optimal: The path found by PRM is only as good as the roadmap. It doesn't guarantee finding the shortest or optimal path, although query algorithms like Dijkstra's find the shortest path within the existing roadmap.

Rapidly-exploring Random Tree (RRT)

While PRM builds a static roadmap that can be queried multiple times, the Rapidly-exploring Random Tree (RRT) is an incremental algorithm designed to find a single path from a start configuration to a goal region. RRT works by growing a tree from the start configuration, exploring the C_{free} space by extending towards randomly sampled points. It's particularly effective for single-query pathfinding in complex, high-dimensional spaces.

Algorithm Steps:

RRT starts with a single node representing the initial configuration q_{start}. It then iteratively grows a tree by:

1.
Sampling: A random configuration $q*{rand}$ is chosen from the C-space. For improved exploration, this sampling can be biased towards the goal region. A common strategy is to sample the goal configuration $q*{goal}$ with a certain probability (e.g., 5-10%) and sample uniformly from the C-space otherwise.

2.
Nearest Neighbor: The node in the existing RRT that is closest to $q*{rand}$ is identified. Let this nearest node be $q*{near}$. The "closest" is typically measured using a distance metric like the Euclidean distance in the C-space.

3.
Steering (Extension): A new configuration, $q*{new}$, is generated by "steering" from $q*{near}$ towards $q*{rand}$. This steering operation moves a small step (defined by a `stepsize` parameter) from $q*{near}$ in the direction of $q*{rand}$. The distance moved is limited by `stepsize` to ensure that extensions are incremental. The steering function needs to be able to compute the next configuration in the direction of $q*{rand}$.

4.
Collision Checking: The path segment from $q*{near}$ to $q*{new}$ is checked for collisions. If the segment is collision-free, then $q*{new}$ is added to the RRT as a new node, and an edge is created connecting $q*{near}$ to q_{new}.

5.
Goal Check: After adding $q*{new}$, the algorithm checks if $q*{new}$ is sufficiently close to the goal configuration $q*{goal}$ (or if it has reached the goal region). If it has, a path has been found. The path can be reconstructed by backtracking from $q*{new}$ to the root q_{start} using the parent pointers stored in the tree.

The process repeats until a path to the goal is found or a time limit is reached.

Key Features and Variations:

Rapid Exploration: The "rapidly-exploring" aspect comes from the fact that RRT's structure allows it to expand quickly into unexplored regions of the C-space. By always extending from the nearest node to a random sample, it biases exploration towards areas that are further away from the existing tree.

Probabilistic Completeness: Like PRM, RRT is probabilistically complete. As the number of samples and extensions increases, the probability of finding a path to the goal, if one exists, approaches 1.

C-space Dimensionality: RRT excels in high-dimensional C-spaces because it avoids discretizing the space entirely. Collision checking for each segment is still necessary, but the overall approach scales much better than grid-based methods.

Path Smoothing: The path generated by a basic RRT is often jagged and consists of many small, straight-line segments. This is because the tree grows by taking small steps towards random samples. To obtain a smoother and more kinematically feasible path, post-processing steps like shortcutting or path smoothing using splines are typically applied.

RRT (RRT-star): A significant improvement upon the basic RRT is RRT. *RRT* incorporates a "rewiring" step that aims to find asymptotically optimal paths. When a new node $q{*new*}$ *is added to the tree, RRT looks at its neighbors within a certain radius. If a path through $q*{new}$ offers a shorter path to any of these neighbors than their current path from the root, the neighbor's parent is updated to q_{new} (rewiring). This process ensures that as more samples are added, the tree converges towards the optimal path. RRT has demonstrated remarkable performance in finding near-optimal paths in complex environments.

Biased Sampling: The effectiveness of RRT is heavily influenced by its sampling strategy. Sampling the goal configuration directly with a certain probability is a common heuristic to accelerate convergence. Other intelligent sampling strategies exist, such as sampling in regions of C_{free} identified as difficult to reach, or using information from previous planning attempts.

Step Size: The `step*size` parameter controls how far the tree extends in each iteration. A larger `step*size` leads to faster exploration but might miss narrow passages. A smaller `step_size` results in a more detailed tree but takes longer to explore.

Advantages of RRT:

Handles High Dimensions Effectively: RRT is one of the primary algorithms for motion planning in high-dimensional C-spaces (e.g., for robotic arms).
Single-Query Planner: It is designed to find a path for a specific start-goal pair, making it efficient when only one pathfinding instance is needed.
Probabilistic Completeness and Asymptotic Optimality (RRT): Guarantees that a path will eventually be found if one exists, and RRT converges to the optimal path.
Handles Complex Geometries: Can navigate around intricate obstacles.

Disadvantages of RRT:

Path Quality: Basic RRT often produces jagged paths that require post-processing.
Computational Cost: Can still be computationally intensive, especially in very high-dimensional spaces or when tight tolerances for collision checking are required.
No Guarantee of Optimality (Basic RRT): The basic RRT does not guarantee finding the shortest path.
Sensitivity to Parameters: Performance can be sensitive to parameters like `step_size` and sampling strategy.

Probabilistic Completeness and Optimality Guarantees

Both PRM and RRT are considered **probabilistically complete**. This means that if a path exists in C_{free}, the probability that the algorithm will find a path approaches 1 as the number of samples (for PRM roadmap construction) or the number of tree extensions (for RRT) goes to infinity. This is a strong theoretical guarantee, but it's important to remember that in practice, computation time is finite. Thus, there's always a non-zero chance that a path might not be found within a given time limit, especially in complex environments or high dimensions.

While PRM finds the shortest path within its constructed roadmap (using Dijkstra's on the graph), the roadmap itself might not perfectly represent C_{free}. RRT, in its basic form, makes no guarantees about path optimality. The paths are often suboptimal, a consequence of the greedy nature of extending the tree towards random samples. However, variants like RRT *introduce* **asymptotic optimality**. This means that as the number of samples goes to infinity, the path found by RRT converges to the optimal path. This is achieved through the rewiring mechanism that continually refines the tree to find shorter paths.

Practical Considerations for Sampling-Based Planners

Implementing sampling-based motion planners in real-world robotics applications involves several practical considerations:

Collision Detection Efficiency: Collision detection is the bottleneck for most sampling-based planners. For complex robots with many degrees of freedom and intricate environments, highly optimized collision detection libraries and techniques are essential. This might involve spatial data structures like bounding volume hierarchies (BVHs) or swept-volume representations.

C-space Sampling Strategies: Simple uniform random sampling is often inefficient. For PRM, techniques like Latin Hypercube sampling can ensure better coverage. For RRT, goal-biasing is common, but more advanced strategies might involve sampling in regions that have been identified as difficult to reach (e.g., narrow passages).

Local Planners: The choice of local planner is crucial. For PRM, it defines how nodes are connected. For RRT, it determines how the tree grows. In high dimensions, simple straight-line interpolation might not be sufficient or might lead to many failed collision checks. More sophisticated local planners might consider robot kinematics and dynamics.

Path Smoothing and Simplification: The raw paths generated by RRT are often not usable directly. Path smoothing techniques are necessary to reduce the number of waypoints, create kinematically feasible trajectories, and improve the aesthetic quality of the path. Common techniques include:
Shortcutting: Iteratively attempting to connect non-adjacent waypoints in the path with a direct, collision-free segment.
Spline Fitting: Using B-splines or Bezier curves to generate a smooth curve that interpolates or approximates the waypoints, while ensuring collision avoidance.
Gradient-based Optimization: Treating the path as a curve and optimizing its parameters to minimize length and deviations from a desired shape, while satisfying collision constraints.

Handling Kinematic and Dynamic Constraints: Basic PRM and RRT plans in the kinematic C-space. For robots with significant velocity or acceleration limits, these kinematic paths must be further processed to generate dynamically feasible trajectories. This might involve time-parameterization of the path or using planners that directly incorporate dynamic constraints (e.g., kinodynamic RRT).

Multi-Query vs. Single-Query: The choice between PRM and RRT often depends on whether the application requires answering many queries in a static environment (favoring PRM) or finding a single path in a potentially complex, high-dimensional space (favoring RRT).

Sampling-based motion planning has revolutionized the field of robotics by enabling robots to navigate complex, high-dimensional environments that were previously intractable. The probabilistic nature of these algorithms, coupled with their ability to avoid explicit discretization, has opened up new possibilities for robot autonomy.

From raw path to executable motion: Trajectory Generation and Motion Primitives

The preceding sections have delved into the intricate world of motion planning algorithms, primarily focusing on how robots can find a feasible, collision-free sequence of configurations – a path – through their operational space. These algorithms, such as

Probabilistic Roadmaps (PRM) and Rapidly-exploring Random Trees (RRT), are crucial for navigating complex environments, especially those with high-dimensional configuration spaces. However, a kinematically feasible path, while mathematically valid, is merely an abstract representation of movement. For a robot to actually execute this plan, the path must be translated into a *trajectory*. A trajectory is not just a sequence of positions, but a sequence of positions *parameterized by time*, incorporating crucial elements like velocity and acceleration. This distinction is vital because real-world robots are governed by physical constraints – their motors have limited speed, their actuators have inertia, and their bodies experience forces. Simply commanding a robot to instantaneously jump between waypoints along a path would likely result in failure, damage, or unsafe operation. Therefore, the process of trajectory generation bridges the gap between an abstract path and a concrete, executable robotic motion.

The fundamental difference between a path and a trajectory lies in the temporal dimension. A path can be thought of as a curve in the configuration space, $q(s)$, where s is an abstract parameter ranging from 0 to 1. This parameter s might loosely correspond to progress along the path, but it carries no explicit notion of time. A trajectory, on the other hand, is $q(t)$, where t represents actual time. This means that at any given time t, the robot is at a specific configuration $q(t)$, and importantly, it possesses a certain velocity $\dot{q}(t)$ and acceleration $\ddot{q}(t)$. These time-derivatives are precisely what define the robot's dynamic behavior.

Generating a trajectory from a path typically involves a two-step process: first, time-parameterization of the path, and second, consideration of kinematic and dynamic constraints.

Time-Parameterization of Paths

The first step in converting a path into a trajectory is to assign a time to each configuration along the path. This process, known as time-parameterization, essentially "drives" the robot along the path at a specific rate. The simplest form of time-parameterization would be to assume a constant velocity along the entire path. If the path is represented by a sequence of waypoints $q 0, q1, \ldots, qn$, and the distance between consecutive waypoints $q{i-1}$ and qi is di, then a constant velocity v would imply that the time taken to traverse each segment is $\Delta ti = di / v$. The total time to reach waypoint qi would be the sum of these durations: $ti = \sum{j=1}^{i} \Delta tj$. This approach, however, often leads to discontinuous velocity changes at waypoints, which are usually undesirable.

A more sophisticated approach involves generating smooth velocity and acceleration profiles. This is where mathematical tools like
piecewise polynomial curves become invaluable. Cubic splines are a popular choice for interpolating a set

of waypoints while ensuring continuity of position, velocity, and acceleration. A cubic spline is a curve composed of multiple cubic polynomial segments, where each segment connects two adjacent waypoints. The key characteristic of splines is that they are designed to be smooth at the connection points (knots) between segments.

For a path defined by a sequence of waypoints $q_0, q_1, \ldots, q_n$, a cubic spline can be constructed such that:
1. The spline passes through each waypoint q_i.
2. The first derivative (representing velocity) is continuous across all waypoints.
3. The second derivative (representing acceleration) is continuous across all waypoints.

A cubic polynomial segment connecting two points q_{i-1} and q_i can be represented as:
$p(s) = a s^3 + b s^2 + c s + d$, where s is a local parameter that varies from 0 to 1 along the segment. The coefficients a, b, c, d are determined by the constraints. For a trajectory, we want to time-parameterize this path. Let's consider a single segment of the path represented by $q(s)$, where $s \in [0, 1]$. We want to find a function $s(t)$ that maps time t to the path parameter s, such that $q(t) = q(s(t))$.

The velocity along the path at time t would be:

$\dot{q}(t) = \frac{dq}{dt} = \frac{dq}{ds} \frac{ds}{dt}$.
And the acceleration:

$\ddot{q}(t) = \frac{d^2q}{dt^2} = \frac{d^2q}{ds^2} \left(\frac{ds}{dt}\right)^2 + \frac{dq}{ds} \frac{d^2s}{dt^2}$.

If we assume $q(s)$ is already a smooth curve in configuration space, the challenge is to define $s(t)$ such that $\dot{q}(t)$ and $\ddot{q}(t)$ satisfy the robot's dynamic constraints. A common strategy is to define $s(t)$ as a polynomial in time. For instance, if we need to ensure continuous velocity and acceleration, a cubic polynomial for $s(t)$ over a time interval $[t *{start}, t*{end}]$ could be used. The function $s(t)$ would be defined by the values of s, $\dot{s}$, $\ddot{s}$ at the start and end of the interval.

For example, consider a path segment from $q *{i-1}$ to $q*i$. If we want to reach $q*i$ at time $t*i$ starting from $q*{i-1}$ at time $t*{i-1}$, and we want specific velocities and accelerations at these points, we can formulate a time-parameterization problem. A common approach is to define the trajectory as a piecewise polynomial function in time. For each segment of the path, we can define a polynomial trajectory such that it passes through the specified configurations with specified velocities and accelerations at the segment endpoints.

Let's consider a single degree of freedom for simplicity. Suppose we have a path segment from $q *{i-1}$ to $q*i$. We want to parameterize this segment with time, such that at time $t*{i-1}$, the configuration is $q*{i-1}$ with velocity $\dot{q}*{i-1}$ and acceleration $\ddot{q}*{i-1}$, and at time $t*i$, the configuration is*

qi with velocity $\dot{q}i$ and acceleration $\ddot{q}i$. We can use a quintic (fifth-order) polynomial for the position $q(t)$ over the time interval $[t{i-1}, ti]$. A quintic polynomial has six coefficients, allowing us to satisfy six boundary conditions:

$q(t{i-1}) = q{i-1}$
$\dot{q}(t{i-1}) = \dot{q}{i-1}$
$\ddot{q}(t{i-1}) = \ddot{q}{i-1}$
$q(ti) = qi$
$\dot{q}(ti) = \dot{q}i$
$\ddot{q}(ti) = \ddot{q}i$

By solving for these coefficients, we obtain a smooth trajectory segment that respects the desired start and end conditions. These segments are then concatenated to form the complete trajectory. The selection of target velocities and accelerations at waypoints ($\dot{q} i, \ddot{q}i$) is critical. They must be chosen such that the resulting trajectories are not only smooth but also kinematically and dynamically feasible for the robot. For instance, these values should not exceed the robot's maximum velocity or acceleration limits.

The choice of the degree of the polynomial (cubic, quintic, etc.) depends on the desired level of smoothness and the number of constraints that need to be satisfied. Cubic polynomials are good for position and velocity continuity, while quintic polynomials can ensure continuity of position, velocity, and acceleration. In higher-dimensional configuration spaces, this process is applied independently to each degree of freedom, or using vector-valued polynomials, ensuring that the resulting trajectory is smooth across all joints or states simultaneously.

Motion Primitives

Beyond generating smooth, time-parameterized paths, robotics often leverages the concept of **motion primitives**. A motion primitive is a fundamental, pre-defined movement pattern that a robot can execute reliably. These primitives serve as building blocks for more complex behaviors. Think of them as the basic strokes a painter uses before creating a masterpiece. By combining these basic motions, a robot can perform a wide range of tasks.

Common examples of motion primitives include:

1.
Straight-line motion: Moving the robot's end-effector or base in a straight line. For a mobile robot, this could be moving forward or backward along a specific heading. For a robotic arm, this might translate to a linear motion of the tool center point (TCP) in Cartesian space.

2.
Rotational motion: Turning the robot in place or rotating a specific joint. For a mobile robot, this could

be a pivot turn or a yaw rotation. For an arm, this is the standard joint-space rotation.

3.
Circular or Arced motion: Following a curved path, often used for tasks involving sweeping or manipulating objects along a contour.

4.
Grasping/Releasing primitives: For manipulators, specific primitives might exist for actuating the gripper to pick up or release an object.

5.
Wake-up/Sleep primitives: For systems that need to be powered up or shut down, or brought into a ready state.

The power of motion primitives lies in their simplicity, reliability, and often, their optimized execution. Because these are fundamental movements, they can be pre-programmed and tested thoroughly. When a motion planner generates a path, it might not output a detailed sequence of joint commands for every millisecond. Instead, it might output a sequence of primitive invocations. For example, a path might be described as: "Execute a 'move straight' primitive to point A, then execute a 'turn 90 degrees' primitive, followed by another 'move straight' primitive to point B."

The trajectory generation component then takes these primitive commands and fleshes them out with temporal information. Each primitive has its own set of parameters (e.g., distance for straight line, angle for turn, radius for arc) and internal trajectory generation logic. For a 'move straight' primitive, the robot might generate a trapezoidal velocity profile to move the specified distance, ensuring smooth acceleration and deceleration. Similarly, a 'turn' primitive might generate a profile for angular velocity.

Benefits of using Motion Primitives:

Modularity and Reusability: Primitives can be reused across different tasks and environments.
Simplicity of High-Level Planning: High-level planners can focus on sequencing primitives rather than dealing with the low-level continuous motion details.
Guaranteed Performance: Well-tested primitives offer predictable and reliable execution.
Efficiency: Optimized primitives can be executed faster and more efficiently than arbitrary trajectories generated from scratch.
Compliance with Constraints: Primitives can be designed to inherently satisfy certain kinematic or dynamic constraints, simplifying the overall trajectory generation process.

For instance, a common strategy is to use RRT or PRM to generate a high-level path in the configuration space. This path might be a sequence of waypoints. Then, the system can decompose this path into a sequence of motion primitives. For example, if the path requires moving from point A to point B, and point B is reachable by a straight-line motion primitive with current robot configuration, that primitive is selected. If point B requires a turn before moving, a turning primitive might be selected first.

Consider a robotic arm tasked with picking up an object. A motion planner might generate a path to approach the object, then the system might invoke a 'close gripper' primitive. After successful grasping, it might generate a path to move the object to a new location, possibly using a combination of 'move straight' and 'rotate joint' primitives.

Integrating Kinematic and Dynamic Constraints

The process of translating a path into a trajectory is fundamentally about satisfying the robot's physical limitations. These limitations can be broadly categorized into kinematic and dynamic constraints.

Kinematic Constraints: These constraints relate to the robot's geometry and its allowed motions, independent of forces. They include:

Joint Limits: Each joint in a robotic arm has a limited range of motion.
Workspace Limits: The robot's end-effector cannot reach beyond its physical workspace.
Velocity Limits: The maximum speed at which joints or end-effectors can move.
Acceleration Limits: The maximum rate of change of velocity for joints or end-effectors.
Singularities: Configurations where the robot loses some degrees of freedom, making certain movements impossible or ill-defined.

Dynamic Constraints: These constraints involve forces, torques, inertia, and friction. They are crucial for robots that operate at higher speeds or handle significant payloads.
Torque/Force Limits: The maximum torque or force that actuators can produce.
Inertial Effects: The inertia of robot links and payloads affects how quickly they can be accelerated or decelerated.
Friction: Friction at joints can oppose motion and affect precise control.
Payload Capacity: The maximum weight a robot can safely lift and manipulate.

When generating trajectories, we must ensure that the resulting $q(t)$, $\dot{q}(t)$, and $\ddot{q}(t)$ profiles respect all these constraints. The time-parameterization of a path using piecewise polynomials is a common method to achieve this. By carefully selecting the polynomial coefficients, we can shape the velocity and acceleration profiles to stay within acceptable bounds.

For example, if a path requires a sharp turn, a simple straight-line interpolation might result in extremely high angular velocities or accelerations, exceeding the robot's limits. A sophisticated trajectory generator would recognize this and, during time-parameterization, introduce a gentler velocity profile for the turn, potentially extending the time taken for that maneuver or subtly altering the path itself if strict time limits are not imposed.

Trapezoidal and S-curve Velocity Profiles:

To ensure smooth transitions and respect velocity/acceleration limits, common velocity profiles are employed during time-parameterization:

Trapezoidal Velocity Profile: This is a fundamental profile where the velocity ramps up linearly from zero to a maximum velocity, stays constant for a period, and then ramps down linearly to zero. This profile ensures continuous acceleration and deceleration. However, it results in discontinuous jerk (the third derivative of position).

S-curve Velocity Profile: To further improve smoothness and reduce vibration, the S-curve profile is often used. It's an extension of the trapezoidal profile that includes smooth transitions for acceleration and deceleration. Instead of instantaneous changes in acceleration, the acceleration itself changes smoothly, resulting in a continuous jerk profile. This is achieved by using cubic polynomials for the acceleration ramps, leading to a quintic polynomial for the position itself over time.

These profiles are applied to each degree of freedom independently or in a coupled manner, depending on the robot's kinematics and the complexity of the trajectory generation. For instance, to move from waypoint $q
*i$ to $q*{i+1}$, a trajectory generator might define $q(t)$ as a quintic polynomial within the time interval $[t*i, t*{i+1}]$ such that the boundary conditions for position, velocity, and acceleration at $t*i$ and $t*{i+1}$ are met, while ensuring that the velocity and acceleration profiles stay within the robot's kinematic limits.

Incorporating Dynamics:

Handling dynamic constraints is more complex and often requires an understanding of the robot's equations of motion. If dynamic feasibility is critical (e.g., for high-speed maneuvers, robots with significant inertia, or tasks involving interaction forces), trajectory generation becomes a more involved optimization problem. This might involve:

Time-Optimal Trajectory Generation: Finding the shortest time to complete a task while respecting all kinematic and dynamic constraints.
Dynamic Models: Explicitly using the robot's dynamic model (e.g., mass, inertia, friction parameters) to predict and control its behavior.
Feedback Control: Employing robust feedback control laws to ensure the robot accurately follows the generated trajectory, compensating for modeling errors and external disturbances.

In many practical applications, especially in industrial settings with slower movements and controlled environments, trajectory generation focuses heavily on kinematic feasibility, using smooth polynomial splines and optimized velocity profiles. For more dynamic tasks or robots operating in less predictable conditions, incorporating dynamic models and advanced control strategies becomes paramount.

The output of the trajectory generation phase is a time-stamped sequence of desired configurations, velocities, and accelerations that can be fed to the robot's low-level controller. This controller then attempts to command the robot's actuators to follow this trajectory as closely as possible, closing the loop between the abstract plan and the physical execution of motion.

In summary, while motion planning algorithms like PRM and RRT provide the fundamental roadmap and path, it is the process of trajectory generation that breathes life into these plans. By parameterizing paths with time and carefully considering the kinematic and dynamic limitations of the robot, we transform abstract geometric sequences into executable, smooth, and safe motions. The use of mathematical tools like piecewise polynomials and concepts like motion primitives are key to this translation, ensuring that robots can not only navigate complex spaces but also do so effectively and reliably in the physical world.

Chapter 9: Advanced Motion Planning and Control

Moving beyond the realm of merely finding *a* path, the field of motion planning increasingly grapples with the question of finding the *best* path. This pursuit of excellence is encapsulated in the concept of **optimal motion planning**. While a standard motion planner might successfully navigate a robot around obstacles, an optimal planner strives to find a path that not only avoids collisions but also optimizes certain desirable characteristics. These characteristics are typically quantified by a **cost function**, which the planner aims to minimize. The choice of cost function is paramount, as it dictates what "optimal" truly means for a given task. Common objectives include minimizing path length, reducing the time taken for the maneuver, conserving energy, or ensuring the smoothest possible motion to minimize wear and tear on the robot or improve the quality of an operation.

The fundamental shift from feasibility to optimality introduces a significant computational challenge. Finding *any* path can often be achieved through heuristic search or random sampling. However, finding the *optimal* path typically requires more rigorous methods that explore the search space more thoroughly and evaluate paths against a specific metric. This often leads to a trade-off: the more optimal the desired solution, the higher the computational resources and time required to find it. Understanding this trade-off is crucial for deploying optimal motion planning in real-time applications.

One of the most influential algorithms in the domain of

optimal pathfinding, and a cornerstone for many optimal motion planning strategies, is the
A *(A-star) search algorithm*. While originally conceived for grid-based pathfinding in 2D or 3D environments, A has been extended and adapted for continuous configuration spaces commonly encountered in robotics. At its core, A is an informed search algorithm that finds the shortest path from a start node to a goal node in a graph. It achieves this by maintaining a priority queue of paths to explore, prioritizing those that appear most promising. The "promise" of a path is evaluated using a cost function, typically denoted as \$f(n)\$, which is the sum of two components: \$g(n)\$ and \$h(n)\$.

Here, \$g(n)\$ represents the actual cost incurred to reach node \$n\$ from the start node. This could be the actual path length traversed, the time elapsed, or any other cumulative cost. The second component, \$h(n)\$, is a heuristic estimate of the cost from node \$n\$ to the goal node. A
is guaranteed to find the shortest path (optimal solution) if the heuristic function \$h(n)\$ is admissible, meaning it never overestimates the true cost to reach the goal. For example, in a 2D environment, the straight-line Euclidean distance from a point to the goal is an admissible heuristic because the actual path cannot be shorter than a straight line.

Adapting A for robotics involves representing the robot's configuration space as a graph. This is often done by discretizing the continuous space into a grid or by sampling points in the configuration space and connecting them to form a graph, similar to the way Probabilistic Roadmaps (PRM) are constructed. When using a discretized grid, each cell can be considered a node, and edges connect adjacent cells that are reachable without collision. The cost $g(n)$ would be the sum of costs of the edges traversed from the start, and $h(n)$ would be the heuristic estimate (e.g., Euclidean distance in joint space or task space) from the current cell to the goal cell.

In the context of
optimal A, *the algorithm is applied to these graph representations. For instance, if we discretize the configuration space into a grid, an optimal A* search will identify the path with the minimum cumulative cost (e.g., minimum total distance) from the start configuration to the goal configuration. The advantage of A is its ability to efficiently find optimal paths in such discrete spaces, especially when combined with a good heuristic.

However, discretizing the configuration space can lead to two primary issues. Firstly, the resolution of the discretization directly impacts the optimality. A finer grid leads to a more accurate representation of the space and potentially a more optimal path, but it drastically increases the number of nodes and edges, leading to higher computational complexity and memory requirements. A coarser grid is computationally cheaper but may result in suboptimal paths or even fail to find a path if the resolution is too low to represent narrow passages. Secondly, a path found on a discrete grid might not be smooth or dynamically feasible for a real robot. It often consists of a sequence of straight-line segments connecting the centers of grid cells, which can lead to jerky movements and high accelerations when translated into robot commands.

To address these limitations, researchers have developed methods for applying optimal search in continuous configuration spaces or for optimizing paths generated by other planners. For continuous spaces, techniques like
optimal sampling-based planners are employed. These methods build upon the principles of planners like RRT and PRM but incorporate optimization. For example, an **RRT**-*like algorithm (RRT-star)* enhances the basic RRT by introducing a rewiring step. As RRT grows its tree towards the goal, RRT also considers rewiring nodes. When a new node is added to the tree, RRT *checks its neighbors. If a newly added node provides a shorter path to an existing neighbor, that neighbor's parent is updated. Crucially, it also checks if a neighbor can be reached with a lower cost* through *the newly added node. This continuous "rewiring" process ensures that as the tree expands, it converges towards an optimal path. The key is that RRT* considers not just the direct connection to the parent but also connections through nearby nodes already in the tree, effectively performing a localized optimization.

The theoretical guarantees of RRT

are significant: under certain conditions (e.g., with an infinite number of samples), RRT converges to the optimal path. However, in practice, with a finite number of samples, RRT *provides a near-optimal path. The degree of optimality improves as the number of samples increases, again highlighting the computational cost associated with achieving higher optimality. The rewiring step in RRT* can be computationally intensive, especially in high-dimensional spaces, as it requires searching for neighbors and evaluating potential path improvements.

Another approach for achieving optimality, particularly when dealing with paths generated by less optimal planners (like basic RRT or PRM), is **post-processing optimization**. Once a feasible path is found, optimization algorithms can be applied to refine it. A common technique is **path smoothing**. This involves iteratively perturbing the path points to reduce curvature, shorten the path, or improve its smoothness, while ensuring that the perturbed path remains collision-free. Algorithms like gradient descent on the path's cost function, or more specialized smoothing techniques like Chaikin's algorithm or spline fitting, can be employed. However, ensuring collision avoidance during smoothing can be challenging. Collision checking needs to be integrated into the optimization loop, potentially turning it into a constrained optimization problem.

Optimal control techniques also play a significant role in optimal motion planning, especially when dynamic constraints are critical. While path planning often focuses on the geometric path, optimal control focuses on finding the control inputs (e.g., joint torques, velocities) over time that steer the robot from an initial state to a final state while minimizing a cost function. Techniques like **Model Predictive Control (MPC)** can be seen as a form of optimal trajectory generation. MPC repeatedly solves an optimization problem over a finite horizon, predicting the robot's future behavior and calculating the optimal control inputs to follow a desired trajectory or achieve a goal.

The cost function in optimal control can be very sophisticated, incorporating terms for:

Deviation from a reference trajectory: Penalizing the robot for not following a desired path.
Control effort: Minimizing energy consumption or actuator wear.
Time: Encouraging faster maneuvers.
State deviations: Keeping the robot within certain operational bounds.

When dynamic constraints are central, such as for high-speed maneuvers or robots with significant inertia, traditional sampling-based or grid-based planners that primarily consider kinematic feasibility might not suffice. Optimal control methods, often rooted in calculus of variations or optimal control theory, can directly generate dynamically feasible and optimal trajectories. For example, algorithms like **Differential Dynamic Programming (DDP)** or **Iterative Linear Quadratic Regulator (iLQR)** can find optimal control policies by iteratively improving an initial trajectory.

The optimization problem for dynamic feasibility can be formulated as follows: Given a robot's dynamic model $f(\mathbf{x}, \mathbf{u})$ (where $\mathbf{x}$ is the state and $\mathbf{u}$ is the control input) and a cost function J, find the control sequence $\mathbf{u}(t)$ that minimizes J subject to the system dynamics and constraints. The cost function might include terms like:

$J = \int{t0}^{tf} L(\mathbf{x}(t), \mathbf{u}(t)) dt + \Phi(\mathbf{x}(tf))$
where L is the running cost and Φ is the terminal cost.

For tasks that require the shortest possible time to complete,
time-optimal trajectory generation is a specialized area. This involves finding the trajectory that reaches the goal state in the minimum amount of time, subject to all kinematic and dynamic constraints. This is often a challenging problem, particularly for systems with non-holonomic constraints or complex dynamics. Techniques from optimal control theory, such as Pontryagin's Maximum Principle, can be used to derive solutions for time-optimal problems, often resulting in bang-bang control strategies where actuators operate at their maximum or minimum limits.

Consider a common scenario: a robotic arm needs to move a payload from point A to point B. A simple kinematic planner might find a short geometric path. However, if the payload is heavy, accelerating and decelerating too quickly along this path could exceed the torque limits of the arm's joints, or even cause instability. An optimal motion planner, considering dynamics, would generate a trajectory that might be geometrically longer or take more time, but is specifically shaped to respect torque limits, minimize energy expenditure, and ensure smooth, stable motion. This often involves generating profiles for joint velocities and accelerations that are well within the robot's capabilities, even if it means taking a less direct route in terms of path length.

The practical implementation of optimal motion planning often involves a hierarchical approach. A high-level planner might use a sampling-based method like RRT to quickly find a near-optimal, collision-free path in the configuration space. This path is then passed to a lower-level trajectory optimizer that refines it, ensuring dynamic feasibility and possibly optimizing for energy or time using methods from optimal control. Collision checking remains a persistent challenge throughout these optimization processes. Techniques like continuous collision checking along the optimized trajectory, or optimization algorithms that incorporate collision constraints directly, are essential.

The concept of
smoothness as an optimization objective is also critical. While not always explicitly a primary goal, smooth

trajectories are often implicitly desired for reasons of efficiency, reduced wear, and better system performance. A trajectory that has low jerk (rate of change of acceleration) tends to be smoother and less taxing on the robot's actuators. Quintic and higher-order polynomials are often used to generate trajectories with continuous acceleration and jerk, which can be considered a form of optimization for smoothness.

Furthermore, for robots operating in uncertain environments or dealing with imprecise models, **robust optimal motion planning** becomes important. This involves finding a trajectory that is optimal not just for a nominal model but also performs well or remains safe under a range of possible uncertainties (e.g., sensor noise, model inaccuracies, external disturbances). This often leads to solutions that are more conservative, perhaps taking longer or being less "optimal" in a deterministic sense, but are significantly more reliable in practice.

In summary, optimal motion planning is a vital extension of basic motion planning, shifting the focus from mere feasibility to achieving superior performance according to defined criteria. Algorithms like A
and its kin, along with enhancements to sampling-based methods such as RRT, provide frameworks for finding paths that minimize metrics like length. Post-processing techniques like path smoothing and advanced optimal control methods further refine these paths to satisfy dynamic constraints, optimize for time or energy, and ensure smooth, robust execution. The perpetual

challenge lies in balancing the computational demands of optimization with the need for real-time applicability, making the choice of algorithm and cost function a critical decision for any robotic system. The pursuit of optimality transforms a robot from a simple navigator into a highly efficient and performant agent.
The discussion thus far has largely assumed a static world, where the robot's environment remains constant throughout its operation. However, in many real-world applications, the environment is anything but static. Robots are often deployed in spaces shared with humans, other robots, or dynamic objects like conveyor belts, forklifts, or even the very products they are manipulating. Planning for such **dynamic environments** introduces a fundamentally new layer of complexity. The presence of moving obstacles means that a path planned in a static environment may quickly become invalid, leading to collisions or an inability to reach the goal.

To cope with the unpredictability and constant change inherent in dynamic environments, robots must adopt strategies that go beyond pre-computed, static paths. This necessitates a shift towards **reactive planning** and **adaptive control**. Reactive planning focuses on making immediate decisions based on the current sensory information, modifying the robot's actions on-the-fly to avoid impending collisions or to seize opportunities presented by the changing scene. While global planning provides a general direction or a rough roadmap, reactive methods handle the fine-grained, real-time adjustments required for safe and efficient navigation amidst motion.

One of the core challenges in dynamic environments is the inherent trade-off between reaction speed and look-ahead capability. A purely reactive system might respond instantly to detected obstacles but could be prone to oscillations or inefficient movements if it lacks an understanding of the broader context or future intent of the moving elements. Conversely, a system that tries to predict the future too far ahead might be too slow to react to sudden changes or might make incorrect predictions, leading to errors. The ideal solution often lies in a hybrid approach, combining a robust global plan with sophisticated local reactive controllers.

When considering local obstacle avoidance in dynamic settings, especially for mobile robots operating in cluttered spaces, algorithms that can quickly assess the immediate danger and determine a safe, executable velocity command are paramount. The **Dynamic Window Approach (DWA)** is a prominent example of such a reactive method. DWA operates by discretizing the robot's possible velocities into a "dynamic window" - a set of reachable velocities within a short time horizon, considering the robot's current velocity and its physical constraints (e.g., maximum acceleration and deceleration). For each velocity within this dynamic window, DWA simulates the robot's trajectory for a brief period, checking for collisions with any detected obstacles.

The core idea behind DWA is to project the robot's

future path for a small time interval, using its current kinematic state and the available velocity commands. This projection generates a set of possible trajectories. Each of these trajectories is then evaluated based on a set of criteria. Typically, these criteria include:

1. **Collision Avoidance:** The most critical factor is ensuring that the projected trajectory does not intersect with any obstacles. Trajectories that lead to collisions are immediately discarded.
2. **Goal Reaching:** Among the collision-free trajectories, those that move the robot closer to its global target position or heading are preferred. This is often quantified by a "heading" score, measuring how well the trajectory aligns with the direction towards the goal.
3. **Velocity Maximization:** To maintain forward progress and avoid becoming stuck, trajectories that utilize higher velocities are generally favored, provided they are safe and directed towards the goal.

These objectives are typically combined into a single evaluation function, which assigns a score to each feasible velocity command. The velocity command that yields the highest score is then selected, and the robot executes a motion corresponding to that velocity for a short duration. At the next control cycle, the process is repeated, re-evaluating the dynamic window and potential trajectories based on updated sensor readings and the robot's new state. This continuous loop of sensing, planning, and acting allows DWA to effectively navigate through environments with moving obstacles.

The "dynamic window" itself is crucial. It represents the set of velocities (v, ω) (linear and angular velocity) that the robot can achieve from its current state $(\nu
c, \omegac)$ within a small time interval Δt, respecting the robot's maximum linear and angular velocities, as well as its maximum linear and angular accelerations. Mathematically, the dynamic window is defined by:
$\nu \in [\nu{min}, \nu{max}]$ and $\omega \in [\omega{min}, \omega{max}]$
where:
$\nu{min} = \nuc - a_v \Delta t$
$\nu{max} = \nuc + a_v \Delta t$
$\omega{min} = \omegac - a_\omega \Delta t$
$\omega{max} = \omegac + a_\omega \Delta t$
subject to additional constraints like $\nu \ge 0$ if the robot cannot move backward, or ν and ω within the robot's operational limits.

Within this dynamic window, a sampling of discrete velocity pairs is considered. For each sampled pair (v, ω), a short trajectory is simulated. The simulation typically involves integrating the robot's kinematic equations:

$\dot{x} = v \cos(\theta)$
$\dot{y} = v \sin(\theta)$
$\dot{\theta} = \omega$
over a predefined time horizon T_{sim}. The resulting path $(x(t), y(t), \theta(t))$ is then checked for collisions. If a trajectory is collision-free, its score is

calculated. A common scoring function might be a weighted sum of the heading towards the goal and the achieved velocity:
$Score(v, \omega) = \alpha \cdot Heading(v, \omega) + \beta \cdot Velocity(v, \omega)$
where $Heading(v, \omega)$ measures the alignment of the robot's future direction with the goal and $Velocity(v, \omega)$ is the linear velocity achieved. The pair (v, ω) that maximizes this score is chosen.

Another influential local planning technique for dynamic environments is the
Vector Field Histogram (VFH) and its subsequent variations like VFH+. VFH works by constructing a local obstacle map, typically a grid-based representation centered around the robot, which is updated continuously using sensor data (e.g., from a laser scanner). This local map is then transformed into a polar histogram grid, which represents obstacle densities in different angular sectors around the robot.

The VFH algorithm then proceeds in two main steps:

1. **Obstacle Representation:** It first builds a local occupancy grid based on sensor readings. For each obstacle point detected, it determines the cells it occupies and marks them. Obstacles detected further away contribute to cells that are "less occupied" than nearby obstacles.
2. **Voting Scheme:** From this local map, a polar histogram grid is constructed. Each angular sector in this grid accumulates "votes" based on the density of

obstacles within that sector. Sectors with a high density of obstacles are considered blocked, while sectors with fewer obstacles are considered free.

Once the histogram grid is built, VFH searches for a clear path. It identifies "free" angular ranges (bins with a low obstacle vote count). It then considers the robot's current direction of motion and the direction towards the global goal. The algorithm looks for the "least obstructed" direction that is also reasonably aligned with the goal direction. This involves finding an angular sector that is free and has an orientation close to the desired heading. If multiple such directions exist, it might choose the one that offers the most clearance. The robot then attempts to steer towards this chosen direction by setting its angular velocity.

VFH+ enhances the basic VFH by incorporating information about the robot's shape and its current velocity to provide more robust obstacle avoidance and smoother motion. It refines the histogram representation and the selection criteria to better handle dynamic obstacles and to produce more fluid movements. The critical aspect of VFH and VFH+ is their ability to quickly process sensor data and make decisions about steering without needing to explicitly predict future trajectories over long horizons, making them highly suitable for real-time operation in dynamic environments.

The interplay between global and local planning is

crucial for achieving robust navigation in dynamic scenarios. A pure reactive approach, like DWA or VFH, might get the robot out of immediate danger but may lead to inefficient, meandering paths or even getting stuck in local minima if the global goal is consistently obstructed. Conversely, a purely global planner will fail as soon as the environment changes.

A common and effective strategy is to use a global planner to generate a general path or a series of waypoints in the static or slowly changing environment. This global plan provides a strategic direction. Then, a local planner (like DWA or VFH) takes over to execute the motion between waypoints, or to follow the global path while continuously adjusting to avoid dynamic obstacles detected by the sensors.

When the local planner encounters an unresolvable situation – for example, if all immediate velocities lead to collisions, or if it consistently fails to make progress towards the next waypoint due to persistent dynamic obstacles – it can signal a need for replanning at a higher level. This might involve the global planner re-evaluating the path from its current position, potentially considering the dynamic nature of the environment more explicitly or seeking alternative routes that are less prone to congestion.

The process can be envisioned as follows:

1. **Global Planning:** A path planner (e.g., A, RRT, or a grid-based planner on a known map) computes a collision-free path from the robot's start to its goal, assuming a static environment. This path might be a sequence of configurations or waypoints.
2. **Local Planning/Execution:** The robot starts executing the global path. A local planner continuously monitors sensor data.
If the environment remains static and the global path is clear, the local planner directs the robot to follow the global path precisely.
If dynamic obstacles are detected, the local planner (e.g., DWA, VFH) takes control. It identifies safe and efficient immediate velocities that avoid collisions while attempting to stay close to the global path or move towards the next waypoint.
3. **Re-planning Trigger:** If the local planner cannot find any safe action for an extended period, or if it gets stuck in a loop, or if the global path becomes permanently blocked, a replanning trigger is activated.
4. **Global Replanning:** The global planner is invoked again. It might re-plan from the robot's current position, potentially using updated environmental information or a more sophisticated model of dynamic obstacles.

This hierarchical structure provides robustness. The global planner handles the long-term strategic navigation, while the local planner manages the short-term, reactive control necessary for survival and progress in a dynamic world. The key is the seamless handoff and communication between these two levels of planning.

The sophistication of the global and local planners themselves can vary greatly depending on the application. For robots navigating in highly dynamic and unpredictable environments, such as warehouses with numerous moving AGVs or human-robot collaborative spaces, more advanced techniques might be employed. These could include:

Predictive Local Planning: Instead of simply evaluating immediate trajectories, local planners can incorporate short-term predictions of obstacle movements. If a laser scanner detects an object moving predictably, the local planner can anticipate its future position and plan accordingly, often leading to smoother avoidance maneuvers and less reactive "swerving." This prediction can be as simple as assuming constant velocity or using more complex Kalman filters or particle filters to estimate future states.

Multi-Robot Coordination: In scenarios with multiple robots, planning must account for the actions and intentions of other agents. This might involve centralized coordination, where a central system allocates paths and velocities to all robots to prevent conflicts, or decentralized approaches where robots communicate their intended paths or restrictions to each other.

Risk-Aware Planning: In safety-critical applications, planning might explicitly incorporate risk assessment. Instead of simply aiming for a collision-free path, the planner might aim for paths that have a low probability of collision, even in the presence of uncertainty about obstacle behavior. This often leads to more conservative, but safer, trajectories.

Reinforcement Learning for Reactive Control: Increasingly, reinforcement learning (RL) techniques are being explored for reactive navigation in dynamic environments. An RL agent can learn a policy for selecting actions (velocities) based on sensor inputs through trial and error, potentially discovering highly effective reactive behaviors that are difficult to hand-engineer. The agent is trained in simulation or in controlled real-world scenarios to maximize a reward function that penalizes collisions and rewards progress towards the goal.

Consider a robotic arm tasked with picking parts from a moving conveyor belt. A static path planner would fail miserably. The robot needs a local controller that can track the moving belt, predict where the next part will be, and intercept it. DWA or VFH could be used to maintain the robot's position relative to the belt, while a more specialized trajectory generation module might predict the part's arrival and generate a smooth, fast trajectory to grasp it, all while ensuring the arm doesn't collide with itself or its surroundings.

Another example is a mobile robot navigating a busy factory floor. A global planner might provide a route from Station A to Station B. However, this route might be blocked by forklifts, other robots, or human workers. The local planner's role is paramount here. It must detect these moving obstacles, assess their speed and direction, and find a way to either wait, go around, or adjust its speed to maintain a safe distance. If the local planner continuously gets blocked, it might request a new global path that avoids the currently congested area.

The challenge of planning in dynamic environments is not just about avoiding collisions; it's also about maintaining efficiency and achieving the task objectives. A robot that is constantly stopping and starting or taking unnecessarily circuitous routes to avoid moving objects will be slow and unproductive. Therefore, the local planning algorithms need to be designed to be as decisive and efficient as possible, making minimum deviations from the intended global path while ensuring safety.

The integration of sensor modalities also plays a role. While laser scanners are excellent for detecting obstacles in a plane, depth cameras and vision-based systems can provide richer information about the environment, including the classification of detected objects (e.g., distinguishing a stationary wall from a person). This richer understanding can feed into more intelligent planning decisions, allowing the robot to anticipate the behavior of certain types of dynamic agents more effectively.

Ultimately, successful motion planning in dynamic

environments hinges on the ability of the robot to perceive its surroundings in real-time, understand the motion of other entities, and make rapid, intelligent decisions about its own motion. This involves a tightly coupled system of sensing, perception, global strategic planning, and local reactive control, all working in concert to ensure safe, efficient, and task-oriented navigation. The ongoing research in this area continues to push the boundaries, enabling robots to operate more autonomously and effectively in the complex, ever-changing world around us.

The preceding discussion has underscored the critical need for robots to navigate dynamic environments by intelligently reacting to immediate threats and adapting their trajectories on the fly. Techniques such as the Dynamic Window Approach (DWA) and Vector Field Histogram (VFH) excel at this local, reactive planning, ensuring that robots can maneuver safely amidst unforeseen obstacles and moving elements. However, these reactive methods, while vital for real-time obstacle avoidance, often operate under the implicit assumption that the robot's dynamics—its physical limitations concerning speed, acceleration, and inertia—can be adequately handled by simply selecting feasible velocities within a short time horizon. This assumption, while practical for many simpler robots like wheeled mobile platforms at moderate speeds, begins to falter when dealing with robots exhibiting more complex dynamic behaviors, or when operating at the edge of their performance envelopes. This is precisely where the domain of **kinodynamic planning** becomes indispensable.

Kinodynamic planning moves beyond the traditional geometric considerations of motion planning to explicitly incorporate the robot's dynamic model. In essence, it seeks to generate trajectories that are not only collision-free in the robot's configuration space and workspace but are also physically realizable by the robot, respecting its inherent limitations in velocity, acceleration, and other dynamic parameters like inertia and torque. For many robotic systems, particularly those operating at higher speeds or in three-dimensional space, neglecting these dynamics can lead to the generation of paths that, while geometrically feasible, are impossible for the robot to follow. Such paths would result in jerky, erratic movements, potential instability, or outright failure to execute, potentially leading to collisions or system damage.

Consider, for instance, an unmanned aerial vehicle (UAV). A UAV's motion is governed by complex aerodynamics, motor speeds, battery limitations, and wind disturbances. Simply planning a straight-line path between two points in space, even if clear of obstacles, might be dynamically impossible if the required turns are too sharp or the ascent/descent rates exceed the drone's capabilities. Similarly, a high-speed ground robot attempting to execute a rapid evasive maneuver must account for its inertia and the grip of its wheels on the surface; a trajectory that requires instantaneous changes in direction or speed will likely result in the robot skidding or even flipping over. For robotic manipulators, especially those designed for high-speed pick-and-place operations, the joint accelerations and velocities must be carefully controlled to avoid excessive wear and tear, vibration, or loss of precision. Kinodynamic planning provides the framework to address these challenges by synthesizing trajectories that are both geometrically and dynamically valid.

The fundamental difference between traditional motion planning and kinodynamic planning lies in the state space and the types of constraints considered. In standard motion planning, the state of the robot is typically represented by its configuration (e.g., position and orientation for a mobile robot, joint angles for a manipulator). The planning process then searches for a path in this configuration space. Kinodynamic planning, however, must consider a higher-dimensional state space that includes not only the robot's configuration but also its velocity and potentially acceleration. For example, the state of a mobile robot in kinodynamic planning might be $(x, y, \theta, \dot{x}, \dot{y}, \dot{\theta})$, where (x, y, θ) are the position and orientation, and $(\dot{x}, \dot{y}, \dot{\theta})$ are the corresponding velocities.

This expanded state space necessitates a re-evaluation of planning algorithms. Traditional algorithms like A or Dijkstra's, which rely on discrete grid representations or graph search in configuration space, become computationally prohibitive when the state space includes continuous velocity and acceleration variables. Furthermore, the "neighbors" of a given state are no longer simply adjacent configurations but rather states reachable within a small time interval, dictated by the robot's dynamic model and control inputs.

Sampling-based motion planning algorithms, such as the Rapidly-exploring Random Trees (RRT) and its variants, have proven particularly adept at handling the complexities of kinodynamic planning. The core idea of RRT is to incrementally build a tree of reachable states by randomly sampling states and connecting them to the existing tree. To adapt RRT for kinodynamic planning, the sampling and expansion process must be modified to incorporate the robot's dynamics.

In a kinodynamic RRT, the tree's nodes represent states in the full state space, including position, orientation, and velocities. When extending the tree, instead of simply interpolating between two configurations, a control input (or a sequence of control inputs) is applied to the robot's dynamic model from a randomly sampled state. This application of control inputs drives the robot's state forward in time according to its dynamics, generating a trajectory segment. The key innovation is that this trajectory generation process explicitly respects the robot's dynamic constraints.

Let's delve deeper into how this adaptation works. Consider a kinodynamic RRT aiming to find a path for a car-like robot. The state of the robot is $(x, y, \theta, v, \omega)$, where v is its linear velocity and ω

is its angular velocity. The dynamic model might be described by:

$\dot{x} = v \cos(\theta)$
$\dot{y} = v \sin(\theta)$
$\dot{\theta} = \omega$
$\dot{v} = a$ (linear acceleration)
$\dot{\omega} = \alpha$ (angular acceleration)

The control inputs are the accelerations (a, α). However, the actual control inputs are typically bounded, e.g., $|a| \le a
{max}$ and $|\alpha| \le \alpha{max}$. To perform the expansion step in kinodynamic RRT, we select a state $S{near}$ from the existing tree. Then, we sample a random state $S{rand}$. Instead of directly connecting $S{near}$ to $S{rand}$, we try to drive the robot from $S{near}$ towards $S{rand}$ using feasible control inputs. This can be achieved by:

1.
Sampling Control Inputs: Randomly sample control inputs (a, α) within their allowed bounds.
2. **Simulating Trajectory:** Integrate the robot's dynamic equations from $S{near}$ using the sampled control inputs for a fixed time duration, Δt. This generates a trajectory segment and a new state $S{new}$.
$S{new} = \text{integrate}(S{near}, (a, \alpha), \Delta t)$
3. **Checking Feasibility:** The generated trajectory segment and the resulting state $S{new}$ must be collision-free. Furthermore, if $S{new}$ is sufficiently

close to $S{rand}$ *(in some metric), it can be added to the tree. More commonly, $S*{new}$ is added as a child of $S*{near}$, and the algorithm continues to explore from $S*{new}$.

A more sophisticated approach involves a **control-based extension**. Instead of randomly sampling control inputs, we can use a controller to steer the robot from $S*{near}$ towards a target state or a randomly sampled state $S*{rand}$. For example, a simple proportional controller could be used to adjust a and α to reduce the error between the current state and the target. This control-based approach is often more efficient in exploring the state space towards the goal.

The process of integrating the dynamic model needs careful consideration. Numerical integration methods, such as Euler integration or Runge-Kutta methods, are typically employed. The choice of Δt is critical: a smaller Δt leads to more accurate trajectory generation but increases the computational cost due to more frequent integration steps. A larger Δt can lead to inaccuracies and potentially explore states that are not truly reachable.

A significant challenge in kinodynamic planning is the computational cost. The state space is significantly larger than in geometric planning, and each expansion step involves simulating the robot's dynamics. To mitigate this, several strategies are employed:

Bi-directional RRT: Grow two trees simultaneously, one from the start state and one from the goal state. When the trees meet, a path has been found. This can significantly speed up convergence, especially in high-dimensional spaces.
RRT (RRT-star): An asymptotically optimal variant of RRT that rewires the tree to find increasingly better paths as more samples are added. This involves checking if connecting to a newly sampled node can improve the path to existing nodes, and if re-routing from an existing node through the new node can yield a shorter, dynamically feasible path.
Trajectory Optimization: Once an initial kinodynamically feasible trajectory is found, optimization techniques can be applied to further refine it, making it smoother, faster, or more energy-efficient, while still respecting dynamic constraints. This often involves casting the problem as an optimal control problem.

Let's consider a specific example: planning for a quadrotor drone to fly through a series of obstacles in a 3D environment. The state space would be $(x, y, z, \phi, \theta, \psi, \dot{x}, \dot{y}, \dot{z}, \dot{\phi}, \dot{\theta}, \dot{\psi})$, where (x, y, z) is the position, (ϕ, θ, ψ) are the Euler angles (roll, pitch, yaw), and the subsequent terms are their respective velocities. The dynamics of a quadrotor are non-linear and coupled, involving forces generated by the rotors and torques. Planning here requires generating a sequence of thrust commands and desired attitude changes that are within the drone's physical limits. A kinodynamic RRT would sample states and apply control signals (e.g., desired rates of change in thrust and torques) to simulate the drone's movement according to its physics model. The tree expansion would involve integrating the differential equations of motion for the quadrotor. Collision checks would be performed against the 3D environment. The goal would be to connect a start state (e.g., hovering at a certain altitude) to a goal state (e.g., reaching a target position with a specific velocity profile).

Another critical aspect is the formulation of the "goal biasing" strategy in kinodynamic RRT. Purely random sampling might be inefficient if the goal is in a specific region. Goal biasing involves increasing the probability of sampling states near the goal or sampling control inputs that drive the robot towards the goal. For kinodynamic planning, this can mean sampling target states that are close to the goal configuration *and* have velocities that are consistent with reaching the goal.

The concept of "kinodynamic planning" is also closely related to optimal control. Optimal control problems aim to find a control input sequence that minimizes a cost function (e.g., time, energy, deviation from a reference trajectory) while satisfying system dynamics and constraints. Kinodynamic planning can be viewed as a specific type of optimal control problem where the primary objective is to find *any* feasible, collision-free trajectory, and secondary objectives might include optimality. Algorithms like Differential Dynamic Programming (DDP) or Model Predictive Control (MPC) are often used in conjunction with or as alternatives to sampling-based methods for kinodynamic trajectory generation, particularly when optimality is a key concern or when dealing with highly predictable dynamics. MPC, for instance, repeatedly solves a finite-horizon optimal control problem, applying only the first control action and then re-planning based on the new state. This makes it inherently reactive and suitable for dynamic environments, while still respecting the system's dynamics.

The integration of kinodynamic planning into a robot's overall motion planning system, especially in complex and dynamic environments, often involves a hierarchical approach. A global planner might first generate a rough, dynamically plausible path or a sequence of waypoints in a simplified model of the environment. Then, a kinodynamic trajectory generator, perhaps using an RRT variant or an optimal control solver, takes these waypoints and generates a smooth, dynamically feasible, and collision-free trajectory segment for each waypoint. This segment respects the robot's dynamics and any local obstacles. This hierarchical decomposition allows for the leveraging of different algorithmic strengths: global planners for broad strategic coverage and kinodynamic planners for detailed, physically realizable motion execution.

The challenge of generating kinodynamically feasible trajectories is further amplified when considering uncertainties in the robot's model or the environment. For instance, if the robot's inertial properties change due to payload variations, or if environmental factors like friction are uncertain, a nominal kinodynamic plan might become infeasible. Robust kinodynamic planning aims to generate trajectories that are resilient to such uncertainties, often by planning for worst-case scenarios or by incorporating a safety margin that accounts for potential deviations. This can involve extending the state space to include uncertainty parameters or using probabilistic methods to ensure a high likelihood of successful execution.

In summary, kinodynamic planning is a critical advancement in motion planning that explicitly accounts for the robot's dynamic limitations. By extending the state space to include velocities and potentially accelerations, and by integrating the robot's dynamic model into the planning process, it enables the generation of trajectories that are not only collision-free but also physically achievable. Sampling-based algorithms, particularly RRT variants adapted for kinodynamic domains, have emerged as powerful tools for tackling these complex problems. The ability to generate dynamically feasible paths is paramount for the safe, efficient, and reliable operation of robots with sophisticated dynamics, paving the way for their deployment in more challenging and high-performance applications.

The preceding discussion on kinodynamic planning has highlighted the sophisticated methods required to generate physically realizable trajectories for robots, moving beyond simple geometric paths to incorporate the inherent dynamic constraints of the system. This focus on dynamic feasibility sets the stage for understanding how these sophisticated trajectories are actually executed and managed within a broader robotic system. The generation of a kinodynamically optimal path is only one piece of the puzzle; integrating this plan into a coherent, responsive, and intelligent system necessitates robust control architectures. These architectures provide the framework for decision-making, task execution, and real-time adaptation, ensuring that the robot can effectively navigate its environment and achieve its objectives.

At the heart of enabling autonomous systems to perform complex tasks lies the concept of control architectures. These architectures define how different modules of the robot's intelligence—perception, planning, decision-making, and actuation—interact and cooperate. They are the blueprints that dictate the flow of information and commands, shaping the robot's overall behavior from high-level goal setting to low-level motor control. The choice of control architecture profoundly impacts a robot's capabilities, its adaptability, and its suitability for particular applications. Broadly, these architectures can be categorized into hierarchical, behavior-based, and hybrid approaches, each offering distinct advantages and facing unique challenges.

One of the most intuitive and widely adopted control architectures is the **hierarchical control architecture**. This paradigm breaks down complex tasks into a series of simpler, manageable subtasks, arranged in a layered structure. At the highest level resides the strategic planner, responsible for defining overarching goals and long-term objectives. This might involve tasks like "navigate to the destination," "inspect the area," or "perform a specific assembly operation." The output of this strategic layer is then passed down to a tactical layer, which decomposes these high-level goals into more concrete plans and sequences of actions. For instance, "navigate to the destination" might be broken down into "plan a global path," "avoid static obstacles," and "follow a specific road."

Further down the hierarchy, a **kinodynamic planner**, as discussed previously, would take the tactical instructions and generate a dynamically feasible trajectory. This trajectory specifies the robot's desired states (position, velocity, acceleration) over time. Finally, at the lowest level, the **executive or control layer** receives this detailed trajectory and translates it into actual actuator commands (e.g., motor torques, steering angles, throttle adjustments). This layer is responsible for real-time execution, ensuring that the robot's physical components follow the planned trajectory as closely as possible, often using low-level controllers like PID

(Proportional-Integral-Derivative) loops to regulate joint positions, velocities, or forces.

A significant advantage of hierarchical architectures is their clarity and modularity. Each layer has a well-defined role and can be developed and tested relatively independently. This makes debugging and maintenance easier, and it facilitates the integration of different planning and control algorithms. For example, a sophisticated kinodynamic planner could be swapped out for a simpler one, or a new perception module could be integrated, without necessarily disrupting the entire system, provided the interfaces between layers are maintained. This layered approach also allows for different levels of abstraction. The strategic planner might operate on a simplified map of the environment, while the executive layer deals with high-frequency sensor data and actuator dynamics.

However, hierarchical architectures can suffer from a lack of flexibility and responsiveness. Information flows primarily in one direction: top-down. If an unexpected event occurs that requires a significant deviation from the current plan, the lowest layers might not have the autonomy to respond effectively. They would typically need to report the issue back up the hierarchy, where a new plan would be generated. This can lead to delays and potentially catastrophic failures in rapidly changing environments. For instance, if a robot following a kinodynamically planned path suddenly encounters a moving obstacle directly in its way, the executive layer might have limited options beyond an emergency stop or a reactive maneuver that could violate the globally planned trajectory. The process of reporting this anomaly, re-planning, and then re-executing can be too slow for critical situations. Moreover, the strict decomposition can sometimes lead to sub-optimal overall behavior, as local decisions at lower levels might not always align perfectly with the global objective.

In contrast to the structured, top-down approach of hierarchical control,
behavior-based architectures offer a more decentralized and reactive paradigm. Instead of decomposing tasks, this architecture focuses on creating a collection of independent, concurrent "behaviors" or "modules," each responsible for a specific, relatively simple capability. Examples of behaviors might include "obstacle avoidance," "follow wall," "seek light," "go to goal," or "maintain speed." Each behavior typically has access to sensor information and generates motor commands directly or influences the commands generated by other behaviors.

The key to behavior-based systems lies in their **arbitration mechanism**. Since multiple behaviors might be active simultaneously and generate conflicting motor commands, a mechanism is needed to decide which behavior's output should ultimately control the robot's actuators at any given moment. Arbitration can be implemented in various ways:

Fixed-priority arbitration: Behaviors are assigned a priority level, and the highest-priority active behavior's commands are executed. For example, "emergency stop" would have the highest priority, followed by "obstacle avoidance," then "go to goal."
Weighted-sum arbitration: The outputs of multiple active behaviors are combined using weights. These weights might be static or dynamically adjusted based on the current situation. For instance, the "go to goal" behavior might have a high weight when the path is clear, but its weight would decrease significantly as an obstacle gets closer, allowing the "obstacle avoidance" behavior to dominate.
Flocking or emergent behavior: In some systems, behaviors interact in a way that leads to emergent, coordinated actions without explicit arbitration. This is common in multi-robot systems where individual robots exhibit simple behaviors that, when combined, lead to complex group movements.

Behavior-based architectures excel in environments that are dynamic and unpredictable. Their inherent reactivity allows robots to respond quickly to changes, as each behavior can directly process sensor data and generate immediate motor commands. For instance, an obstacle avoidance behavior can react instantly to a new obstacle without waiting for a global re-plan. This makes them well-suited for tasks like navigating cluttered spaces, responding to unexpected events, or operating in unstructured environments. The modular nature also makes it relatively easy to add new behaviors or modify existing ones.

However, behavior-based systems can struggle with complex, long-term, or sequential tasks. Planning and reasoning about sequences of actions can be difficult when there is no explicit high-level planner dictating the order. For example, achieving a complex goal like "pick up an object, bring it to a charging station, and then return to the work area" might require careful sequencing that is not easily managed by independent behaviors. Coordinating multiple behaviors to achieve a compound objective can become challenging, and ensuring that the overall system behavior is predictable and robust can be a significant design challenge. The arbitration mechanism itself can become complex and difficult to tune. It's also harder to guarantee safety or optimality in the same way that a kinodynamic planner might aim to do, as the system is more focused on immediate reactions.

Recognizing the limitations of purely hierarchical or purely behavior-based approaches, many modern autonomous systems employ
hybrid control architectures. These architectures seek to combine the strengths of both paradigms, leveraging layered planning for complex tasks while retaining reactive capabilities for immediate environmental interaction. A common hybrid model is the "subsumption architecture" in an advanced form, or more generally, a layered approach where distinct functionalities are managed at different levels of abstraction and reactivity.

In a typical hybrid architecture, a **global planner** (which might be hierarchical) operates at a strategic level, generating a long-term path or a sequence of goals. This path might be a series of waypoints or a more detailed kinodynamically feasible trajectory. This global plan serves as a guideline for the robot's behavior. Concurrently, a **local planner** or a set of reactive behaviors operate at a lower level. This local component is responsible for real-time obstacle avoidance, path following, and immediate environmental adjustments. The local planner receives the current state from perception, the desired trajectory or waypoint from the global planner, and dynamically adjusts the robot's motion to ensure collision avoidance and adherence to the general direction of the global plan.

For instance, a robot tasked with navigating a factory floor might use a hierarchical planner to determine the most efficient route from its current location to a target workstation, taking into account known static obstacles and workflow constraints. This global plan could be a kinodynamically feasible trajectory. As the robot moves, a reactive obstacle avoidance module, perhaps using methods like DWA or VFH, continuously monitors the environment for unexpected moving obstacles (e.g., forklifts, personnel). If an obstacle is detected, the reactive module overrides the global plan's velocity commands, steering the robot to safely maneuver around the obstruction. Once the obstacle is cleared, the reactive module relinquishes control, and the robot resumes following the global plan, potentially recalculating its trajectory to account for the detour.

The integration of perception, planning, and control is seamless within these hybrid architectures. **Perception** modules provide the raw sensor data (e.g., from cameras, LiDAR, sonar) and process it to identify objects, estimate their positions and velocities, and build or update a representation of the environment. This perceived information is then fed into the **planning** modules. The global planner uses this information for long-term navigation, while the local planner and reactive behaviors use it for immediate decision-making. Finally, the **control** layer takes the output from the planning modules—whether a detailed trajectory or immediate actuator commands—and executes them, often using feedback loops to ensure accurate execution and to correct for discrepancies between the desired and actual state of the robot.

This tight integration is crucial. A perception system might identify a dynamic obstacle. This information is relayed to both the global and local planners. The local planner immediately begins generating reactive commands to avoid the obstacle. Meanwhile, the global planner might assess if the obstacle's presence necessitates a significant deviation from the original global path, potentially triggering a re-planning of the overall route. The control layer then executes the commands from whichever module is currently in charge, ensuring that the robot's actions are safe and effective.

To illustrate with an example, consider an autonomous delivery robot operating in a busy urban environment. The
global planning layer might utilize a map-based planner to generate a route to the delivery address, considering road networks, traffic rules, and known pedestrian areas. This route is then translated into a sequence of desired states, potentially a kinodynamically feasible trajectory for smooth driving. The **perception system** continuously streams data from its cameras and LiDAR, identifying pedestrians, vehicles, cyclists, and road boundaries.

When the robot is on an open street, the
local planner or **reactive control module** primarily focuses on following the global trajectory and maintaining safe distances. However, if a pedestrian unexpectedly steps onto the road, the perception system detects this. This information is immediately passed to the reactive control module, which might override the global plan's velocity command. The reactive module, perhaps employing a variant of the Dynamic Window Approach, calculates a safe and feasible evasive maneuver—a slight steering and deceleration—that respects the robot's dynamics. Once the pedestrian has passed and the path is clear, the reactive module signals that the global plan can be resumed. The robot then smoothly transitions back to following its original kinodynamically planned trajectory, potentially with a slight adjustment to account for the time lost.

This hybrid approach provides a robust framework for autonomous systems. It allows for the deliberate, strategic planning of complex tasks, ensuring that the robot progresses towards its ultimate goals, while simultaneously providing the agility to react to the unpredictable nature of real-world environments. The ability to handle sequences of actions is inherent in the global planning component, which can define steps like "approach charging station," "dock with station," "charge," and "depart station." Within each of these steps, reactive behaviors ensure safe execution. For instance, during the "dock with station" phase, precise control and short-range obstacle avoidance are critical, and a hybrid system can seamlessly transition between the global instruction and the local, high-precision control required for docking.

The design of these architectures also influences how robots learn and adapt. While purely reactive systems might learn specific responses to stimuli, and purely hierarchical systems might benefit from planners that can adapt their search strategies, hybrid systems offer a rich environment for learning. Reinforcement learning, for example, can be applied to tune the arbitration weights in a behavior-based component, or to learn optimal parameters for the local planner, or even to refine the global planning strategies based on experience.

In essence, control architectures are the nervous

systems of autonomous robots. They dictate how a robot perceives its world, makes decisions, and acts upon those decisions. Hierarchical architectures provide structure and planning depth, behavior-based architectures offer reactivity and robustness in dynamic conditions, and hybrid architectures judiciously blend these strengths. By carefully integrating perception, planning, and control within a well-defined architectural framework, engineers can design autonomous systems capable of performing complex sequences of actions, adapting to unforeseen events, and navigating the complexities of the real world with increasing levels of sophistication and reliability. The ability to decompose tasks, manage multiple concurrent objectives, and react to immediate stimuli is paramount, and the evolution of control architectures continues to drive the progress of robotics towards ever-greater autonomy and capability.

The successful execution of a kinodynamically generated trajectory is not merely about generating a sequence of desired states; it hinges on the robot's ability to accurately track that trajectory in the real world. This is where feedback control plays a pivotal role, transforming a theoretically optimal path into tangible, precise robot motion. Without effective feedback control, even the most sophisticated trajectory planner would struggle to compensate for external disturbances, unmodeled dynamics, actuator limitations, and sensor noise, leading to significant deviations from the intended path. The core principle of feedback control is to continuously monitor the robot's actual state (e.g., position, velocity, orientation) and compare it to the desired state specified by the trajectory. The difference, or error, between these two states is then used to compute corrective commands that are sent to the robot's actuators, thereby driving the actual state closer to the desired state.

One of the most widely adopted and foundational feedback control strategies is the **Proportional-Integral-Derivative (PID) controller**. This elegant yet powerful technique has found extensive application across a vast array of engineering disciplines, including robotics, due to its simplicity, robustness, and ease of implementation. A PID controller operates by calculating an output signal that is a weighted sum of three terms: the proportional error, the integral of the error, and the derivative of the error.

Let $e(t)$ represent the error at time t, defined as the difference between the desired value (from the trajectory) and the actual measured value: $e(t) = \text{desired\ value}(t) - \text{actual\ value}(t)$. The PID control law is then given by:

$$u(t) = K_p e(t) + K_i \int_0^t e(\tau) d\tau + K_d \frac{de(t)}{dt}$$

where:

$u(t)$ is the control output signal (e.g., a motor torque command or a joint velocity command).

$K*p$ is the proportional gain. This term provides a control output proportional to the current error. A larger $K*p$ will cause the system to react more strongly to the current error, leading to faster response times but potentially increasing overshoot and oscillations.
$K*i$ is the integral gain. This term integrates the error over time. It helps to eliminate steady-state errors that might persist with a purely proportional controller. If there is a persistent error, the integral term will grow, increasing the control output until the error is reduced to zero. However, a large $K*i$ can lead to integral windup (where the integral term becomes excessively large) and can make the system sluggish or prone to oscillations.
$K*d$ is the derivative gain. This term is proportional to the rate of change of the error. It acts as a damping mechanism, anticipating future error by looking at its trend. A larger $K*d$ can reduce overshoot and oscillations, leading to a more stable system. However, it can also amplify noise in the error signal, potentially causing erratic control actions if the sensor measurements are noisy.

In the context of trajectory tracking, PID controllers can be applied at various levels. For a robot with multiple joints, a cascaded PID control structure is common. The outer loop might control the joint position, taking the desired joint position from the trajectory and producing a desired joint velocity. The inner loop then uses another PID controller to track this desired joint velocity by generating the appropriate motor torque or voltage command. For systems where position and velocity are directly controlled, individual PID loops can be implemented for each degree of freedom, ensuring that the robot's end-effector or each joint follows its specified path in space and time.

Consider a simple example of a mobile robot tasked with following a straight-line trajectory. The trajectory specifies a constant forward velocity and a constant heading. The robot's actual forward velocity and heading are measured by its odometry or other sensors. A PID controller can be used to regulate the forward velocity: the error is the difference between the desired forward velocity and the actual forward velocity, and the output adjusts the motor commands to correct this error. Simultaneously, another PID controller can manage the heading: the error is the difference between the desired heading and the actual heading, and its output can control a steering mechanism or the differential speeds of the drive wheels. By tuning the K_p, K_i, and K_d gains for both controllers, one can achieve precise and smooth tracking of the straight-line path, even in the presence of minor disturbances like surface irregularities or slight wheel slippage.

The tuning of PID gains is a critical aspect of their successful implementation. This process, often referred to as PID tuning, can be done empirically through trial and error, or using more systematic methods like the Ziegler-Nichols tuning rules, or through optimization techniques based on system models. The goal is to achieve a balance: a fast response to reduce tracking errors quickly, minimal overshoot and oscillations for smooth motion, and good disturbance rejection for robustness. Poorly tuned PID controllers can lead to instability, excessive vibrations, or sluggish performance, severely compromising the robot's ability to follow trajectories accurately.

While PID controllers are excellent for regulating individual states or simple combinations of states, they have limitations when dealing with highly complex systems or when precise prediction of future behavior is required. For instance, they do not explicitly account for the system's future states or constraints over an extended horizon, relying solely on the current and past error. This is where more advanced feedback control techniques, such as
Model Predictive Control (MPC), become invaluable.

Model Predictive Control is an advanced control strategy that leverages a mathematical model of the robot's dynamics to predict its future behavior over a finite time horizon. At each control step, MPC solves an optimization problem to determine a sequence of control inputs that will minimize a cost function, subject to system dynamics and constraints. The key idea is to "predict" the system's trajectory and optimize the control actions over this prediction horizon, typically by considering factors like tracking accuracy, energy consumption, and actuator limits.

The core components of an MPC system are:

1. **System Model:** A dynamic model that describes how the robot's state evolves over time in response to control inputs and disturbances. This model can be based on physics, empirical data, or a combination of both.
2. **Prediction Horizon (N):** The time horizon into the future over which the system's behavior is predicted and optimized.
3. **Control Horizon (M):** The time horizon over which control inputs are optimized. Often, $M \le N$. For simplicity, in many MPC formulations, the control input is assumed to remain constant over a certain period (e.g., one control step), and then this sequence of optimal inputs is applied.
4. **Cost Function:** A function that quantifies the performance of the system over the prediction horizon. It typically penalizes deviations from the desired trajectory, control effort, and constraint violations. For trajectory tracking, the cost function would likely include terms that measure the difference between the predicted robot states and the desired trajectory states.
5. **Constraints:** Limitations on the system's states and control inputs, such as actuator saturation limits (e.g., maximum motor torque), velocity limits, and safety constraints (e.g., avoiding collisions).

At each sampling instant t, the MPC controller:

Measures the current state of the robot.
Uses the system model to predict the robot's future states over the prediction horizon, assuming a sequence of control inputs.

Solves an optimization problem to find the sequence of control inputs (over the control horizon) that minimizes the cost function while satisfying all constraints.
Applies only the first control input from the optimal sequence to the robot's actuators.
Discards the rest of the predicted control sequence and repeats the process at the next sampling instant, using the new measured state. This "receding horizon" principle is crucial for adaptability and robustness.

For trajectory tracking, the desired trajectory provides the reference points over the prediction horizon. The MPC controller aims to generate control commands that drive the robot's predicted states as close as possible to these reference points, while respecting the robot's physical limitations. For example, if a robot is following a curved trajectory, the MPC would predict how the robot's position and orientation will evolve based on current velocities and steering commands. It would then optimize the steering angle and speed over the next few seconds to minimize the error from the planned curve, ensuring that the robot does not exceed its maximum turning rate or velocity limits, and that it smoothly transitions through the desired path.

The advantages of MPC for trajectory tracking are manifold:

Explicit handling of constraints: MPC can directly incorporate actuator limits, kinematic constraints, and even dynamic constraints into the optimization

problem, leading to safer and more feasible control actions. This is particularly beneficial for robots with complex dynamics or tight operational envelopes.
Predictive capability: By looking ahead, MPC can anticipate future deviations and take proactive measures, leading to smoother and more accurate tracking compared to purely reactive controllers.
Optimization of multiple objectives: MPC allows for the simultaneous optimization of various performance criteria (e.g., tracking error, energy consumption, comfort) through the design of the cost function.
Robustness to disturbances: The receding horizon nature of MPC, combined with the system model, allows it to adapt to external disturbances and model uncertainties, re-optimizing control actions at each step.

However, MPC also comes with significant challenges:

Computational complexity: Solving an optimization problem at every control step can be computationally intensive, requiring powerful onboard processors, especially for systems with long prediction horizons, complex models, or many states and inputs.
Model accuracy: The performance of MPC heavily relies on the accuracy of the system model. Inaccurate models can lead to suboptimal or even unstable control.
Tuning: Tuning the prediction horizon, control horizon, cost function weights, and constraint parameters can be complex and requires a good understanding of the system's behavior.

Despite these challenges, MPC has become a powerful

tool for high-performance trajectory tracking in many robotic applications, including autonomous driving, industrial automation, and aerial robotics, where precise and constrained motion is paramount.

The choice between PID and MPC, or other feedback control strategies, often depends on the specific application requirements, the complexity of the robot's dynamics, the availability of computational resources, and the need to handle constraints. For simpler systems or less demanding tracking tasks, well-tuned PID controllers often provide a cost-effective and highly capable solution. For applications demanding high precision, adherence to strict constraints, and proactive handling of future states, MPC offers a more sophisticated and powerful approach.

Regardless of the specific controller chosen, the fundamental importance of robust feedback control for trajectory tracking cannot be overstated. It is the bridge between the abstract, generated plan and the physical reality of robot motion. Without it, robots would be incapable of reliably performing tasks that require accurate navigation, manipulation, or interaction with their environment. The continuous monitoring of the robot's state, the calculation of errors, and the generation of corrective commands form the bedrock of dynamic motion control, ensuring that the robot not only attempts to follow a trajectory but reliably and accurately executes it, fulfilling the promise of advanced motion planning. The interplay between planning and control is a symbiotic one; planning provides the "what" and "when," while feedback control ensures the "how" – the precise execution that brings robotic capabilities to life.

Chapter 10: Robot Manipulation and Grasping

Robotic manipulation is the branch of robotics concerned with the design, construction, and operation of robots that can interact with their physical environment to perform tasks, typically involving the handling and repositioning of objects. At the heart of most robotic manipulation systems lies the **robotic manipulator**, often referred to as a **robot arm**. These mechanical structures are designed to mimic the capabilities of a human arm and hand, enabling a robot to reach, grasp, and manipulate objects within its workspace. The primary function of a robotic manipulator is to provide a controllable and configurable means for a robot to exert forces and torques on objects, thereby altering their state—be it position, orientation, or even shape.

A typical robotic manipulator is composed of several key elements:
joints and **links**. Links are the rigid segments that form the structure of the arm, connecting the base of the robot to its end-effector. These links are interconnected by joints, which are the points of articulation that allow relative motion between adjacent links. The type of joint dictates the nature of this motion. The most common types of joints used in robotic manipulators are **revolute joints** and **prismatic joints**.

A
revolute joint allows for rotational motion around a fixed axis, much like the elbow or shoulder joint in a human arm. This type of joint typically introduces a single degree of freedom, allowing for rotation within a

specified range. For example, a joint that rotates along the Z-axis would change the angular position of the connected link relative to the previous one.

A
prismatic joint, on the other hand, allows for linear or sliding motion along an axis. This is analogous to the extension and retraction of a telescope or the movement of a drawer. A prismatic joint also typically introduces a single degree of freedom, enabling linear displacement. For instance, a joint that slides along the X-axis would change the linear position of the connected link.

The combination of these links and joints forms a kinematic chain, which defines the structure and the possible configurations of the manipulator. The end of this chain is equipped with an
end-effector. The end-effector is the tool or device that directly interacts with the environment. This could be a simple gripper designed to grasp objects, a vacuum suction cup, a welding torch, a paint sprayer, a cutting tool, or a complex multi-fingered hand. The choice of end-effector is dictated by the specific task the robot is intended to perform. For instance, a robot arm designed for pick-and-place operations in an electronics assembly line might have a vacuum gripper, while a robot on a car manufacturing assembly line might be equipped with a welding gun or a paint sprayer. The end-effector is crucial because it is the interface through which the robot exerts its influence on the world.

The capability of a robotic manipulator to reach and orient objects within its workspace is fundamentally determined by its
degrees of freedom (DOF). The degrees of freedom refer to the number of independent parameters that can uniquely define the configuration of the robot arm. In simpler terms, it is the number of independent joints that can be actuated. Each revolute or prismatic joint typically contributes one DOF. For example, a simple robot arm with one rotating joint and one sliding joint has two degrees of freedom.

The number of DOFs directly influences the manipulator's dexterity and its ability to perform complex motions. A human arm, for instance, has a very high number of DOFs, allowing for a wide range of movements and precise positioning of the hand in three-dimensional space. Similarly, robotic manipulators with more DOFs generally have greater flexibility and reachability. A common configuration for industrial robots is a six-DOF arm, which is often considered the minimum required to achieve arbitrary positioning (x, y, z coordinates) and arbitrary orientation (roll, pitch, yaw) of the end-effector in 3D space. This is often referred to as being "fully-articulated."

To understand and control the motion of a robotic manipulator, it is essential to study its
kinematics. Kinematics is the study of motion without considering the forces that cause it. In the context of

robotics, kinematics deals with the geometric relationships between the manipulator's joints and the position and orientation of its end-effector. There are two primary branches of manipulator kinematics: forward kinematics and inverse kinematics.

Forward kinematics is concerned with determining the position and orientation of the end-effector given the values of the joint variables (i.e., the angles for revolute joints and the displacements for prismatic joints). In essence, if you know how much each joint is rotated or translated, forward kinematics tells you exactly where the end-effector is and how it is oriented in space. This is a relatively straightforward calculation, typically performed using transformation matrices (e.g., Denavit-Hartenberg parameters) that describe the relationship between consecutive links. For a manipulator with n joints, forward kinematics is a mapping from the n-dimensional joint space to the 6-dimensional task space (3 for position, 3 for orientation). The result of forward kinematics is a unique end-effector pose (position and orientation) for each valid joint configuration. For a manipulator with n DOFs, the forward kinematics equations can be written as:

$$(x, y, z, \alpha, \beta, \gamma) = F(q_1, q_2, \dots, q_n)$$

where (x, y, z) are the coordinates of the end-effector in Cartesian space, (α, β, γ) represent its orientation (e.g., using Euler angles or a rotation matrix), and q_i are the joint variables. This calculation is fundamental for simulating robot motion, visualizing the robot's configuration, and understanding its reachable workspace.

Inverse kinematics, on the other hand, is the more challenging and often more critical problem. It is concerned with determining the required joint variables (angles or displacements) that will place the end-effector at a desired position and orientation in space. In other words, if you know where you want the end-effector to be and how you want it oriented, inverse kinematics tells you how to set each joint angle to achieve that pose. This is the problem that needs to be solved for most manipulation tasks, such as picking up an object at a specific location or performing a welding seam along a particular path.

The inverse kinematics problem can be significantly more complex than forward kinematics for several reasons:

1. **Multiple Solutions:** For a given end-effector pose, there may be multiple valid sets of joint angles that can achieve it. Think of reaching for an object with your elbow bent in different ways; you can often achieve the same hand position with different arm configurations. These different configurations are called "kinematic alternatives" or "multiple solutions." The robot's control system must then choose among these solutions, often based on factors like avoiding joint

limits, minimizing joint velocities, or staying away from singular configurations.

2. **No Solution:** It is also possible that a desired end-effector pose is outside the robot's reachable workspace, in which case there is no solution to the inverse kinematics problem.

3. **Singularities:** Certain configurations of the robot arm, known as singular configurations, can lead to infinite joint velocities for finite end-effector velocities, or a loss of controllability in certain directions. These configurations must be avoided or handled with care during motion planning and control.

4. **Computational Complexity:** Unlike forward kinematics, which usually has a closed-form analytical solution, inverse kinematics often requires iterative numerical methods to solve, especially for manipulators with many degrees of freedom. These numerical methods involve solving systems of non-linear equations, which can be computationally intensive and may not always converge to a solution.

The methods for solving inverse kinematics problems can be broadly categorized into analytical and numerical approaches.

Analytical solutions are possible for simpler robot structures, such as those with fewer DOFs or specific joint configurations (e.g., a spherical wrist). These methods involve deriving explicit mathematical equations that directly compute the joint variables from the desired end-effector pose. While computationally efficient and guaranteed to find all possible solutions, analytical solutions are not available for all robot geometries.

Numerical solutions, on the other hand, are more general and can be applied to almost any robot manipulator. These methods typically involve an iterative process where an initial guess of the joint variables is refined over successive steps until the end-effector pose converges to the desired target. Common numerical techniques include:

Jacobian-based methods: These methods use the Jacobian matrix, which relates joint velocities to end-effector velocities. The Jacobian is the derivative of the forward kinematics equations. By inverting or pseudo-inverting the Jacobian, one can compute the required joint velocity changes to achieve a desired end-effector velocity. This process is then integrated over time to find the joint positions.

Optimization-based methods: These methods formulate the inverse kinematics problem as an optimization problem, where the objective is to minimize the error between the current end-effector pose and the desired pose, subject to joint limits and other constraints. Techniques like gradient descent or Newton's method can be employed.

The choice of method for inverse kinematics depends on the specific robot design, the required accuracy, and the available computational resources. For many industrial robots, especially those with six DOFs and a common configuration like the PUMA robot (a Stanford arm derivative), analytical solutions are well-established. For more complex or custom-designed manipulators, numerical methods are often necessary.

Understanding both forward and inverse kinematics is fundamental to robot manipulation. Forward kinematics allows us to predict where the robot's end-effector will be given the joint commands, which is essential for simulation and for understanding the robot's capabilities. Inverse kinematics allows us to command the robot to move its end-effector to a specific location and orientation, which is the basis for performing tasks. Together, these kinematic relationships form the foundation upon which all subsequent motion planning, control, and manipulation algorithms are built. They provide the geometric framework for translating desired end-effector actions into actuator commands for the robot's joints, enabling the robot to physically interact with and modify its environment. The ability to precisely control the position and orientation of the end-effector through these kinematic transformations is what allows a robot arm to execute tasks ranging from simple pick-and-place operations to intricate assembly and intricate surgical procedures. The study of these geometric relationships, therefore, is not just an academic exercise but a practical necessity for anyone designing, programming, or operating robotic manipulation systems.

The geometric description of a robot manipulator's configuration and its relationship to the end-effector's position and orientation is a cornerstone of robotic manipulation. This geometric understanding is formalized through the study of kinematics, which, as previously introduced, is divided into two fundamental problem domains: forward kinematics and inverse kinematics. While forward kinematics is generally a more straightforward computational task, inverse kinematics presents significant challenges that are central to enabling robots to perform meaningful tasks in the physical world. This section will elaborate on the mathematical underpinnings and practical implications of both forward and inverse kinematics, highlighting the methods used to tackle these problems and their importance in robot control.

Forward Kinematics: From Joint Angles to End-Effector Pose

Forward kinematics provides a direct mapping from the robot's joint configuration to the pose (position and orientation) of its end-effector in Cartesian space. For a manipulator with n degrees of freedom, denoted by the vector of joint variables $\mathbf{q} = [q_1, q_2, \dots, q_n]^T$, *the forward kinematics problem is to compute the end-effector's pose, typically represented as a transformation matrix* $\mathbf{T}_{end\_effector}$ or a set of Cartesian coordinates (x, y, z) and orientation parameters (e.g., Euler angles or a rotation matrix). This relationship can be expressed as:

$$

\text{Pose}*{end*\effector} = \mathcal{F}(\mathbf{q})
$$

where $\mathcal{F}$ is the forward kinematics function. For a serial manipulator, this function is typically derived by chaining together the individual transformations between consecutive links. The Denavit-Hartenberg (D-H) convention is a widely adopted method for systematically defining the coordinate frames attached to each link and deriving these transformations.

The D-H convention assigns four parameters to the relationship between consecutive joint coordinate frames: a link length ($a
*i$), a link twist ($\alpha*i$), a link offset ($d*i$), and a joint angle ($\theta*i$). For revolute joints, $d*i$ and $\theta*i$ are variable, while $a*i$ and $\alpha*i$ are fixed link parameters. For prismatic joints, $a*i$ and $\alpha*i$ are fixed, while $d*i$ is variable and $\theta*i$ is fixed. Using these parameters, a standard homogeneous transformation matrix $\mathbf{A}_i$ can be constructed that describes the transformation from the coordinate frame of link i to the coordinate frame of link $i-1$.

For a manipulator with n links and n joints, the transformation from the base frame (link 0) to the end-effector frame (link n) is obtained by multiplying these individual transformation matrices in sequence:

$$

\mathbf{T}{base}^{end\effector} = \mathbf{A}1
\mathbf{A}2 \dots \mathbf{A}_n
$$

Each $\mathbf{A}_i$ is a 4×4 homogeneous transformation matrix of the form:

$$

\mathbf{A}_i = \begin{bmatrix}
\cos(\thetai) & -\sin(\thetai)\cos(\alphai) &
\sin(\thetai)\sin(\alphai) & ai\cos(\theta_i) \\
\sin(\thetai) & \cos(\thetai)\cos(\alphai) & -
\cos(\thetai)\sin(\alphai) & ai\sin(\theta_i) \\
0 & \sin(\alphai) & \cos(\alphai) & d_i \\
0 & 0 & 0 & 1
\end{bmatrix}
$$

If the D-H parameters are defined such that $\theta
i$ is the variable for a revolute joint and di is the variable for a prismatic joint, then the i-th matrix $\mathbf{A}i$ will incorporate the current value of the i-th joint variable. By substituting the current joint angles (for revolute joints) and displacements (for prismatic joints) into the $\mathbf{A}i$ matrices and multiplying them together, the resulting homogeneous transformation matrix $\mathbf{T}{base}^{end\effector}$ directly provides

the position (the first three elements of the last column) and orientation (the upper-left 3×3 submatrix) of the end-effector relative to the robot's base frame.

For instance, consider a simple 2-DOF planar manipulator with two revolute joints. Let the first joint be at the origin and the second joint be at the end of the first link. Using the D-H convention, we can define the coordinate frames and parameters. The transformation matrix from the end-effector frame to the base frame would be $\mathbf{T}
{base}^{end\effector} = \mathbf{A}1 \mathbf{A}2$. Expanding this product would yield a 4×4 matrix. The position (x, y) of the end-effector would be given by the first two elements of the last column, and its orientation (angle with respect to the x-axis) would be derived from the rotation submatrix.

The calculation of forward kinematics is computationally efficient and deterministic. For any given set of joint variables $\mathbf{q}$, there is always a unique and well-defined end-effector pose. This is crucial for tasks such as:

Workspace Analysis: Determining the reachable volume and reachable orientations of the end-effector. By sampling various joint configurations and computing their resulting end-effector poses, one can map out the robot's workspace.
Simulation and Visualization: Predicting and displaying how the robot arm will move in response to

joint commands, allowing for virtual testing and debugging of robot programs.
Collision Detection: Checking if any part of the robot arm will collide with obstacles or itself given a specific configuration.

Inverse Kinematics: From End-Effector Pose to Joint Angles

In contrast to forward kinematics, inverse kinematics (IK) addresses the more complex problem of finding the joint variables $\mathbf{q}$ required to achieve a desired end-effector pose, often specified in the Cartesian workspace. Given a target pose $\text{Pose}{target} = (x{des}, y{des}, z{des}, \mathbf{R}{des})$, where $\mathbf{R}{des}$ is the desired orientation, the IK problem is to find $\mathbf{q}$ such that:

$$
\mathcal{F}(\mathbf{q}) = \text{Pose}_{target}
$$

This problem is significantly more challenging for several reasons:

1.
Non-linearity: The relationship between joint variables and end-effector pose is inherently non-linear. This

means that the IK problem often involves solving systems of non-linear equations.
2. **Multiple Solutions:** For a given end-effector pose, there can be multiple distinct configurations of the robot's joints that achieve it. Imagine reaching for a cup on a table; you can do so with your elbow bent in various ways, or even by using your other arm. For a robot arm, different combinations of joint angles can result in the same end-effector position and orientation. These are often referred to as "reachable configurations" or "kinematic alternatives." The choice among these solutions can depend on factors such as energy consumption, joint limits, proximity to singularities, or desired end-effector motion path.
3. **No Solution:** The desired end-effector pose might be unreachable. This occurs if the target pose lies outside the robot's workspace, or if it requires the robot to exceed its joint limits.
4. **Singularities:** Certain joint configurations, known as singularities, pose particular difficulties. At a singularity, the robot loses one or more degrees of freedom, meaning that certain end-effector motions become impossible, or infinitesimally small end-effector movements require infinitely large joint velocities. This can lead to numerical instability and control issues.

Given these complexities, inverse kinematics problems are typically solved using either analytical methods (when possible) or numerical iterative methods.

Analytical Solutions

Analytical solutions are derived through direct algebraic manipulation of the forward kinematics equations. They involve solving the system of non-linear equations for the joint variables in closed form. This approach is only feasible for robot manipulators with specific kinematic structures, often those with fewer degrees of freedom or particular arrangements of joints, such as a spherical wrist. A spherical wrist, for instance, consists of three consecutive revolute joints whose axes intersect at a single point. This geometric property simplifies the problem significantly, allowing for the decoupling of the position and orientation problems.

For a common 6-DOF industrial robot with a spherical wrist, the IK problem can often be decomposed into two parts:

1. **Position Problem:** Determining the first three joint angles (q_1, q_2, q_3) that position the wrist center at the desired Cartesian coordinates (x, y, z). This is often solvable using geometric approaches or trigonometry.
2. **Orientation Problem:** Once the wrist center is positioned, the remaining three joint angles (q_4, q_5, q_6) determine the orientation of the end-effector. This part of the problem is essentially finding the orientation of the wrist's coordinate frame relative to

the frame of the last link, which can be solved by comparing the desired end-effector orientation with the orientation of the base frame relative to the wrist.

The advantage of analytical solutions is their computational efficiency and their ability to find all possible solutions directly. However, they are highly dependent on the specific robot geometry and are not generally applicable to all manipulators. Developing analytical solutions for arbitrary robot structures can be a very complex and time-consuming task.

Numerical Solutions

When analytical solutions are not available or are too difficult to derive, numerical methods are employed. These methods typically involve an iterative process that starts with an initial guess for the joint variables and refines this guess over successive steps until the end-effector pose converges to the desired target pose within a specified tolerance.

Jacobian-Based Methods

One of the most common and powerful numerical techniques for solving inverse kinematics is based on the

Jacobian matrix. The Jacobian matrix, denoted by $\mathbf{J}(\mathbf{q})$, is a matrix that relates the angular and linear velocities of the joints to the linear and angular velocities of the end-effector. Specifically, for an n-DOF manipulator, the Jacobian is an $m \times n$ matrix (where m is typically 6 for 3D space, representing 3 translational and 3 rotational degrees of freedom) such that:

$$
\dot{\mathbf{x}}{end\effector} = \mathbf{J}(\mathbf{q}) \dot{\mathbf{q}}
$$

Here, $\dot{\mathbf{x}}{end\effector}$ is a vector of end-effector velocities (e.g., $[\dot{x}, \dot{y}, \dot{z}, \omegax, \omegay, \omega_z]^T$), and $\dot{\mathbf{q}}$ is the vector of joint velocities. The Jacobian matrix is computed by taking the partial derivatives of the forward kinematics equations with respect to each joint variable.

The inverse kinematics problem can be approached using the Jacobian by considering the relationship between desired changes in end-effector pose and the required joint velocities. If we have a desired end-effector velocity $\dot{\mathbf{x}}_{des}$, we can attempt to compute the required joint velocities $\dot{\mathbf{q}}$ as:

$$

\dot{\mathbf{q}} = \mathbf{J}(\mathbf{q})^{-1}
\dot{\mathbf{x}}_{des}
$$

This requires the Jacobian to be invertible, which is true when the robot is not in a singular configuration and has full rank. When the Jacobian is not square or is singular, the
pseudo-inverse $\mathbf{J}^+$ is used:

$$

\dot{\mathbf{q}} = \mathbf{J}^+(\mathbf{q})
\dot{\mathbf{x}}_{des}
$$

The pseudo-inverse provides a solution that minimizes the joint velocities while still achieving the desired end-effector velocity.

To solve the IK problem for a specific target pose rather than just velocity, an iterative scheme is used. Starting with an initial guess $\mathbf{q}
0$, the algorithm iteratively updates the joint angles. At each step k, the current end-effector pose $\mathcal{F}(\mathbf{q}*k)$ is computed. The error between the current pose and the desired pose, $\Delta \mathbf{x} = \text{Pose}{target}* -
*\mathcal{F}(\mathbf{q}*k)$, is calculated. Then, the required joint changes $\Delta \mathbf{q}$ are estimated using the Jacobian:

$$
\Delta \mathbf{q}_k = \mathbf{J}^+(\mathbf{q}_k) \Delta \mathbf{x}
$$

The next set of joint angles is then computed as:

$$
\mathbf{q}_{k+1} = \mathbf{q}_k + \Delta \mathbf{q}_k
$$

This process is repeated until the error $\Delta \mathbf{x}$ falls below a predefined tolerance. The choice of an initial guess is important for convergence. Often, the previous joint configuration or a configuration that places the end-effector near the target can serve as a good starting point.

Handling singularities is a critical aspect of Jacobian-based methods. When the robot approaches a singularity, the Jacobian becomes ill-conditioned (its determinant approaches zero), and its inverse or pseudo-inverse can lead to very large joint velocities or unstable behavior. Techniques to mitigate this include damping the pseudo-inverse, limiting joint velocities, or actively avoiding singular regions during motion planning.

Optimization-Based Methods

Another class of numerical IK solvers treats the problem as an optimization task. The goal is to find the joint configuration $\mathbf{q}$ that minimizes the error between the current end-effector pose and the desired pose, subject to various constraints, such as joint limits and obstacle avoidance. The objective function might be:

$$
\text{Minimize } E(\mathbf{q}) = \frac{1}{2} ||\mathcal{F}(\mathbf{q}) - \text{Pose}_{target}||^2
$$

This is a non-linear least-squares problem. Various optimization algorithms, such as gradient descent, Newton's method, or Levenberg-Marquardt, can be used to find the minimum of this error function. These methods can be more robust than simple Jacobian inversion, especially in the presence of singularities or when dealing with complex constraints.

These numerical methods, while general and widely applicable, come with the caveat that they may not always converge to a solution, or they might converge to a local minimum that does not correspond to the desired pose. Furthermore, they can be computationally more expensive than analytical solutions, which is a consideration for real-time control applications.

The importance of inverse kinematics cannot be overstated. It is the bridge between task-level commands (e.g., "pick up the object at location X") and the low-level control signals that drive the robot's actuators. Without effective IK solvers, robots would be unable to perform most manipulation tasks that require precise positioning and orientation of their end-effectors in the physical world. The ongoing research in IK focuses on improving convergence speed, robustness, handling of singularities, and incorporating additional constraints for more complex and safer robotic operations.

The interplay between forward and inverse kinematics forms the fundamental basis for robot motion planning and control. Forward kinematics tells us where the robot
is based on its joint states, providing essential feedback for control and awareness. Inverse kinematics tells us how to get the robot *where we want it to be*, translating high-level goals into actionable joint commands. Mastering both of these kinematic transformations is therefore an essential step in understanding and implementing sophisticated robotic manipulation capabilities, paving the way for robots to interact intelligently and effectively with their environments.

The previous discussion has laid the groundwork for understanding how robots move and position their end-effectors through kinematics. However, for a robot to truly interact with the physical world, simply knowing its position and orientation is insufficient. Many critical tasks require the robot not only to reach a certain pose but also to apply specific forces, or to absorb unexpected forces, during its interaction. This is where the concepts of force control and compliance become paramount. Without these capabilities, robots would be limited to highly structured environments, unable to perform delicate manipulations, assemble intricate parts, or safely operate alongside humans.

Force control, in its essence, is about enabling a robot to apply a desired force to its environment, or to maintain a specific force against a surface. Consider the task of inserting a peg into a hole. If the robot simply moves to the hole's location and pushes, it might jam the peg or damage the hole if misaligned. A force-controlled robot, however, can actively sense the forces it is exerting. It can apply a gentle, consistent pushing force while searching for the correct alignment. Once the peg is seated, it can reduce the force or signal that the insertion is complete. Similarly, tasks like polishing, grinding, or even wiping a surface require the robot to maintain a constant contact force. This allows for uniform material removal or application, which would be impossible with simple position control alone.

Achieving controlled force interaction necessitates a different approach than pure position control. In position control, the robot is commanded to move to a specific Cartesian coordinate or joint angle configuration. In force control, the robot is commanded to exert a specific force, or to maintain a certain force profile. This often involves feedback from force/torque sensors mounted at the robot's wrist or within its end-effector. These sensors provide crucial information about the external forces acting on the robot, allowing for corrective actions.

There are several fundamental strategies for force control. One straightforward approach is **direct force control**, often referred to as hybrid position-force control. In this paradigm, the task space is decomposed into directions where position is controlled and directions where force is controlled. For example, in the peg-in-hole insertion task, the robot might control its position along the insertion axis (z-axis) while controlling its force in the perpendicular directions (x-y plane). This allows the robot to maintain a consistent depth while being able to "slide" or adjust its position to find the correct alignment when encountering resistance. The challenge here lies in accurately defining and implementing this decomposition, especially for complex geometries and unknown environments.

A more sophisticated and widely adopted approach is **impedance control**. Impedance control doesn't directly command forces or positions; instead, it regulates the *dynamic relationship* between the forces exerted by the robot and its resulting motion. It aims to make the robot behave like a mechanical impedance – a spring, a damper, or a combination of both – with desired properties. The core idea is to control the robot's interaction dynamics such that it exhibits a specified impedance, typically relating end-effector force and velocity.

The fundamental equation for impedance control can be expressed as:

$$
\mathbf{M} \ddot{\mathbf{x}} + \mathbf{D} \dot{\mathbf{x}} + \mathbf{K} \mathbf{x} = \mathbf{F}_{ext}
$$

Here, $\mathbf{M}$, $\mathbf{D}$, and $\mathbf{K}$ represent the desired inertial, damping, and stiffness properties of the robot's interaction, respectively. $\mathbf{x}$ denotes the deviation of the end-effector from a reference position (which can be dynamically updated), $\dot{\mathbf{x}}$ and $\ddot{\mathbf{x}}$ are its velocity and acceleration, and $\mathbf{F}_{ext}$ is the external force acting on the end-effector. This equation essentially defines how the robot *should* move in response to external forces, or how it *should* apply forces in response to its own motion.

By selecting appropriate values for $\mathbf{M}$, $\mathbf{D}$, and $\mathbf{K}$, the robot can be made to behave in various compliant or stiff ways. For instance:

A high stiffness ($\mathbf{K}$) coupled with low damping ($\mathbf{D}$) would make the robot behave like a stiff spring, reacting strongly to any displacement. This might be useful for tasks requiring precise positioning against a hard surface but would be very unforgiving of errors.

A low stiffness ($\mathbf{K}$) and high damping ($\mathbf{D}$) would make the robot behave like a soft damper, absorbing shocks and smoothly yielding to external forces. This is ideal for tasks where the robot might encounter unexpected obstacles or needs to move in a very gentle manner.

A purely inertial behavior ($\mathbf{M} \neq 0$, $\mathbf{D} = 0$, $\mathbf{K} = 0$) would mean the robot's motion is primarily governed by its inertia, making it less responsive to immediate external forces but potentially useful for specific dynamic tasks.

The challenge in implementing impedance control lies in the fact that robots are not ideal mechanical systems. The robot's own dynamics, including friction, motor inertia, and flexibility in its links and joints, are significant. Therefore, the controller must actively compensate for these inherent dynamics to achieve the desired impedance. This often involves complex feedback loops that utilize both joint-space and Cartesian-space information, along with force/torque sensor readings. The control signals generated by the impedance controller are then translated into joint torques or motor commands that drive the robot's actuators.

Compliance, a closely related concept, can be thought of as the robot's ability to "give way" when subjected to external forces. A compliant robot is less likely to break itself or the object it is interacting with when unexpected forces arise. It's a crucial safety feature and an enabler of delicate manipulation.

There are two primary ways to achieve compliance in robots:

1. **Passive Compliance:** This is inherent to the robot's mechanical design. For example, a robot arm with flexible links will naturally bend under load, providing a degree of passive compliance. Grippers with compliant fingers are another example. While simpler to achieve, passive compliance is difficult to control and is often insufficient for complex tasks. The compliance is fixed by the material properties and geometry.
2. **Active Compliance:** This is achieved through control strategies like impedance control or hybrid force/position control. The robot's controller actively adjusts its joint commands in response to force sensor feedback, effectively creating a dynamically adjustable level of compliance. This allows the robot to be stiff when needed for precise positioning and compliant when interaction forces are encountered.

The benefits of active compliance are profound. It

allows robots to perform tasks that would otherwise be impossible or dangerous. Imagine a robot assembling electronics: it needs to be precise enough to place tiny components but also compliant enough to accommodate slight variations in part dimensions or seating positions without applying excessive force. Similarly, a robot assisting a human operator needs to be highly compliant to ensure safety, yielding immediately if its path is blocked or if the human makes contact.

The implementation of force control and compliance often relies on accurate and sensitive force/torque sensing. These sensors, typically mounted at the robot's wrist, measure the six components of the wrench (three forces and three torques) applied to the end-effector. The quality of this sensing directly impacts the performance of force control algorithms. Noise in the sensor readings, calibration errors, or limitations in bandwidth can all degrade the effectiveness of the control system.

Moreover, the integration of force control and compliance with higher-level motion planning and task execution is an active area of research. For tasks like object manipulation, a robot might need to switch between different control modes – for instance, moving in free space with position control, then engaging with an object using force control for regrasping, and finally placing it with compliant insertion. This requires a robust framework for managing these transitions and ensuring smooth, safe behavior.

The ability of robots to understand and control forces is also crucial for human-robot interaction (HRI). As robots move from controlled industrial settings into more dynamic and collaborative environments, their capacity to interact physically with humans without causing harm is paramount. A compliant robot can detect human presence through force feedback and adjust its behavior accordingly, perhaps slowing down or stopping. This "gentle" interaction is a hallmark of modern collaborative robots, or "cobots."

In summary, force control and compliance are not mere extensions of position control; they represent a fundamental shift in how robots interact with their physical environment. They enable robots to move beyond simply executing pre-programmed paths and instead allow them to actively, adaptively, and safely engage with objects and humans. The development of advanced algorithms, coupled with improved sensing technologies, continues to push the boundaries of what robots can achieve in the realm of physical manipulation, making them more versatile, capable, and integrated partners in a wide array of applications. This opens up possibilities for tasks that require fine motor skills, adaptability to uncertainties, and a high degree of safety in physical interactions, ultimately leading to more intelligent and capable robotic systems.

The ability of a robot to successfully interact with its environment hinges critically on its capacity to acquire and manipulate objects. While concepts like kinematics and force control address how a robot positions itself and manages its interactions, the fundamental act of holding and controlling an object falls under the domain of grasping. Grasping is not merely about closing a gripper around an object; it is a complex process that involves understanding object properties, selecting an appropriate manipulation strategy, and executing a stable hold that allows for subsequent actions. The success or failure of a robotic task often begins or ends with a grasp.

The inherent properties of an object play a pivotal role in determining whether a stable and functional grasp can be achieved. These properties can be broadly categorized into geometric features, material characteristics, and dynamic attributes.

Geometric properties, such as an object's shape, size, and surface features, are primary determinants of graspability. Consider the difference between grasping a smooth, spherical ball and a cubical block with well-defined edges. The ball presents a continuous, smooth surface that offers limited points of contact and is susceptible to slippage. In contrast, the cubical block provides distinct edges and flat surfaces that can be leveraged for more secure grasps, such as an edge grasp or a planar grasp. The complexity of the object's geometry also matters. Objects with intricate shapes, undercuts, or thin protrusions pose significant challenges. A grasp that might be stable on a simple convex shape could become unstable if it relies on a contact point that is not robust against small perturbations. The presence of concavities, holes, or re-entrant features can either facilitate specific grasp types (e.g., a ring grasp around a handle) or, conversely, obstruct them by preventing the gripper fingers from making adequate contact. The overall size and scale of the object, relative to the robot's end-effector, also dictate feasibility. An object that is too large may not fit within the gripper's aperture, while one that is too small might be difficult for the gripper to isolate and grasp without accidentally interacting with surrounding objects.

Material properties, particularly friction and surface texture, are equally crucial. Friction is the force that opposes relative motion between two surfaces in contact. A high-friction surface will resist slippage more effectively than a low-friction surface. For instance, grasping a metallic object with a polished, low-friction surface requires a different strategy than grasping a rubberized object. The coefficient of friction between the gripper fingers and the object's surface directly impacts the forces that can be transmitted. Without sufficient friction, an object can easily slide out of the gripper, even if the gripper is exerting considerable closing force. This is why grippers often employ specialized materials like high-friction rubber or textured pads on their fingers. Surface texture, while related to friction, also plays a role. Rough surfaces can provide mechanical interlocks that enhance stability, whereas very smooth surfaces offer little resistance. The weight of the object is another critical material-related property. Heavier objects require stronger grasps and place greater demands on the robot's arm and gripper actuators. Moreover, the center of mass of the object is important; if the center of mass is not properly supported by the grasp, the object can pivot and fall.

Dynamic attributes, such as an object's inertia and its potential for deformation or fragility, also influence grasp selection. A heavy object with high inertia will resist changes in motion, meaning that applying forces to move it will require more effort. Conversely, a light object with low inertia can be easily accelerated, but also easily knocked off course. Fragile objects, like glassware or delicate electronic components, require a very different approach to grasping. The gripper must exert just enough force to prevent slippage but not so much as to cause damage. This often necessitates the use of sophisticated force sensing and control strategies to modulate the grasping force precisely. Deformable objects, such as fabrics or soft plastics, introduce an additional layer of complexity. They can change their shape under the influence of the gripper, making it difficult to maintain a stable grasp. The weight distribution and how it might shift during manipulation are also important considerations.

Understanding these object properties is fundamental to devising effective grasping strategies. Without this knowledge, a robot is left to trial and error, which is often impractical and can lead to damage to the object or the robot itself.

Robotic grasps can be broadly categorized into two main types, distinguished by their objective: power grasps and precision grasps. These categories, inspired by human hand capabilities, highlight different ways in which an end-effector can secure an object.

Power Grasps are designed to exert significant holding forces, often enveloping the object to maximize contact and stability. The primary goal of a power grasp is to resist large external forces that might try to dislodge the object. Think of a human firmly gripping a hammer handle or a heavy tool. In a robotic context, power grasps typically involve multiple fingers or gripper jaws closing around the object. A common form of power grasp is the **enveloping grasp**, where the object is fully or partially surrounded by the gripper. This can be achieved with a two-jaw parallel gripper that closes tightly on the sides of an object, or with a multi-fingered hand that conforms to the object's shape. The key characteristic of a power grasp is its emphasis on **force closure**, a concept we will explore further. This means that the grasp can resist any external force applied to the object, regardless of its direction, by generating sufficient internal friction and normal forces. Examples include using a three-jaw chuck to hold a cylindrical workpiece in a lathe, or a multi-fingered hand closing around a bottle. These grasps are generally robust but may not allow for fine manipulation or orientation changes of the object while it is held. They are often used when the object needs to be lifted, moved, or subjected to significant forces during a task, such as in heavy lifting or assembly operations where precise dexterity is not paramount.

Precision Grasps, on the other hand, are focused on enabling fine manipulation and dexterous control of an object. The goal is not necessarily to exert maximum holding force, but rather to achieve a stable hold that allows for delicate movements, rotations, and precise placement. A human picking up a pen, a needle, or a coin exemplifies a precision grasp. In robotics, these grasps often involve fewer points of contact and rely on a combination of friction and the geometry of the object and the gripper. A common type of precision grasp is the **pinch grasp**, where an object is held between two opposing digits or surfaces. This can be a fingertip-to-fingertip pinch, a fingertip-to-palm pinch, or a two-jaw pinch. Another precision grasp is the **tripod grasp**, where three points of contact are used to stabilize an object, much like holding a small ball. This type of grasp offers greater stability than a simple two-point pinch and is often used for smaller, irregularly shaped objects. Precision grasps often aim for **form closure**, which means the object is geometrically trapped by the gripper, preventing it from escaping its confinement. This can be achieved even with minimal friction. However, precision grasps can be less resistant to external forces compared to power grasps. They are essential for tasks requiring intricate manipulation, such as inserting a small component into a tight-fitting hole, picking and placing small electronic parts, or handling delicate surgical instruments.

Beyond these two broad categories, numerous specific grasp types exist, often described by the way the fingers or jaws interact with the object's geometry:

Cylindrical Grasp: The gripper fingers wrap around the cylindrical surface of an object. This can be a full wrap (like a chuck) or a partial wrap.
Spherical Grasp: The gripper fingers conform to the curved surface of a spherical or near-spherical object.
Planar Grasp: The gripper makes contact with a flat surface of the object. This is often used for thin, planar objects like credit cards or sheets of metal.
Edge Grasp: The gripper grasps the object along one of its edges. This is common for objects like books or plates.
Point Grasp: The gripper makes contact with a single point on the object, often a concave surface or a protruding feature.

The choice between a power grasp and a precision grasp, and indeed the specific type of grasp within these categories, depends heavily on the task requirements, the object's properties, and the capabilities of the robotic end-effector.

The ultimate goal in grasping is to achieve a **stable grasp**. A stable grasp is one that can resist external disturbances and maintain its hold on the object without requiring continuous active control to prevent slippage or rotation. Stability is often analyzed using concepts from mechanics, primarily **form-closure** and **force-closure**.

Form-closure is a geometric condition. An object is said to be in form-closure with respect to a gripper if its movement is completely constrained by the gripper geometry alone. In other words, no matter how the object is moved, it cannot escape the gripper without the gripper itself changing its configuration. For a 2D case, this typically requires at least three contact points that form a polygon enclosing the object. In 3D, it requires more complex arrangements of contact points and surfaces. Form-closure grasps are inherently stable against any external force and moment, as long as the contact points remain within the gripper's configuration. However, achieving true form-closure can be difficult in practice. It often requires a very specific object shape and gripper design, and it may not allow for any relative motion between the object and the gripper, which is often needed for manipulation. Many precision grasps, especially those with multiple contact points that geometrically "trap" an object, aim for form-closure.

Force-closure is a condition related to forces. An object is in force-closure if the gripper can exert forces and torques that can balance any possible external wrench (a combination of forces and moments) that might be applied to the object. This means that even if the object can move freely within the gripper, the gripper can still generate sufficient internal friction and normal forces to counteract any external disturbance. Unlike form-closure, force-closure does not necessarily require the object to be geometrically trapped. A grasp between two compliant fingers on a smooth ball can be in force-closure if the friction between the fingers and the ball is high enough.

Mathematically, for a grasp to be in force-closure, the set of grasp wrenches (the wrenches that the gripper can exert on the object through its contact points) must span the entire space of possible wrenches. In 2D, this generally requires at least three points of contact. In 3D, it typically requires at least four non-coplanar points of contact, or a combination of points and surfaces that can generate forces in all directions. Force-closure is a more practical condition than form-closure for many robotic applications because it can be achieved with a wider range of object shapes and gripper designs, and it allows for movement. However, achieving force-closure relies heavily on friction, and thus the friction coefficients at the contact points are critical. If the friction is insufficient, a grasp that would otherwise be in force-closure can fail.

The process of determining whether a grasp is stable,

and how stable it is, often involves analyzing the grasp wrench space (GWS). The GWS is the set of all possible wrenches that a gripper can exert on an object. A grasp is stable if the GWS can resist any expected external wrench. Metrics like the force closure wrench space metric (FCWMS) are used to quantify the degree of force closure, indicating how well the grasp can resist disturbances. Higher values of such metrics generally correspond to more stable grasps.

Grasp planning, the process of determining how and where to grasp an object, is a complex problem, especially in real-world scenarios. Several challenges arise:

One of the most significant challenges is **grasping in cluttered environments**. When an object is surrounded by other objects, the available space for the gripper to approach and make contact is limited. This means that a grasp that might be ideal in isolation may become impossible due to collisions with neighboring objects. The robot's planner must not only identify a stable grasp for the target object but also ensure that the approach and withdrawal trajectories of the gripper are collision-free. This often requires sophisticated path planning algorithms that consider both the robot's kinematics and the geometry of the environment. The gripper itself can also be an obstacle. If a large gripper is used to grasp a small object in a confined space, the gripper may collide with the environment before its fingers can reach the object.

This highlights the importance of gripper design and selection for specific task environments. Multi-fingered hands offer greater dexterity and can often reach objects in more constrained spaces than simple parallel jaw grippers, but they also present a more complex planning problem due to their increased degrees of freedom.

Another major hurdle is **grasping unknown objects**. In many industrial settings, robots are programmed to grasp objects whose precise geometry, size, and properties are known in advance. However, in more dynamic or less structured environments, robots may encounter objects that have not been previously modeled. This necessitates the ability to *perceive* and *infer* graspable properties from sensor data. Computer vision, using cameras, is a primary tool for this. Robots can use depth sensors (like LiDAR or structured light) to create 3D point clouds of objects, from which their shape can be reconstructed. Machine learning, particularly deep learning, has shown great promise in enabling robots to predict stable grasps directly from sensor data, often bypassing the need for explicit geometric modeling. These systems learn to associate visual features with grasp affordances – the potential ways an object can be interacted with. However, challenges remain in achieving robustness across a wide variety of objects and lighting conditions, and in accurately estimating grasp stability from visual data alone. The tactile sensing capabilities of grippers can also provide valuable information about unknown objects, such as surface texture, hardness, and the presence of edges or corners. Integrating visual and tactile information is a promising avenue for overcoming the challenge of grasping unknown objects.

The problem of grasp planning becomes even more intricate when considering **dynamic environments** where objects might be moving. Grasping a moving object requires real-time perception, prediction of the object's trajectory, and a highly responsive grasping system. This is a significant area of research in fields like autonomous driving and high-speed robotic assembly.

Furthermore, the selection of an appropriate grasp is not just about stability, but also about the subsequent manipulation task. A grasp that is stable for picking up an object might not be suitable for orienting it, inserting it, or performing any other operation. This leads to the concept of **task-aware grasping**, where the grasp is chosen not only for its inherent stability but also for its ability to facilitate the required manipulation. For example, when inserting a peg into a hole, the initial grasp might be oriented to allow for easy insertion, even if it's not the most stable grasp for lifting.

Finally, the computational complexity of grasp planning can be substantial. If a robot needs to consider thousands of potential grasp configurations on an object, and then simulate their stability and collision-free trajectories, the computation time can become prohibitive for real-time applications. This has driven the development of efficient grasp planning algorithms, including methods that use probabilistic sampling, grasp synthesis based on learned models, and heuristics that prune the search space of possible grasps. The challenge lies in developing algorithms that are both computationally efficient and capable of finding high-quality, stable, and task-appropriate grasps in complex scenarios.

In summary, grasping is a fundamental yet challenging aspect of robotic manipulation. It requires a deep understanding of object properties, the selection of appropriate grasp types (power vs. precision), and the rigorous analysis of grasp stability through concepts like form-closure and force-closure. Overcoming challenges posed by cluttered environments, unknown objects, and the need for task-aware manipulation is crucial for enabling robots to interact with the physical world in a truly versatile and intelligent manner. This forms the bedrock upon which advanced robotic manipulation strategies are built.

The culmination of understanding object properties, grasp types, and stability metrics is the practical realization of a robot interacting with an object: grasp planning and execution. This phase transforms theoretical possibilities into tangible actions, moving from "what is a stable grasp?" to "how do we achieve it?" The process involves a series of critical steps, beginning with selecting the right tool for the job and progressing through sophisticated sensing and motion control.

The initial and often crucial decision in grasp planning

is the selection of the appropriate end-effector or gripper. The vast diversity of objects and tasks necessitates a corresponding diversity in gripper designs. For instance, tasks involving the manipulation of cylindrical components, such as those found in manufacturing lines or assembly processes, often benefit from the use of two-jaw parallel grippers. These grippers are straightforward to control, can exert significant closing forces, and are well-suited for grasping objects by their sides. Their simplicity makes them a workhorse in industrial automation. However, when dealing with objects of irregular shapes, or when the task demands a more delicate or dexterous hold, a multi-fingered hand might be a more suitable choice. Robotic hands, mimicking aspects of the human hand with multiple degrees of freedom in their fingers, offer a far greater range of possible grasp configurations. They can conform to complex object geometries, achieve precision grasps, and even perform in-hand manipulation, such as reorienting an object while it is still held. The choice of gripper is not merely about mechanical form but also about sensing capabilities. Grippers equipped with force sensors can provide feedback on the contact forces exerted, which is vital for tasks involving fragile objects or for achieving precise force control during the grasp. Tactile sensors embedded in gripper fingers can offer rich information about surface texture, slip detection, and object shape, which are invaluable for grasping unknown objects or for adapting to uncertainties.

Once the end-effector is chosen, the next step is to

determine the optimal grasp points on the object. This is where perception and geometric reasoning intersect. For known objects with pre-defined CAD models, grasp planning algorithms can computationally search for stable grasp configurations. These algorithms often discretize the object's surface and evaluate potential contact points based on their proximity, orientation, and the predicted stability of the resulting grasp. Techniques like analyzing the object's geometry to identify graspable features—edges, planar surfaces, concavities—are employed. For example, if a robot is tasked with picking up a cubical box, algorithms might identify the edges as potential grasp locations for an edge grasp, or the flat surfaces for a planar grasp. The selection is then guided by stability metrics. The objective is to find a set of contact points and associated gripper finger poses that result in a grasp with a high degree of force closure or form closure, as discussed previously.

In scenarios where objects are not pre-modeled, or when grasping in cluttered environments, vision plays a pivotal role. Sophisticated computer vision algorithms are used to recognize the object, estimate its pose (its position and orientation in 3D space), and sometimes even infer its shape and graspable features. Techniques such as template matching, feature-based recognition (e.g., using SIFT or SURF descriptors), and more recently, deep learning-based object detection and segmentation models, are employed. Once the object is localized and its pose is known, the grasp planning can proceed, often using generalized grasp affordance models that can predict stable grasps for a class of objects or even for previously unseen objects based on their visual appearance. For instance, a robot might be trained on a dataset of successful grasps for various objects, and a convolutional neural network (CNN) can then learn to predict grasp parameters (like gripper width, orientation, and finger positions) directly from an image of a target object.

The actual execution of the grasp is a tightly coupled process involving both grasp planning and motion planning for the robot arm. The motion planner must ensure that the robot's arm and end-effector can reach the planned grasp configuration without colliding with the environment or the object itself. This involves generating a collision-free trajectory from the robot's current configuration to the grasp approach pose. The approach path is critical: it must be designed to avoid any premature contact that could disturb the object or lead to an unstable grasp. For example, when approaching an object on a flat surface, the gripper might be positioned above the object and then lowered vertically, or it might approach at an angle that avoids collision with the object's sides before reaching the grasp points.

During the execution of the grasp, real-time feedback from sensors is often indispensable. Force sensors in the gripper can monitor the applied forces, allowing the robot to adjust its grip to avoid crushing delicate objects or to ensure sufficient contact force for a secure hold. Tactile sensors can detect slippage, prompting the robot to re-adjust its grip or even abort the grasp and replan. Vision systems can also be used to monitor the grasp execution online, for example, by verifying that the gripper fingers are positioned as intended relative to the object. This closed-loop control is essential for robust grasping in dynamic or uncertain environments.

Let us consider a typical industrial scenario: an automated assembly line where a robot is tasked with picking small electronic components from a feeder tray and placing them onto a printed circuit board. Here, grasp planning is highly refined. The components are typically known, with precise CAD models available. Vision systems, often integrated with the robot's end-effector, are used for accurate localization and pose estimation of the components within the tray. This is crucial because even slight misalignments in the tray or component orientation can render a pre-programmed grasp unsuccessful. The planning algorithm, leveraging the 3D model and the perceived pose, identifies optimal grasp points. For small surface-mount components, this might involve a vacuum gripper that applies suction to a flat surface, or a micro-gripper that pinches a specific edge or feature. The motion planner generates a trajectory for the robot arm to move from its previous position, approach the component without collision, execute the grasp by activating the gripper, and then withdraw from the tray. During withdrawal, force feedback might be used to ensure the component is securely held and not dislodged. On the assembly line, the entire process happens at high speed, requiring efficient and reliable grasp planning and execution.

In research laboratories, the challenges often extend beyond simple pick-and-place operations. Consider a robot tasked with grasping an object from a bin of randomly oriented items. This is a classic bin-picking problem. Here, the robot's vision system, often a combination of a 2D camera and a depth sensor (like a structured light or time-of-flight camera), captures a point cloud of the bin's contents. Algorithms then process this point cloud to segment individual objects, identify their 3D poses, and crucially, predict stable grasps. Since the objects are randomly oriented, the planner must consider a wide range of potential grasps. Often, a "sense-and-adapt" approach is employed. The vision system identifies potential grasp locations, and the robot attempts to grasp at one of these locations. If the grasp fails (detected by force sensors or lack of object acquisition), the robot might try a different grasp, perhaps from a different angle, or ask for human assistance. Advanced methods involve exploring a probability distribution of grasps and learning to identify the most promising ones from partial sensory data.

Another research frontier is grasping deformable objects, such as fabrics or cables. Planning a grasp for a flexible object requires understanding its mechanical properties and how it will deform under manipulation. Grasping a cable to plug it into a port, for instance, necessitates not just finding a stable grip but also controlling the cable's shape and flexibility to guide it into the connector. This often involves complex feedback control strategies that combine visual servoing with force sensing to maintain appropriate tension and alignment.

The integration of grasp planning with task-level planning is also an active area. Instead of simply picking up an object, the robot might need to perform a sequence of manipulations. For example, in a screwing task, the robot must first grasp the screw, then insert it into a hole, and then rotate it. The initial grasp choice will significantly influence the feasibility and efficiency of subsequent operations. A grasp that provides good torque transmission might be preferred over one that is merely stable for picking. This "task-aware grasping" requires a more holistic approach where grasp selection is optimized in the context of the entire task.

In summary, grasp planning and execution represent the bridge between the intelligence of grasp selection and the physical reality of robotic manipulation. It encompasses the critical steps of choosing the right end-effector, identifying optimal grasp points using perception and geometric reasoning, planning collision-free trajectories for the robot arm, and executing the grasp with sensor feedback to ensure success. From the highly optimized pick-and-place operations in industrial automation to the more adaptive and intelligent grasping required in research settings for unknown or deformable objects, the principles of efficient and robust grasp planning and execution are fundamental to the advancement of robotic capabilities. The continuous evolution of sensing technologies, computational power, and algorithmic sophistication is steadily enhancing the robot's ability to confidently and skillfully interact with the physical world.

Chapter 11: Multi-Agent Systems and Coordination

The exploration of robotic systems has, thus far, largely focused on the capabilities and complexities of individual robotic agents. We have delved into the intricate details of how a single robot perceives its environment, plans its actions, and executes tasks, from sophisticated grasp planning to the nuanced understanding of object properties and stability. However, the frontiers of robotics are increasingly being defined not by the prowess of solitary machines, but by the collective intelligence and coordinated efforts of multiple robots working in concert. This shift ushers us into the domain of Multi-Robot Systems (MRS), a field dedicated to understanding, designing, and implementing systems composed of two or more robots that collaborate to achieve objectives that would be difficult, inefficient, or impossible for a single robot to accomplish.

The concept of multi-robot systems is as old as the idea of teamwork itself, and it is finding ever more sophisticated manifestations in the realm of robotics. At its core, an MRS is a collection of intelligent agents, each possessing its own sensing, processing, and actuation capabilities, that are networked and programmed to interact in a shared environment to solve a common problem. This shared problem can range from mapping an unknown territory to assembling complex structures, from exploring hazardous environments to performing intricate logistical operations. The fundamental premise is that by pooling resources, sharing information, and coordinating actions, a group of robots can achieve a higher level of performance, robustness, and adaptability than any individual robot acting alone. This collaborative paradigm is not merely an academic pursuit; it is a critical enabler for many of the most ambitious applications of robotics envisioned for the future, from autonomous fleets of delivery drones to coordinated teams of planetary rovers, and from advanced manufacturing assembly lines to large-scale environmental monitoring.

To better understand the landscape of multi-robot systems, it is useful to categorize them based on the nature of their constituent agents. One primary distinction lies between
homogeneous and **heterogeneous** systems. In a homogeneous multi-robot system, all the robots are identical. They possess the same hardware, the same sensing capabilities, and the same processing power. Think of a swarm of identical drones performing aerial surveillance or a fleet of identical autonomous vehicles navigating a warehouse. The advantage of homogeneity lies in its simplicity of design, deployment, and control. Algorithms can often be developed and tested assuming uniform robot capabilities, and spare parts or maintenance procedures are standardized. This uniformity also simplifies fault tolerance; if one robot fails, another identical robot can seamlessly take over its role without significant adaptation. However, homogeneous systems can be less versatile. If a task requires a specific capability that none of the robots possess, the entire system may be unable to perform it.

In contrast, a heterogeneous multi-robot system

comprises robots with diverse capabilities. This diversity can manifest in various ways: robots might differ in their physical form factors (e.g., a wheeled robot and a flying drone), their sensing modalities (e.g., one robot with a high-resolution camera, another with a thermal sensor), their manipulation abilities (e.g., a robot with a powerful gripper and another with delicate manipulators), or their computational resources. An example could be a team of robots designed for disaster response: a ground robot equipped with sensors for detecting hazardous materials, a flying drone for aerial reconnaissance and providing an overhead view, and a robot with a heavy-lifting capacity for clearing debris. Heterogeneous systems offer significant advantages in terms of task specialization and overall system performance. By assigning tasks to the robot best suited for them, efficiency and effectiveness can be greatly enhanced. The overall system can tackle a wider range of complex problems because the combined capabilities of its diverse members exceed the limitations of any single robot. However, designing, controlling, and coordinating heterogeneous systems is inherently more complex. Task allocation algorithms must account for varying robot skills and resource availabilities, and communication protocols need to be robust enough to handle potentially different data formats and processing capabilities among the robots. Furthermore, ensuring interoperability and smooth collaboration between agents with disparate characteristics presents a substantial engineering challenge.

The appeal of multi-robot systems stems from a

confluence of compelling advantages that extend beyond mere redundancy. Perhaps the most intuitive benefit is
increased efficiency and speed. When a task can be broken down into smaller, parallelizable sub-tasks, multiple robots can work on these sub-tasks simultaneously, drastically reducing the overall completion time. Consider the task of mapping a large area. A single robot would have to traverse the entire area systematically. However, if multiple robots are deployed, the area can be divided among them, and they can map their respective zones concurrently, leading to a significantly faster overall mapping process.

Another critical advantage is
enhanced robustness and fault tolerance. In any complex system, failures are inevitable. In a single-robot system, the failure of that robot means the complete cessation of the task. In a multi-robot system, especially one with some degree of redundancy, the failure of one or even a few robots does not necessarily bring the entire mission to a halt. Other robots can often pick up the slack, adapt their roles, and continue the operation, albeit perhaps with reduced efficiency. This resilience is invaluable in applications where mission success is paramount, such as in military operations, search and rescue, or critical infrastructure monitoring. This redundancy can also extend to critical components within a robot; if one robot has a backup, it's more robust.

Multi-robot systems also offer significant

scalability. As the scope of a task increases, or as the requirements for coverage or capacity grow, additional robots can be added to the system to meet these demands. This "plug-and-play" scalability allows systems to adapt to evolving mission requirements without needing to redesign the entire architecture or replace individual robots with more powerful, but potentially more expensive, single units. For example, a fleet of cleaning robots in a large facility can be scaled up by simply adding more robots as the floor area increases.

Furthermore, MRS can enable
new capabilities that are simply not possible with single robots. This can include cooperative manipulation, where multiple robots must work together to lift or maneuver a large or unwieldy object. Imagine two robotic arms working in tandem to assemble a car chassis or a team of robots coordinating to form a temporary bridge. Another example is distributed sensing, where the collective information from multiple sensors, possibly of different types and strategically placed, provides a more comprehensive and accurate understanding of the environment than any single sensor could. Cooperative exploration, where robots coordinate their movements to map unknown terrain efficiently and safely, is another such capability.

However, the power of multi-robot systems comes with a unique set of formidable challenges, primarily centered around
coordination. Unlike a single robot that operates autonomously, multiple robots must coordinate their

actions to avoid conflicts, optimize performance, and achieve their shared goals. This coordination can be broadly broken down into several key areas:

Communication is the lifeblood of any multi-robot system. Robots need to exchange information about their state, their intentions, the environment, and the tasks they are performing. This can range from simple status updates to complex negotiation protocols. The design of effective communication protocols is critical. Issues such as bandwidth limitations, communication latency, reliability of the communication links (especially in environments with obstacles or electromagnetic interference), and the sheer number of robots that need to communicate can pose significant hurdles. For instance, in a swarm of hundreds or thousands of small, simple robots, broadcasting detailed information from every robot to every other robot would be infeasible. Instead, communication might need to be localized (robots only communicate with their neighbors) or hierarchical. The choice between centralized communication (where a base station coordinates all information flow) and decentralized communication (where robots communicate directly with each other) also has profound implications for robustness and scalability. A centralized system is easier to manage but is a single point of failure, while a decentralized system is more robust but significantly more complex to design and control.

Task Allocation is the process of assigning specific tasks or sub-tasks to individual robots within the system. This is particularly critical in heterogeneous systems, where different robots have different capabilities, but it is also relevant in homogeneous systems to ensure efficient workload distribution. An effective task allocation mechanism must consider the capabilities of each robot, its current workload, its proximity to the task, and the overall system objective. For example, if a robot is already performing a high-priority task, it should not be assigned another demanding task. Conversely, if a robot is idle, it should be prioritized for available tasks. Algorithms for task allocation can range from simple greedy approaches (assigning the task to the first available robot that can perform it) to more sophisticated optimization techniques that aim to minimize overall completion time or resource consumption. Issues like dynamic reassignment (when a robot fails or a new task emerges) and fair distribution of work also need to be addressed.

Collision Avoidance is paramount for the safe and effective operation of any MRS, especially in dynamic environments or when robots are performing tasks in close proximity. Robots must be able to sense each other and the static environment and then plan their trajectories to avoid collisions. This is not just about preventing physical impact; it also involves avoiding situations where robots might impede each other's progress or block access to critical areas. Cooperative collision avoidance strategies, where robots actively coordinate their movements to ensure deconfliction, are often necessary. This can involve robots signaling their intended paths, negotiating right-of-way, or dynamically adjusting their trajectories based on the predicted movements of other agents. The complexity increases significantly with the number of robots and the density of the environment.

Beyond these core challenges, other critical aspects of multi-robot coordination include
localization and mapping (especially when robots need to build a common map of an unknown environment or maintain accurate positions relative to each other), **formation control** (maintaining specific geometric arrangements of robots, such as lines or formations for coordinated movement or coverage), and **behavioral coordination** (ensuring that individual robot behaviors aggregate into coherent and effective system-level actions). The field of multi-agent systems and coordination is thus a rich area of research, exploring how to imbue these collections of robots with the intelligence and ability to act as a unified, effective team.
The successful operation of any multi-robot system hinges on the seamless flow of information between its constituent agents. Without effective communication, robots would be little more than independent entities, unable to leverage the synergistic potential that defines the very essence of multi-robot collaboration. This section embarks on a detailed exploration of the multifaceted landscape of communication within multi-robot systems, examining the diverse architectural paradigms, the inherent challenges, and the

sophisticated solutions that enable these systems to function cohesively.

At the heart of multi-robot communication lies the need for agents to share critical information. This information can encompass a wide spectrum, from the mundane yet vital status updates - such as battery levels, operational modes, and sensor readings - to the strategic intelligence required for collaborative task execution. Robots must be able to convey their perceived environment, their intended actions, and their progress towards shared objectives. The efficacy of this information exchange directly dictates the system's ability to adapt to dynamic environments, resolve ambiguities, and ultimately achieve its goals with optimal performance.

Several fundamental architectural paradigms govern how robots communicate within a multi-robot system. Each paradigm presents a distinct trade-off between simplicity, robustness, scalability, and efficiency.

One of the most straightforward approaches is **centralized communication**. In this model, a single, designated entity - often referred to as a base station, mission planner, or control node - acts as the central hub for all information flow. Robots exclusively communicate with this central entity, sending their data and receiving instructions from it. The central node is responsible for aggregating information from all agents, performing global analysis, and disseminating coordinated commands. This architecture offers several advantages. It simplifies the design of individual robot communication modules, as they only need to interface with one other entity. It also provides a single point for sophisticated global decision-making and task allocation, leveraging the potentially greater computational power of the central node. Furthermore, it can simplify debugging and monitoring, as all system activity is channeled through a single point.

However, the centralized model is fraught with significant drawbacks. The most critical is its inherent **single point of failure**. If the central communication node fails, the entire multi-robot system can become paralyzed, unable to coordinate its actions or receive new directives. This vulnerability is unacceptable in many critical applications where system continuity is paramount. Another major concern is **scalability**. As the number of robots increases, the central node can become overwhelmed by the sheer volume of incoming data and the demand for outgoing commands. This can lead to significant communication bottlenecks, increased latency, and reduced system responsiveness. The limited bandwidth of the central node's communication link can also become a significant constraint, especially when robots are generating large amounts of sensor data. In essence, the centralized architecture struggles to scale beyond a moderate number of agents or to handle highly data-intensive operations without substantial investment in high-capacity communication infrastructure for the central node. Examples of this paradigm might be found in early drone swarm demonstrations where a single operator's computer dictated every drone's movement, or in warehouse logistics systems where a central server manages the precise movements and task assignments of a fleet of autonomous mobile robots.

In stark contrast, **decentralized communication** architectures distribute communication and decision-making responsibilities among the individual robots. In a purely decentralized system, there is no single point of control or communication hub. Instead, robots communicate directly with each other, often in a peer-to-peer fashion. This can involve robots broadcasting information to all other agents within range, or more targeted communication to specific neighbors. Decision-making authority is also distributed, with each robot potentially capable of making local decisions based on the information it receives from its peers.

The primary advantage of decentralized architectures is their **inherent robustness and fault tolerance**. Since there is no single point of failure, the system can continue to operate even if some individual robots or communication links fail. The failure of one or even several agents does not necessarily cripple the entire system. This makes decentralized systems highly desirable for applications in unpredictable or hazardous environments where system resilience is crucial. Decentralization also offers superior **scalability**. As more robots are added, the communication load is distributed among them, avoiding the bottlenecks associated with centralized systems. Each robot only needs to handle communication with a limited number of its neighbors, rather than the entire swarm. This allows for the graceful scaling of systems to hundreds or even thousands of agents.

However, decentralized communication and control introduce significant complexity. Designing algorithms that enable effective coordination among a large number of autonomous agents, each making local decisions, is a formidable challenge. Ensuring that these local actions aggregate into coherent, system-level behavior requires sophisticated algorithms for consensus building, information sharing, and conflict resolution. Debugging and understanding the emergent behavior of a highly decentralized system can also be far more difficult than in a centralized system. Furthermore, ensuring that all robots possess the necessary computational resources and communication capabilities to participate effectively in a decentralized network is critical. Examples of decentralized communication are prevalent in many bio-inspired swarm robotics research, where simple robots interact with their immediate neighbors to achieve emergent global patterns, such as flocking or foraging. Another example is in formations of autonomous vehicles where each vehicle communicates with its adjacent vehicles to maintain spacing and avoid collisions without a central commander.

A middle ground, often referred to as

distributed communication, seeks to balance the benefits of both centralized and decentralized approaches. In a distributed architecture, there may not be a single monolithic control node, but rather a hierarchy or network of interconnected control entities. Alternatively, robots might communicate in a more structured, network-based fashion where information is relayed through a series of intermediate agents or within defined communication neighborhoods. This can involve concepts like multi-hop communication, where data is passed from one robot to another to reach its destination, or cluster-based communication, where robots are organized into groups, each with a representative agent responsible for inter-group communication.

Distributed architectures can offer a more nuanced approach to scalability and robustness. For instance, in a clustered system, the failure of a single robot might only affect its cluster, while the rest of the system continues to operate. The system can be made more robust by having redundant cluster heads or by allowing robots to dynamically re-cluster. Scalability can be improved by increasing the number of clusters or by allowing clusters to merge or split as needed. This paradigm can also facilitate more efficient information dissemination. Instead of a single point broadcasting to everyone, information can be routed more intelligently through the network, potentially reducing bandwidth requirements and latency.

The practical implementation of any of these

communication architectures faces a host of real-world challenges. The nature of the operating environment can profoundly impact communication reliability. In indoor environments with structured layouts and Wi-Fi infrastructure, communication might be relatively stable. However, in outdoor environments, or those with significant physical obstacles, electromagnetic interference, or dynamic changes (like moving vehicles or fluctuating foliage), communication links can become intermittent, degraded, or completely lost. This unreliability necessitates robust communication protocols that can handle packet loss, corrupted data, and temporary disconnections.

Limited bandwidth is another pervasive issue, especially in systems with a large number of agents or those generating high-resolution sensor data (e.g., from cameras or LiDAR). Robots may need to transmit detailed environmental maps, precise state information, or video feeds, all of which consume significant bandwidth. Efficient data compression, intelligent data filtering (only sending what is necessary), and prioritizing critical information are essential strategies for managing bandwidth constraints.

Latency, the time delay between sending a message and receiving a response, is a critical factor, particularly for real-time control applications. If robots are reacting to each other's movements, a significant communication latency can lead to outdated information, resulting in misjudgments and potential collisions or task failures. For instance, in a cooperative manipulation task where two robots are lifting a heavy object, a delay in one robot sensing the other's movement could cause it to apply force at the wrong time, leading to instability or damage. Minimizing latency often involves using high-speed communication hardware, optimizing network topology, and employing predictive control strategies that anticipate future states based on current information, even if slightly delayed.

To address these challenges, researchers and engineers have developed a variety of communication protocols and architectural patterns. One crucial aspect is the definition of the
communication protocol stack, which dictates how data is formatted, addressed, transmitted, and received. This typically involves several layers, from the physical layer (how bits are transmitted over the air or wires) to the application layer (how robots interpret the exchanged information). Standardized protocols like TCP/IP are often adapted, but specialized protocols tailored for multi-robot systems are also common, especially in resource-constrained or highly dynamic environments.

For swarm robotics, which often involves a large number of simple, low-cost robots, communication protocols are designed with scalability and energy efficiency in mind.
Probabilistic communication or **stochastic protocols** are sometimes employed, where information is not guaranteed to reach every agent, but rather with a

certain probability. This can be sufficient for collective behaviors where local interactions propagate information throughout the swarm. **Neighborhood-based communication** is another common strategy, where robots only exchange information with their immediate neighbors. This limits the range of communication and the number of messages each robot must process, contributing to scalability. Techniques like **event-triggered communication** can also reduce overhead; instead of robots continuously broadcasting their status, they only send messages when a significant event occurs (e.g., detecting a new obstacle, completing a sub-task).

In the context of fleet management for autonomous vehicles or industrial robots, communication architectures often prioritize reliability and efficient task management.
Message Queuing Telemetry Transport (MQTT), a lightweight publish-subscribe messaging protocol, is frequently used. It allows robots (publishers) to send messages to specific "topics," and other robots or the central system (subscribers) can receive these messages without needing direct connections. This decouples the sender and receiver, improving flexibility and scalability. **Data serialization formats** are also critical for efficient data exchange. Formats like Protocol Buffers or Avro can be more efficient than plain text formats like JSON or XML, reducing message size and parsing time, which is crucial for limited bandwidth and low-latency applications.

Beyond simple data exchange, multi-robot systems often require mechanisms for **data sharing and consensus building**. When multiple robots are building a map of the same environment, they need to share their partial maps and integrate them into a coherent, unified map. This requires algorithms for map merging and the resolution of conflicting information. Similarly, in collaborative decision-making, robots may need to reach a consensus on the best course of action. This can involve iterative message passing, voting mechanisms, or more complex negotiation protocols, especially when robots have different perspectives or preferences.

For example, consider a team of robots tasked with searching a disaster zone. Each robot might be equipped with different sensors (e.g., thermal cameras, audio sensors, visual cameras). To effectively coordinate the search, they need to share their sensor data. A robot with a thermal camera might detect a heat signature, and this information needs to be communicated to other robots, perhaps a robot with a visual camera that can attempt to confirm the source. A consensus might then be needed on the priority of investigating this potential survivor. The communication architecture must support the efficient sharing of diverse sensor data and enable the team to agree on a prioritized action plan, all while dealing with potential communication dropouts and the need for real-time decision-making.

In applications like aerial surveillance using drones, **swarm communication protocols** play a pivotal role. Drones might operate in formations, sharing their flight paths and observed targets. A common strategy is to use a combination of local and global communication. Each drone broadcasts its position and status to its immediate neighbors. This local communication is crucial for maintaining formation integrity and avoiding collisions. Periodically, or when a significant event occurs (e.g., detecting a target of interest), drones might report this information to a central command and control system or relay it through a designated "leader" drone to a wider network. This tiered approach allows for both real-time local coordination and the aggregation of high-level information for mission command.

The design of communication architectures must also consider the
resource constraints of individual robots. Small, low-power robots, like those found in some swarm applications, may have limited processing power and battery life. Their communication modules might be simple, low-bandwidth transceivers. In such cases, the communication protocol must be designed to minimize computational load and energy consumption. This often means relying on simple, local interactions rather than complex, long-range communication or heavy data processing.

Furthermore, as robots become more autonomous and

operate in increasingly complex environments, the need for
semantic communication is emerging. This goes beyond simply transmitting raw sensor data or state information. Semantic communication aims to exchange information in a more abstract, context-aware manner. For instance, instead of sending a raw image of a door, a robot might communicate "door detected, status: closed, requires opening." This richer form of communication can reduce the amount of data transmitted and enable more intelligent collaborative decision-making, as robots can understand the *meaning* of the information being shared.

In summary, communication in multi-robot systems is a cornerstone that enables collaboration, coordination, and collective intelligence. While centralized architectures offer simplicity, they suffer from a lack of robustness and scalability. Decentralized architectures excel in robustness and scalability but introduce significant design complexity. Distributed architectures attempt to strike a balance, offering a more nuanced approach. The practical realization of these architectures is fraught with challenges, including unreliable links, limited bandwidth, and latency, all of which necessitate the development of sophisticated communication protocols, efficient data management strategies, and robust consensus-building mechanisms. The ongoing evolution of multi-robot systems will undoubtedly continue to drive innovation in communication technologies and paradigms, enabling ever more complex and ambitious collective robotic endeavors.

The effective deployment of multi-robot systems hinges not only on their ability to communicate and coordinate their actions but crucially on how tasks are assigned and distributed amongst them. In a collaborative environment, where multiple agents work towards a common objective, the process of allocating responsibilities becomes paramount for achieving efficiency, optimizing performance, and ensuring the successful completion of the mission. This subsection delves into the intricate problem of task allocation and assignment, exploring various methodologies that enable robotic teams to judiciously distribute work, thereby maximizing their collective capabilities.

At its core, task allocation in multi-robot systems can be framed as an optimization problem. The objective is typically to achieve the best possible system-wide outcome, which can be defined in several ways: minimizing the overall time to complete all assigned tasks (makespan minimization), maximizing the total value of completed tasks, minimizing the cost of task execution, or ensuring an equitable distribution of workload among the robots to prevent overburdening any single agent. The complexity arises from the fact that robots possess heterogeneous capabilities, tasks may have dependencies, and the environment itself can be dynamic, leading to the need for sophisticated allocation strategies.

One of the foundational approaches to static task allocation is the

Assignment Problem. This classic combinatorial optimization problem seeks to find a one-to-one mapping between a set of tasks and a set of agents (robots) such that the total cost (or value, depending on the formulation) is minimized (or maximized). In the context of robotics, if we have m tasks and n robots, and we can define a cost matrix C, where C_{ij} represents the cost (e.g., time, energy, risk) of robot i performing task j, the goal is to select one task for each robot (or vice versa, if $m \neq n$) such that the sum of the costs of the assigned tasks is minimized.

The most renowned algorithm for solving the assignment problem is the
Hungarian Algorithm. This algorithm efficiently finds an optimal assignment in polynomial time. It works by transforming the cost matrix through a series of row and column reductions and then identifying a maximum matching in a bipartite graph constructed from the reduced matrix. For instance, consider a scenario where a team of four robots needs to perform four distinct reconnaissance missions in different zones. Each robot has varying travel times and sensor capabilities, leading to different costs (estimated mission duration) for each robot to cover each zone. The cost matrix might look like this:

$$
C = \begin{pmatrix}
c{11} & c{12} & c{13} & c{14} \\
c{21} & c{22} & c{23} & c{24} \\

c{31} & c{32} & c{33} & c{34} \\
c{41} & c{42} & c{43} & c{44}
\end{pmatrix}
$$

Where $c
*{ij}$ is the estimated time for robot i to complete mission j. The Hungarian algorithm would process this matrix to determine the assignment of missions to robots that results in the shortest total mission completion time for the entire team. For example, if the optimal assignment is robot 1 to mission 2, robot 2 to mission 4, robot 3 to mission 1, and robot 4 to mission 3, the total completion time would be $c*{12} + c{24} + c{31} + c_{43}$. The beauty of the Hungarian algorithm is that it guarantees finding the absolute minimum total time.

However, the classical assignment problem assumes a static environment and fixed task requirements. In many real-world robotic applications, tasks arrive dynamically, robots become unavailable due to faults or battery depletion, and task priorities can change. This necessitates more flexible and adaptive task allocation mechanisms.

Auction-based task allocation provides a powerful framework for dynamic task assignment. In this paradigm, tasks are treated as "items" to be auctioned, and robots are "bidders." The auction process typically proceeds in rounds:

1.
Task Announcement: Available tasks are announced to the robot team.
2. **Bidding**: Each robot evaluates the tasks it can perform and submits bids based on its own cost or utility for completing each task. The bid usually reflects the robot's estimated cost to perform the task, its current availability, and potentially its priorization for certain task types.
3. **Awarding**: The task is awarded to the robot that submits the highest bid (in a value maximization scenario) or the lowest bid (in a cost minimization scenario), subject to certain rules to ensure fairness and efficiency. Often, a robot winning an auction commits to performing that task, temporarily removing it from its available bandwidth.

This auction mechanism can be implemented in various ways. A common approach is a **simultaneous, single-item auction**, where each task is auctioned independently. Robots can bid on multiple tasks, but they typically can only win one task at a time. Alternatively, a **multi-item auction** allows robots to bid on and potentially win multiple tasks. More sophisticated mechanisms involve **sequential auctions**, where tasks are auctioned off one by one, or **contract net protocols**, which generalize the auction concept by allowing robots to bid on complex task bundles and negotiate terms.

For example, consider a fleet of delivery robots. New delivery requests (tasks) continuously arrive. When a new request is made, it's announced. Each robot, based on its current location, battery level, and route, calculates the "cost" of taking on this delivery (e.g., time to reach the pickup location, travel time to the destination, energy consumed). Robots then submit bids representing this cost. The system (or a designated auctioneer robot) awards the task to the robot with the lowest bid. If the winning robot has already committed to other tasks, it might withdraw its bid, and the auction for that task might be re-run among the remaining bidders. This dynamic process ensures that tasks are assigned to robots that can complete them most efficiently at any given moment.

A key advantage of auction-based mechanisms is their ability to handle dynamic environments and heterogeneous robot capabilities. Robots can bid based on their individual strengths and limitations, leading to efficient assignments that might not be obvious through static optimization. Furthermore, these mechanisms can be decentralized, with robots communicating directly to bid and negotiate, enhancing robustness.

Another important dimension of task allocation is the distinction between
centralized and **decentralized** approaches. In a **centralized task allocation** system, a single entity (e.g., a base station or a designated robot) is responsible for collecting task information and assigning them to all robots in the team. This approach can leverage global knowledge of all tasks and robot states to make optimal

decisions, similar to how the Hungarian algorithm operates on a global cost matrix. However, it suffers from the same drawbacks as centralized communication: it can be a single point of failure, and it may become a bottleneck as the number of robots and tasks grows.

Decentralized task allocation, on the other hand, distributes the decision-making process among the robots themselves. Each robot makes independent decisions about which tasks to undertake, often based on local information and communication with its neighbors. This can be achieved through various protocols. One method is **market-based allocation**, where robots form a market for tasks, exchanging services and payment (or bids) for task completion. Another approach is **distributed constraint optimization (DCOP)**, where each robot is an agent trying to satisfy its own objectives while respecting constraints imposed by other agents and the overall system goals. DCOPs are often solved iteratively, with agents communicating their decisions and preferences to reach a joint solution.

Consider a scenario of search and rescue robots in a large, unknown area. A centralized approach would require all robots to report their findings and receive task assignments from a central controller. This could be slow and prone to failure if the controller is compromised. In a decentralized system, robots might explore their local areas and, upon detecting something of interest, broadcast this information. Other robots in proximity could then bid to investigate further, perhaps based on their sensor capabilities or proximity. This self-organization allows the team to adapt quickly to new discoveries and operate even if communication with a central authority is lost.

When designing decentralized task allocation strategies, several factors come into play.
Task dependencies are crucial: if task B can only be performed after task A is completed, this dependency must be accounted for in the allocation process. This can lead to complex planning and coordination, especially when dealing with multiple interdependent tasks. **Robot capabilities** are also a major consideration. A robot equipped with a manipulator arm might be best suited for assembly tasks, while a robot with advanced sensing might be assigned reconnaissance missions. Allocation algorithms must be able to match tasks to robots that possess the necessary skills or can acquire them.

Moreover, the concept of
dynamic re-tasking is essential for robust systems. If a robot fails to complete its assigned task, or if a higher-priority task emerges, the allocation system must be able to adapt. This might involve re-assigning the failed task to another robot or preempting lower-priority tasks to address the new urgent need. Auction-based mechanisms are particularly well-suited for this, as failed tasks can be re-auctioned, and new tasks can be introduced into the bidding process.

Let's consider an example of dynamic re-tasking in a team of autonomous mobile robots in a warehouse. Suppose robots are assigned to pick and deliver inventory items. If a robot assigned to retrieve item X breaks down midway, the task of retrieving item X needs to be reassigned. The system could detect this failure and immediately re-auction item X to other available robots. Alternatively, if a critical order with a very short deadline arrives, the system might need to re-evaluate current assignments. Robots might be instructed to pause their current, lower-priority tasks and bid on the critical order, potentially leading to a temporary shift in the entire team's focus. This adaptability is vital for maintaining operational efficiency in real-world logistics.

Task partitioning is another aspect of task allocation. For very large or complex tasks, it might be beneficial to divide the task into smaller sub-tasks that can be performed concurrently by different robots or by the same robot in sequence. For instance, building a large structure might involve sub-tasks like material transport, component assembly, and structural welding. A sophisticated allocation system would partition the overall objective into these manageable sub-tasks and then assign them optimally to the available robots. This is closely related to hierarchical task networks, where complex tasks are decomposed into a tree of simpler actions.

The choice of allocation algorithm often depends on the specific application domain and the desired trade-offs. For simple, static assignments, the Hungarian algorithm might suffice. For dynamic environments with continuous task arrivals, auction-based or market-based mechanisms are more appropriate. For highly collaborative tasks requiring tight synchronization, more complex coordination protocols might be needed.

Furthermore, the integration of task allocation with other multi-agent system components is crucial. The information gathered through communication, such as environmental perceptions and robot states, directly feeds into the task allocation process. Conversely, the outcomes of task allocation – which robot is assigned which task – inform the coordination and execution phases. This intricate interplay highlights the holistic nature of multi-agent system design.

The problem of task allocation and assignment is not merely about distributing work; it is about orchestrating the collective intelligence and physical capabilities of a robotic team to achieve a common goal more effectively than any individual robot could alone. As multi-robot systems become increasingly prevalent in diverse applications ranging from space exploration and disaster response to manufacturing and logistics, the development of efficient, robust, and scalable task allocation mechanisms will remain a critical area of research and engineering. The ongoing evolution of algorithms, from classic optimization techniques to sophisticated multi-agent learning and negotiation strategies, continues to push the boundaries of what is possible for collaborative robotic systems.

The ability of multiple robots to work together seamlessly is foundational to many advanced multi-agent system applications. Beyond simply assigning tasks, a critical aspect of collaboration lies in how these agents move and position themselves relative to one another. This subsection delves into the realm of **Cooperative Navigation and Formation Control**, exploring the principles and techniques that enable robotic teams to navigate and maintain spatial arrangements for collective efficiency and effectiveness. Such coordination is vital for tasks ranging from distributed sensing and environmental mapping to collaborative manipulation and the physical transportation of large objects.

At its core, formation control aims to ensure that a group of robots maintains a specific geometric configuration, often referred to as a "formation." This formation can be as simple as a line or a grid, or as complex as a dynamic, adaptable shape dictated by the ongoing task. The primary goal is to achieve synchronized movement while preserving these relative positions, thereby enabling a suite of capabilities that would be impossible for individual robots acting in isolation. For instance, a swarm of drones might form a wide perimeter to collectively scan a large area for search and rescue operations, or a group of ground robots might arrange themselves into a tight formation to move a bulky object that none could move alone. The success of these endeavors hinges on robust methods for maintaining the desired spatial relationships, even in the face of external disturbances or internal system uncertainties.

One of the most intuitive approaches to formation control is the
leader-follower paradigm. In this setup, one robot is designated as the leader, and its trajectory dictates the overall movement of the team. The other robots, termed followers, are responsible for maintaining their predefined positions relative to the leader and/or their neighboring followers. This can be achieved through various control strategies. A simple implementation might involve each follower robot calculating its desired position based on the leader's current position and a fixed offset defining the formation. The follower then computes a control command (e.g., velocity or acceleration) to move towards this desired relative position.

Consider a scenario where a team of four robots is tasked with traversing a landscape, with one robot leading the way. The leader robot navigates the terrain, perhaps following a pre-planned path or reacting to environmental obstacles. The three follower robots are programmed to maintain a specific formation, for example, a triangular pattern behind the leader. Each follower would continuously monitor the leader's position and its own position. If a follower drifts too far from its designated spot in the formation, its controller would generate corrective commands to steer it back into alignment. This might involve using relative positioning sensors like GPS, lidar, or even visual markers. The control law for a follower might look something like:

$$v_f = K_p (p_{desired} - p_f) + K_d (\frac{d p_{desired}}{dt} - \frac{d p_f}{dt})$$

where v_f is the follower's velocity, $p_{desired}$ is its target position relative to the leader, p_f is the follower's current position, and K_p and K_d are proportional and derivative gains, respectively. The desired position $p_{desired}$ would be a function of the leader's position p_l and the static offset defining the formation. For example, if the formation is a column and the follower is supposed to be 5 meters behind the leader, $p_{desired} = p_l - [0, 5, 0]^T$ (assuming a 3D space and the leader is moving along the x-axis).

While the leader-follower approach is conceptually simple and effective for many applications, it has inherent limitations. The entire formation's behavior is dependent on the leader's performance. If the leader encounters an insurmountable obstacle or experiences a failure, the entire team's movement is compromised. Furthermore, if the leader's path is complex or involves sharp turns, the followers may struggle to keep up, especially if they have kinematic constraints (e.g., limited turning radius). This can lead to the formation breaking apart, particularly in dynamic environments.

To overcome these limitations and enhance robustness, **decentralized formation control** strategies have emerged. In these approaches, there is no single leader whose trajectory is exclusively followed. Instead, each robot's movement is determined by its interactions with its neighbors and potentially a global objective or desired formation shape. This distributed control architecture relies on local sensing and communication, making the system more resilient to individual robot failures and allowing for more agile adjustments to the formation.

One prominent class of decentralized methods is based on
neighbor-based repulsion and attraction forces. Inspired by biological systems like schooling fish or flocking birds (e.g., Reynolds' Boids model), these algorithms define simple rules for each agent:
1. **Separation**: Steer to avoid crowding local flockmates.
2. **Alignment**: Steer towards the average heading of local flockmates.
3. **Cohesion**: Steer to move toward the average position (center of mass) of local flockmates.

When adapted for robotic formations, these principles can be translated into control laws. Each robot adjusts its velocity based on the perceived positions and velocities of its immediate neighbors. For example, a robot might try to maintain a minimum distance from its neighbors (separation), align its velocity with the average velocity of its neighbors (alignment), and converge towards the average position of its neighbors to maintain the overall formation shape (cohesion). The desired formation can be implicitly encoded by the relative desired positions or offsets within these neighbor interactions.

Consider a line formation where robots are meant to be spaced equally along the x-axis. Robot i might have desired positions relative to its left neighbor ($i-1$) and right neighbor ($i+1$). Its control input would then be a combination of forces or velocity commands aimed at achieving these relative positions. For instance, robot i might try to stay d distance away from robot $i-1$ and d distance away from robot $i+1$. Its acceleration command could be a weighted sum of the errors in these distances. This decentralized approach allows for dynamic adjustments: if a robot needs to slow down or speed up, its neighbors will react accordingly, and the formation can adapt smoothly.

A more structured approach to decentralized formation control involves defining a
virtual structure. In this model, the desired formation is represented as a rigid geometric structure, but this structure does not have a physical leader. Instead, the robots are assigned to specific points within this virtual structure. Each robot then controls its motion to maintain its assigned position within the virtual structure, while the virtual structure itself might be

moved or reconfigured based on a global objective. This can be achieved by defining a consensus-based mechanism where robots negotiate to move the entire virtual structure as a collective.

For example, imagine a rectangular formation for a group of robots. Each robot is assigned a corner or a side position within this rectangle. The robots then communicate with their immediate neighbors to ensure they are occupying their designated relative spots. The entire rectangle might then be tasked with moving to a new location or rotating. This rotation or translation of the rectangle could be achieved through a distributed consensus algorithm where each robot attempts to reach a consensus on the desired new pose of the entire formation.

A critical aspect of cooperative navigation is **cooperative path planning**. While individual robots can plan their own paths to reach waypoints, coordinating these paths for a group of robots requires additional considerations, primarily to avoid collisions and ensure efficient traversal of the environment. In a multi-robot system, a collision can occur between any two robots in the team, or between a robot and a static obstacle.

When robots are operating in formation, the problem of collision avoidance becomes partially solved by the formation control itself. If the formation control law enforces a minimum separation distance between robots, then collisions within the team are less likely.

However, robots must also avoid collisions with external obstacles in the environment. This requires integrating the formation control with environmental sensing and obstacle avoidance algorithms.

One common strategy is to use a **hybrid approach**, where formation control dictates the relative positioning, and a separate obstacle avoidance layer manages deviations from the formation to navigate around unexpected barriers. If an obstacle is detected directly in the path of a robot or its intended formation position, that robot might temporarily break formation, maneuver around the obstacle, and then re-establish its position within the formation once the path is clear. This requires intelligent sensing and reactive control.

More sophisticated methods involve **joint path planning**, where the paths for all robots are planned simultaneously. This can be computationally intensive, especially for large teams, but it offers the potential for globally optimal solutions that minimize overall travel time, reduce the likelihood of deadlocks, and maximize the use of available space. Techniques like **Multi-Agent Path Finding (MAPF)** address this problem by formulating it as a combinatorial optimization challenge. MAPF algorithms aim to find paths for all agents such that no two agents occupy the same location at the same time. Many variations exist, including prioritized planning (where agents are assigned priorities to move sequentially), conflict-based search (where conflicts between agents' paths are identified and resolved), and flow-based methods.

For instance, consider a team of robots tasked with collecting samples from different locations in a complex laboratory environment. A centralized MAPF solver could take all the sample locations, robot starting positions, and environment layout as input, and then output a set of collision-free paths for each robot. However, in dynamic environments where new tasks or obstacles appear, such centralized planning might be too slow.

This leads to the concept of **decentralized or distributed path planning** for multi-robot systems. In a decentralized setting, each robot plans its own path while considering the potential paths and presence of its neighbors, often through limited communication. This can be achieved through various methods, such as:

Local Occupancy Grids: Each robot maintains an internal map of the environment, which includes known static obstacles and dynamically updated information about the positions of nearby robots. Path planning is then performed within this local map, with the assumption that other robots will also perform their own local planning and avoid conflicts.
Velocity Obstacles (VOs): This is a reactive approach where each robot determines a set of velocities that would lead to a collision with other agents within a

short time horizon. The robot then chooses a velocity that is not in any of the velocity obstacles, effectively avoiding immediate collisions. This method is particularly effective for dynamic environments. **Cooperative Sensing and Map Merging**: In tasks like exploration, robots may have incomplete knowledge of the environment. Cooperative navigation can involve robots sharing their sensor data and map information to build a common, more complete understanding of the world. This shared map can then be used for more effective path planning and formation control. For example, one robot might detect an obstacle in its vicinity, communicate this to its neighbors, and they can collectively adjust their paths or formation to navigate around it.

The effectiveness of decentralized strategies heavily relies on the communication topology and bandwidth. Robots need to exchange information about their positions, velocities, and intentions. A well-designed communication protocol ensures that this information is timely and accurate enough for robots to make informed decisions. For example, in a line formation, a robot might only need to communicate with its immediate left and right neighbors. In a more complex, dispersed formation, communication might need to extend to more distant robots to maintain overall cohesion.

Let's consider an example of a search and rescue scenario in a collapsed building. A team of small, agile robots equipped with cameras and sensors is deployed.

The objective is to explore the structure and locate any survivors. A decentralized approach would be highly beneficial here:

1.
Exploration: Robots move autonomously, using local obstacle avoidance and their sensing capabilities to explore different parts of the building. They might initially spread out to cover more ground.
2. **Cooperative Mapping**: As they explore, robots build local maps of the areas they visit. They periodically communicate their explored areas and any detected points of interest (e.g., open spaces, potential signs of survivors) to their neighbors. This allows for the merging of local maps into a more comprehensive global map.
3. **Formation for Access**: If a robot discovers a large open area that requires coordinated entry or a narrow passage that can only be navigated by multiple robots working in tandem (e.g., pushing aside debris), it can signal this. Other robots can then converge, form a temporary formation (e.g., a wedge to push debris, a line to transmit information through a narrow tunnel), execute the task collaboratively, and then re-disperse for further exploration.
4. **Collision Avoidance**: Throughout this process, each robot uses its local sensing and communication with neighbors to avoid collisions with both static debris and other robots. If two robots are about to enter the same space, they would communicate and one would defer, or they might adopt a temporary maneuvering formation to pass each other safely.

In such a scenario, maintaining a strict, rigid formation might be impractical due to the unpredictable and often confined nature of the environment. Instead, the "formation" becomes more fluid, dictated by the immediate task requirements and the need for safe passage and exploration. The principles of separation, alignment, and cohesion are still at play, but they are applied more flexibly. For instance, the "cohesion" might be about maintaining proximity within a local neighborhood to facilitate communication and mutual support, rather than adhering to precise geometric offsets.

Furthermore, cooperative navigation and formation control are deeply intertwined with **localization and mapping**. For robots to navigate and maintain formations, they need an accurate understanding of their own positions (localization) and the environment around them (mapping). In a multi-robot system, this can be a cooperative effort. **Cooperative Simultaneous Localization and Mapping (CSLAM)** techniques allow multiple robots to collaboratively build a map of their environment while simultaneously localizing themselves within that map. As robots share their sensor readings and their estimated poses, they can correct each other's errors, leading to a more accurate and robust map and more precise localization for the entire team. This enhanced situational awareness is crucial for effective formation control and path planning, especially in environments where GPS signals are unavailable or unreliable.

The ability to dynamically change formations is also a critical capability. A team of robots might start in a wide, dispersed formation for rapid exploration, then converge into a tighter formation to navigate a bottleneck, and then reconfigure into a line formation for efficient transit across an open area. This requires sophisticated control algorithms that can manage transitions between different formation types. Such transitions can be triggered by mission objectives, environmental conditions, or communication from a central planner.

Consider a scenario involving autonomous underwater vehicles (AUVs) surveying a large area of the seabed. They might initially deploy in a line abreast formation to cover a wide swathe of the ocean floor. As they approach a complex reef structure, they might transition to a more dispersed formation to avoid collisions and allow each AUV to independently navigate the intricate terrain. Once clear of the reef, they might regroup and form a column formation to return to the surface or transit to the next survey area efficiently. The control system must be able to execute these formation changes smoothly and safely, ensuring that no AUV gets too close to another or to the environment during the reconfiguration.

Finally, the robustness of cooperative navigation and formation control systems is often assessed based on their ability to handle

failures and uncertainties. This includes:

Robot Failures: If a robot fails, the remaining robots should be able to continue their mission, perhaps by adapting their formation or re-assigning tasks. Decentralized approaches are naturally more resilient to this.
Sensor Noise and Failures: Real-world sensors are imperfect. Algorithms must be robust to noisy measurements and occasional sensor failures.
Communication Loss: In many environments, wireless communication can be intermittent. Robots must be able to operate autonomously for periods without communication, relying on their last known information and local sensing.
Environmental Dynamics: Unexpected changes in the environment, such as moving obstacles or changing terrain, must be handled gracefully.

In summary, cooperative navigation and formation control are essential for unlocking the full potential of multi-robot systems. By enabling robots to move and position themselves in a coordinated manner, these techniques facilitate tasks that are beyond the capabilities of individual agents. From simple leader-follower arrangements to complex decentralized neighbor-based control, and from collective path planning to dynamic formation reconfigurations, the ongoing development in this field continues to pave the way for more sophisticated and capable robotic teams operating in an increasingly diverse range of applications. The seamless interplay between individual robot intelligence, inter-robot communication, and collective strategic movement is at the heart of achieving true collaborative autonomy.

As the sophistication of multi-agent systems escalates, the need for precisely coordinating the movements of multiple robots becomes paramount. While cooperative navigation and formation control, as discussed previously, focus on maintaining relative spatial arrangements for collective tasks and enhanced maneuverability, a related yet distinct challenge arises when each agent must independently navigate to a specific, often unique, destination while strictly avoiding collisions with all other agents. This is the domain of Multi-Agent Pathfinding (MAPF). MAPF moves beyond maintaining static or dynamic formations; it is fundamentally concerned with the problem of simultaneously finding collision-free paths for a group of agents, each with its own start and goal location, within a shared environment.

The core of the MAPF problem can be succinctly stated: given a set of agents, their initial positions, and their respective goal positions, compute a set of trajectories, one for each agent, such that no two agents occupy the same location at the same time throughout their journey. This constraint extends beyond avoiding static obstacles in the environment; it explicitly addresses the dynamic nature of the system itself – the agents themselves are obstacles to each other. The temporal aspect is crucial here: a collision can occur if agent A is at location (x, y) at time t, and agent B is also at location (x, y) at the same time t. Alternatively, collisions can also occur along edges, meaning if agent A is moving from location (x1, y1) to (x2, y2) between time t and t+1, and agent B is moving from (x2, y2) to (x1, y1) during the same time interval, this constitutes a collision. MAPF algorithms aim to generate paths that respect these spatial-temporal constraints, ensuring that the collective movement of the agents is safe and feasible.

The MAPF problem is typically formulated in discrete spaces, often represented as a grid graph, where agents can move between adjacent cells or remain stationary. The environment can be simple, like an open grid, or complex, featuring static obstacles that agents must also avoid. The goal is to find paths of minimal length (or some other cost metric, such as minimizing the sum of path lengths or the maximum path length) for all agents. The computational complexity of MAPF is notoriously high. In its most general form, it is PSPACE-hard, meaning that the computational resources required to solve it can grow exponentially with the number of agents and the size of the environment. This inherent complexity presents a significant challenge for real-world applications where fast, efficient, and scalable solutions are often required.

Several categories of algorithms have been developed to tackle the MAPF problem, broadly falling into rule-based methods, constraint-based approaches, and optimal search algorithms.

Rule-based methods are often inspired by simple, decentralized heuristics that guide agent behavior. One of the earliest and most intuitive approaches is **Prioritized Planning**. In this method, agents are assigned a fixed priority level, and then paths are planned sequentially. The highest priority agent plans its path first, treating all other agents as dynamic obstacles. Once its path is determined, the next highest priority agent plans its path, now considering the first agent's trajectory as a fixed constraint, and so on. This process continues down the priority list. While conceptually simple and relatively easy to implement, prioritized planning has several drawbacks. It does not guarantee optimality in terms of path length or makespan (the time until the last agent reaches its goal). Furthermore, the choice of priority order can significantly impact the quality of the solution, and a poor ordering can lead to long, suboptimal paths or even deadlocks where lower-priority agents become trapped by the movements of higher-priority agents. For instance, if a low-priority agent is blocked by multiple higher-priority agents who are all converging on a central area, it might be impossible for it to reach its goal without causing a collision.

Another family of rule-based methods involves **Rule-Based Systems** that use local interactions and simple collision avoidance rules. These often resemble some aspects of formation control, but with a focus on reaching individual goals. For example, an agent might move towards its goal, and if it detects another agent on a potential collision course, it might employ a simple maneuver such as waiting, yielding, or taking a slight detour. The **Velocity Obstacle (VO)** method is a prominent example of this reactive approach. For each agent, it computes a set of forbidden velocities that would lead to a collision with another agent within a certain time horizon. The agent then selects a velocity from the set of allowed velocities that is closest to its preferred velocity (e.g., the velocity that moves it most directly towards its goal). VO methods are efficient and can handle dynamic environments, but they typically focus on avoiding immediate collisions and do not explicitly plan for reaching goals in an optimal or globally collision-free manner over longer horizons. They are more suited for dynamic, reactive collision avoidance than for solving the full MAPF problem of finding complete, collision-free paths.

Constraint-based approaches, on the other hand, attempt to represent and resolve conflicts between agent paths more systematically. These methods often involve identifying specific points in time and space where two or more agents are predicted to collide and then imposing constraints to prevent these collisions. A significant advancement in this area is the **Conflict-Based Search (CBS)** algorithm. CBS operates on two levels: a high-level search that reasons about the constraints and a low-level search that finds paths satisfying those constraints.

The low-level search in CBS is typically a standard single-agent pathfinding algorithm, such as A, that finds an optimal path for an individual agent in a given search space, potentially considering a set of constraints. The high-level search, however, is responsible for discovering and resolving conflicts. It maintains a search tree where each node represents a set of constraints that apply to all agents. The root node has no constraints. When a node is expanded, the high-level search selects an agent and a time at which a conflict is detected (i.e., two agents would collide). It then branches the search tree by creating child nodes, each with an additional constraint preventing that specific collision. For example, if agent A and agent B collide at location (x, y) at time t, one child node might have the constraint "agent A cannot be at (x, y) at time t," while another child node might have the constraint "agent B cannot be at (x, y) at time t." The algorithm then re-plans the paths for the affected agents (or all agents, depending on the variant) based on these new constraints and continues the search until a node is found where all agents have collision-free paths.

CBS has shown remarkable success in finding optimal or near-optimal solutions for MAPF problems, often outperforming other methods in terms of solution quality and scalability for a moderate number of agents. Its effectiveness stems from its ability to abstract away individual agent paths and focus on resolving critical conflicts. However, the number of constraints can still grow exponentially in the worst case, leading to a high-level search that becomes computationally intractable for very large numbers of agents or highly congested environments. Variants of CBS, such as Enhanced CBS (ECBS) and Prioritized CBS (PCBS), have been developed to improve its performance and scalability by incorporating priority schemes or more sophisticated conflict detection and resolution strategies.

Beyond CBS, other optimal algorithms for MAPF have been explored, often relying on more sophisticated search techniques. For instance, **Multi-Agent A** *(MAA)* is a generalization of A *that searches in the joint state-space of all agents. The state in MAA* is a tuple representing the positions of all agents. While this approach can find optimal solutions, its state-space grows exponentially with the number of agents ($O(N^d)$, where N is the number of possible locations for a single agent and d is the number of agents), making it infeasible for even moderately sized problems. Techniques like **Reduction to SAT/SMT** solvers or Integer Linear Programming (ILP) can also be used to formulate and solve MAPF problems, leveraging powerful general-purpose solvers. These methods can provide formal guarantees of optimality but often suffer from scalability issues.

The inherent computational complexity of MAPF means that finding a truly optimal solution for a large number of agents in real-time is often impossible. This has led to the development of various **suboptimal algorithms** and **heuristics** that aim to provide good-enough solutions quickly. These might include anytime algorithms that can return a feasible solution and then improve it over time, or approximation algorithms that guarantee a solution within a certain factor of the optimal cost.

Applications of MAPF are widespread and growing. In **warehouse robotics**, where fleets of automated guided vehicles (AGVs) or mobile robots navigate complex warehouse layouts to pick and transport goods, MAPF is crucial for ensuring efficient and collision-free operation. Imagine hundreds of robots moving in close proximity, picking up items, and delivering them to designated locations. Without effective MAPF, congestion, deadlocks, and collisions would quickly bring operations to a halt. MAPF algorithms help to coordinate these movements, optimizing traffic flow and maximizing throughput.

Air traffic control (ATC) can also be viewed as a large-scale MAPF problem, albeit in a continuous 3D space with highly stringent safety requirements. While traditional ATC relies heavily on human controllers and radar, advanced concepts for future air traffic management systems are exploring automated solutions that resemble MAPF. Here, the goal is to manage the trajectories of numerous aircraft to prevent collisions in busy airspace, especially during takeoff, landing, and transit through crowded flight corridors. The challenge is amplified by the need for rapid re-planning in response to weather changes, emergencies, or unexpected aircraft deviations.

Other domains where MAPF plays a significant role include:

Robotic exploration: Guiding multiple drones or ground robots to map an unknown area while ensuring they don't collide with each other or with newly discovered obstacles.
Search and rescue operations: Coordinating a team of robots to search a disaster site, where the environment is dynamic and potentially hazardous.
Autonomous driving: Managing the movement of multiple self-driving vehicles in dense urban environments, coordinating lane changes, intersections, and merging maneuvers.
Surveillance and monitoring: Deploying swarms of robots to patrol a large area, ensuring complete coverage without overlap or mutual obstruction.
Video game AI: Directing non-player characters (NPCs) in video games to navigate complex environments and interact with each other and the player without collision.

The scalability of MAPF algorithms remains a primary research challenge. As the number of agents increases, the state-space and the number of potential conflicts grow dramatically. Researchers are continually exploring new techniques to address this, including:

Decentralized MAPF: Developing methods where each agent plans its own path with limited communication with its neighbors, rather than relying on a central

planner. This can improve scalability but often leads to suboptimal or even incomplete solutions.
Abstraction and Hierarchical Planning: Breaking down the problem into smaller, manageable subproblems or planning at different levels of abstraction.
Machine Learning Approaches: Training neural networks or other machine learning models to predict or generate paths, potentially learning complex collision avoidance strategies from data. Reinforcement learning is particularly promising in this area.
Domain-Specific Optimizations: Tailoring algorithms to specific types of environments or agent behaviors, such as grid worlds, or agents with specific kinematic constraints.

In summary, Multi-Agent Pathfinding is a fundamental problem in multi-agent systems that focuses on generating collision-free, simultaneous trajectories for multiple agents moving to distinct goals. While early methods like prioritized planning offered simple solutions, more advanced techniques like Conflict-Based Search have achieved significant performance gains. Despite these advances, the inherent computational complexity of MAPF continues to be a major hurdle for its application in large-scale, real-time scenarios. Nonetheless, ongoing research in decentralized planning, machine learning, and algorithmic optimization is steadily pushing the boundaries, making MAPF increasingly viable for a wide array of critical applications, from the busy floors of automated warehouses to the complex coordination of future air traffic.

Chapter 12: Human-Robot Interaction (HRI)

The intricate dance between humans and machines, once relegated to the realm of science fiction, is now a palpable reality, shaping industries, augmenting human capabilities, and fundamentally altering our daily lives. As robots transition from rigidly programmed industrial arms to more versatile and adaptable agents capable of operating in shared spaces with people, the field of Human-Robot Interaction (HRI) has emerged as a critical discipline. HRI is not merely about building smarter robots; it is about fostering effective, efficient, and safe collaborations between humans and these artificial entities. It is an interdisciplinary endeavor, drawing upon insights from computer science, psychology, sociology, engineering, and design to bridge the gap between human cognition and robotic action. At its core, HRI seeks to understand, design, and evaluate robotic systems that can work alongside, assist, or even entertain humans in a manner that feels intuitive, natural, and ultimately, beneficial.

The foundational goal of HRI is to create systems that are not only technically proficient but also socially intelligent and psychologically attuned to human users. This involves moving beyond purely functional specifications to consider the broader experience of interacting with a robot. Several key concepts serve as pillars for understanding and achieving successful HRI. **Usability**, for instance, is paramount. A robot, no matter how advanced its capabilities, will fail to achieve its purpose if humans cannot easily and effectively operate or interact with it. This encompasses aspects like intuitive control interfaces, clear communication of the robot's status and intentions, and predictable behavior. A robot designed for a domestic setting, for example, should be simple enough for a layperson to command for tasks like fetching an object or providing information, without requiring extensive training or technical expertise. The design of the user interface, whether physical (buttons, joysticks) or virtual (touchscreens, voice commands), must be aligned with human cognitive models and capabilities.

Equally critical is
safety. In environments where humans and robots coexist, the paramount concern is preventing harm. This extends beyond physical safety to include psychological and social well-being. Robots operating in close proximity to humans must be engineered with robust safety mechanisms, including collision avoidance, force sensing, and fail-safe modes. However, safety in HRI also encompasses the predictability of robot behavior. Humans need to be able to anticipate a robot's actions to feel secure and to plan their own responses effectively. This requires robots to communicate their intentions clearly and to exhibit consistent, understandable behavior, especially in dynamic and potentially hazardous situations. For instance, a robotic assistant in a hospital must not only navigate crowded corridors without bumping into people but also communicate its approach and its intended path to avoid startling patients or staff.

Engagement is another vital aspect, particularly for robots designed for long-term interaction or for specific applications like education, entertainment, or companionship. An engaging robot can hold a user's attention, motivate them, and make the interaction more enjoyable and productive. This can be achieved through aspects like personality, responsiveness, and the ability to adapt to the user's emotional state or preferences. Robots designed for children's education, for example, might employ playful vocalizations, expressive movements, and personalized feedback to maintain a child's interest and facilitate learning. The design of robot appearance and behavior plays a significant role in fostering engagement, often drawing from principles of character design in animation and game development.

Finally,
trust forms the bedrock of any successful human-robot relationship. Humans are more likely to collaborate with, rely on, and accept the assistance of robots they trust. Trust in HRI is built on a combination of factors: perceived competence (the robot performs its tasks reliably), benevolence (the robot acts in the user's best interest), and integrity (the robot's actions are consistent with its stated goals and ethical principles). A robot that consistently malfunctions, misunderstands commands, or operates unpredictably will quickly erode trust. Conversely, a robot that demonstrates reliability, transparency, and a clear understanding of its role and the user's needs can foster a strong sense of trust. This is particularly crucial in high-stakes applications, such as robotic surgery or autonomous driving, where human lives depend on the robot's

performance and the user's confidence in its capabilities.

To achieve these HRI goals, a deep understanding of **human factors** is indispensable. Humans are not simply interchangeable components in a robotic system; they possess complex cognitive, perceptual, and social characteristics that profoundly influence their interaction with robots. **Cognitive load**, for example, refers to the amount of mental effort required to process information and perform tasks. HRI designers must strive to minimize unnecessary cognitive load on users. An overly complex interface, confusing instructions, or a robot that demands constant attention can overwhelm a user, leading to errors and frustration. For instance, a robot assisting in a manufacturing assembly line should present information and instructions in a clear, concise manner, allowing the human worker to focus on the physical task rather than struggling to decipher robotic cues.

User expectations also play a significant role. Humans bring pre-conceived notions and assumptions about how robots should behave, often shaped by past experiences with technology, media portrayals, and their own understanding of intelligence and agency. If a robot's behavior deviates too drastically from these expectations, it can lead to confusion, disappointment, or even rejection. A robot designed to be a helpful household assistant, for instance, might be expected to understand natural language commands and adapt to household routines. If it instead requires highly specific, rigid commands, users will likely find it less useful and intuitive than anticipated. Understanding these expectations allows designers to create robots that align with user mental models and to proactively manage potential discrepancies.

Furthermore, the ability of robots to perceive and respond to
social cues is increasingly important for natural and effective interaction. Humans are inherently social beings, and much of our communication relies on subtle non-verbal signals such as facial expressions, body language, and tone of voice. Robots that can interpret these cues and respond appropriately can foster more nuanced and empathetic interactions. For example, a social robot designed to assist elderly individuals might need to recognize signs of distress or loneliness in a user's voice or posture and offer comforting words or suggest an activity. Conversely, robots can also use their own non-verbal communication, such as gestures, gaze, and vocal intonation, to convey information, manage turn-taking in conversations, and signal their intentions, making the interaction more fluid and human-like.

The design of HRI systems thus requires a holistic approach that integrates technical prowess with an understanding of human psychology and social dynamics. It is about creating robots that are not just tools but partners, assistants, or even companions, capable of seamlessly integrating into human lives and contributing positively to our experiences. The subsequent sections will delve deeper into the specific aspects of designing such systems, exploring various modalities of interaction, ethical considerations, and the future trajectory of this rapidly evolving field.

In order for robots to seamlessly integrate into human environments and collaborate effectively, their ability to communicate using social cues is paramount. This extends far beyond mere functional command and response; it delves into the nuanced realm of human interaction, where meaning is conveyed not just through explicit statements but through a rich tapestry of verbal and non-verbal signals. As robots become more prevalent in our homes, workplaces, and public spaces, their capacity to understand and generate these social cues will dictate the quality and success of the human-robot relationship. This subsection delves into the multifaceted nature of robot social cues and communication, exploring the various modalities through which robots can express themselves and interpret human signals, thereby fostering more natural, intuitive, and trustworthy interactions.

Verbal communication remains a cornerstone of human interaction, and its integration into robotic systems is a key area of development. The ability of a robot to understand spoken language through advanced speech recognition technologies allows users to interact with it in a natural, conversational manner, eliminating the need for complex interfaces or specialized training. This capability is crucial for robots intended for roles such as personal assistants, information kiosks, or educational tools. For instance, a robotic concierge at a hotel could understand a guest's request for directions to the nearest restaurant or for information about local attractions, responding with clear and concise verbal guidance. The sophistication of speech recognition has advanced to a point where robots can now distinguish between different speakers, account for accents and background noise, and even interpret emotional nuances in human speech, although this latter capability is still an active area of research.

Complementing speech recognition is speech synthesis, the ability of robots to generate human-like speech. This is not merely about converting text to audio; it involves creating speech that is intelligible, engaging, and appropriately modulated. The quality of synthesized speech directly impacts user perception and engagement. A robotic voice that is monotonous, robotic, or difficult to understand can quickly lead to frustration and disengagement. Therefore, significant research efforts are focused on developing speech synthesizers that can produce natural-sounding prosody – the rhythm, intonation, and stress patterns of speech – which are critical for conveying meaning and emotion. For example, a robot designed to provide comfort or support to individuals might employ a warmer, more empathetic tone of voice, whereas a robot tasked with issuing safety warnings would likely adopt a more urgent and assertive vocalization. The ability to vary pitch, speed, and volume in response to the context of the interaction is essential for creating a believable and effective communication channel.

Beyond spoken words, non-verbal communication plays an equally, if not more, significant role in human interaction. Robots can leverage a variety of non-verbal cues to convey information, express intent, and build rapport with humans. These can be broadly categorized into gestures, facial expressions, and body language. The specific modalities through which these cues are expressed depend heavily on the robot's physical form and design. For robots equipped with anthropomorphic features, such as a human-like head and arms, the generation of gestures and facial expressions can be quite sophisticated.

Gestures can serve a multitude of purposes in robot communication. A robot might use pointing gestures to direct a user's attention to a specific object or location, akin to how a human would. For example, a robotic assistant in a warehouse could point to a misplaced item for a human colleague to retrieve. Robots can also use gestures to indicate agreement or disagreement, such as nodding or shaking their head. Simple hand gestures, like waving to greet someone or giving a thumbs-up to signify approval, can further enhance the naturalness of the interaction. The design of these gestures must be carefully considered to ensure they are universally understood or at least contextually appropriate for the target user group. For instance, a robot designed to interact with children might employ exaggerated, playful gestures, while a robot in a professional setting would use more subtle and precise movements.

Facial expressions are a powerful channel for conveying emotions and intentions in humans. While replicating the full spectrum of human facial expressions on a robot can be challenging, even rudimentary expressions can significantly enhance communication. Robots with a screen displaying an avatar or those with movable facial features can convey basic emotional states like happiness, sadness, anger, or surprise. For instance, a robot tutor might display a smiling face when a student successfully answers a question, reinforcing the positive outcome. Conversely, a slightly furrowed brow might signal confusion or concern if the student is struggling. The timing and intensity of these expressions are crucial; an improperly timed or overly dramatic expression can be unsettling or even alarming. The goal is to create expressions that are congruent with the robot's verbal communication and the overall context of the interaction, thereby enhancing empathy and understanding.

Body language, encompassing posture, gait, and proxemics (the use of space), also contributes significantly to robot communication. A robot's posture can convey its state of readiness or attentiveness. For example, standing upright and facing a human might indicate that the robot is ready to interact, while slumping or turning away could signal disengagement or a low-power state. The way a robot moves – its gait – can also communicate information. A smooth, deliberate gait might suggest efficiency and control, while a jerky or erratic movement could indicate a malfunction or instability. Proxemics involves how a robot uses space in relation to humans. Maintaining an appropriate distance, not invading personal space, and signaling intentions before entering a shared space are all important aspects of robot body language that contribute to user comfort and safety. For instance, a robot entering a crowded room should slow down, perhaps emit a soft audible cue, and perhaps make a subtle visual signal (like illuminating a light strip) to alert people to its presence and intended path, thus avoiding startling or blocking individuals.

For robots that may not have a physical form, or to augment the communication of physical robots, the use of virtual avatars can be an effective strategy. These avatars, displayed on screens or in augmented reality environments, can convey a wide range of social cues through animated gestures, facial expressions, and body language. This approach offers flexibility in design and can allow for more dynamic and expressive communication than might be possible with a purely physical robot. For example, a virtual assistant on a computer screen could use animated eye movements to simulate gaze following, making the user feel more directly addressed. The avatar could also use hand gestures to emphasize points during a spoken explanation or change its posture to reflect a different level of formality or enthusiasm.

The design principles guiding the generation of these

robot social cues are critical for ensuring effective and intuitive communication. One key principle is **consistency**. The robot's social cues should be consistent with its overall persona, its stated capabilities, and the context of the interaction. A robot that displays a friendly, approachable demeanor through its gestures and expressions should also have a welcoming vocal tone and helpful behaviors. Inconsistency can lead to user confusion and mistrust. For example, a robot that gestures enthusiastically while its synthesized voice sounds bored and monotone will create a dissonance that undermines the intended message.

Another important principle is **transparency**. The robot's social cues should clearly communicate its intentions, its understanding of the situation, and its internal state. If a robot is about to perform an action that might be perceived as intrusive or surprising, it should signal this in advance. For instance, a robot preparing to extend an arm to pick up an object should first subtly indicate its intention, perhaps by shifting its weight, orienting its body, or making a soft vocalization, allowing the human to adjust their position if necessary. Transparency also extends to conveying the robot's limitations. If a robot does not understand a command or is unable to perform a requested task, it should communicate this clearly rather than attempting to guess or providing an incorrect response.

Adaptability is also a crucial design consideration. Humans vary in their communication styles and preferences, and a truly effective social robot should be able to adapt its communication to the individual user and the specific situation. This might involve adjusting the speed of speech, the complexity of gestures, or the level of emotional expression. For example, a robot interacting with a young child might use simpler language and more animated expressions, while one interacting with a scientist might adopt a more formal and technical communication style. Furthermore, robots should ideally be able to learn from user feedback and adjust their social cues over time to better suit individual preferences.

The ability of robots to interpret human social cues is as vital as their ability to generate them. This involves sophisticated perception systems capable of analyzing human facial expressions, body posture, gestures, and vocalizations for emotional and intentional information. A robot that can recognize signs of confusion in a user's facial expression might proactively offer clarification. Similarly, a robot that detects a user's frustration through their tone of voice or body language could de-escalate the situation by offering assistance or taking a break from the task. This reciprocal understanding of social cues is the foundation of effective rapport building.

Rapport, in the context of HRI, refers to a harmonious and comfortable relationship between a human and a robot, characterized by mutual understanding, trust, and a positive emotional connection. Effective social communication is the primary driver for building rapport. When a robot can understand and respond appropriately to human social cues, and when its own social signals are clear and congruent, it fosters a sense of familiarity and connection. This is particularly important for robots designed for roles that involve prolonged interaction or emotional support, such as companion robots for the elderly, educational robots for children, or service robots in healthcare settings.

For instance, a companion robot designed to assist elderly individuals could be programmed to recognize signs of loneliness or sadness, perhaps through changes in speech patterns or facial expressions. Upon detecting such cues, the robot could initiate a conversation, suggest playing a game, or offer to connect the user with a family member. This proactive, empathetic response, communicated through a combination of verbal and non-verbal cues, can significantly improve the user's well-being and create a strong sense of companionship. Similarly, a robot working in a customer service role, like a greeter at a store or a help desk attendant, can use a welcoming smile (if physically capable), an open posture, and a friendly tone of voice to create a positive first impression and make customers feel more at ease.

The prosody of speech – the rhythm, intonation, stress, and pitch variations – is a subtle yet powerful aspect of communication that robots are increasingly being designed to understand and generate. Humans convey a wealth of information through prosody, including their emotional state, their attitude towards the speaker or topic, and the intended emphasis of their words. A robot that can analyze the prosody of human speech can gain deeper insights into the user's feelings and intentions. For example, a sharp, rising intonation might indicate surprise or a question, while a slow, low-pitched delivery could suggest sadness or seriousness.

Conversely, robots that can modulate their own synthesized speech prosody can significantly enhance their expressiveness and naturalness. A robot that uses varied intonation can sound more engaging and less monotonous. It can use rising intonation to signal a question, falling intonation to indicate the end of a statement, and specific stress patterns to emphasize key information. For example, when a robot is providing instructions, it might use a clear, steady pace with emphasis on action verbs. If it is delivering good news, it might use a more upbeat and varied pitch. The ability to convey appropriate emotional tone through prosody – warmth, enthusiasm, concern – is a critical step towards creating robots that can form meaningful social connections with humans.

The development of robot social cues and communication is an ongoing endeavor that is deeply intertwined with advancements in artificial intelligence, cognitive science, and psychology. As our understanding of human social interaction deepens, so too will our ability to imbue robots with the capacity for more sophisticated and meaningful communication. The ultimate goal is to move beyond robots as mere tools and towards creating intelligent agents that can interact with us in ways that are not only efficient and safe but also socially intelligent, empathetic, and ultimately, enriching to the human experience. This continuous evolution promises to redefine the landscape of human-robot collaboration, making our future together more seamless, intuitive, and harmonious.

The integration of robots into our daily lives and professional environments hinges not only on their physical capabilities and artificial intelligence but also, crucially, on how effectively humans can interact with them. This interaction is mediated through the design of User Interfaces (UIs) and the overarching User Experience (UX) they provide. A well-designed UI/UX for robotic systems transforms a complex piece of machinery into an accessible and collaborative partner, enabling users to command, monitor, and provide feedback with ease. Conversely, a poorly designed interface can render even the most advanced robot frustrating and impractical to use, hindering adoption and potentially leading to errors or unsafe operations. Therefore, a deep understanding of UI/UX principles, tailored to the unique challenges and opportunities presented by robotic systems, is paramount.

At its core, user-centered design (UCD) serves as the guiding philosophy for crafting effective robotic interfaces. This approach prioritizes understanding the needs, capabilities, and limitations of the end-users throughout the entire design and development process. It's an iterative cycle that involves research, design, prototyping, and testing, ensuring that the final product is not only functional but also intuitive, efficient, and satisfying to use. For robotic systems, UCD is particularly critical because the stakes can be significantly higher than with traditional software. Errors in interaction can have physical consequences, ranging from damaged property to injury. This necessitates a rigorous approach to understanding how users will perceive, interpret, and respond to the robot's interface and overall behavior.

When considering the spectrum of robotic applications, the nature of the UI/UX design will vary considerably. For instance, a domestic service robot, intended for use in a home environment by individuals with varying levels of technical proficiency, will demand a very different interface than a sophisticated industrial robot operating in a manufacturing plant, typically controlled by trained operators. In the domestic setting, simplicity, ease of learning, and minimal cognitive load are paramount. The interface should allow users to quickly set tasks, adjust parameters, and understand the robot's status without requiring extensive training or technical expertise. Mobile applications, often leveraging familiar paradigms from smartphones and tablets, have become a dominant form of UI for many domestic robots. These apps can provide visual dashboards for monitoring the robot's progress, allow for straightforward task selection (e.g., "clean the living room"), and offer options for scheduling or adjusting cleaning intensity. Gesture-based control, voice commands, and even simple button interfaces integrated directly onto the robot's body can further enhance accessibility. The UX here aims to make the robot feel like a helpful appliance or even a digital assistant, seamlessly integrated into the household routine.

In stark contrast, industrial robots, which often perform repetitive, high-precision tasks in controlled environments, require interfaces designed for expert users. These interfaces need to provide granular control, detailed diagnostic information, and robust programming capabilities. Operators must be able to define complex motion paths, set safety parameters, and troubleshoot issues efficiently. Physical teach pendants, often featuring physical joysticks, programmable buttons, and high-resolution displays, have long been the standard for programming and controlling industrial robots. These pendants allow operators to physically guide the robot through its desired movements and record these points for automated execution. Modern industrial robot UIs are evolving, incorporating touch-screen interfaces with advanced graphical representations of the robot's workspace and kinematics, allowing for more intuitive programming and simulation. The UX goal in this context is to maximize productivity, ensure safety, and minimize downtime through precise control and rapid problem resolution.

Beyond these two extremes, a vast array of intermediate applications exist, each with its own UI/UX considerations. For example, robots used in logistics and warehousing might require interfaces that balance the need for efficient task management with the potential for human-robot collaboration. Here, UIs might involve augmented reality (AR) overlays to guide human workers or robots, providing real-time information about pick-and-place locations, inventory levels, or potential hazards. A warehouse worker might wear AR glasses that highlight the correct items to pick, the optimal path to take, and provide status updates from the robots operating in their vicinity. This creates a shared operational awareness and enhances the collaborative potential.

The concept of
direct control refers to interfaces that allow users to manipulate the robot's movements or actions in real-time. This is common in applications where precise, moment-to-moment guidance is necessary. For example, in surgical robotics, surgeons use highly sensitive joysticks and haptic feedback devices to control robotic arms that perform delicate procedures. The UI must provide a clear visual representation of the surgical field and the robotic instruments, along with precise feedback on forces and movements. In contrast, for a telepresence robot that allows a user to remotely navigate and interact with a physical environment, the UI would focus on providing a live video feed, directional controls, and perhaps controls for the

robot's camera or manipulator arm. The UX here is about creating a sense of embodiment, making the user feel present and in control of the remote robot.

Programming interfaces are designed to enable users to define sequences of actions or behaviors for the robot. This can range from simple "if-then" logic for a basic task robot to complex scripting languages for sophisticated AI-driven robots. Visual programming environments, often using drag-and-drop blocks or flowcharts, have become popular for their ability to abstract away complex coding syntax, making programming more accessible. For example, platforms like Scratch, adapted for robotics, allow children to learn programming concepts by visually assembling code blocks that control robot movement, sensor input, and output actions. For more advanced robotics, languages like Python with specialized libraries (e.g., ROS - Robot Operating System) are commonly used, offering powerful tools for defining complex behaviors, algorithms, and inter-robot communication. The UX of programming interfaces is heavily dependent on the target user's expertise, aiming to reduce the cognitive load associated with translating a desired behavior into executable code.

High-level task specification represents a more abstract form of interaction, where users define what they want the robot to achieve, rather than precisely how to achieve it. This often relies on natural language processing or intuitive graphical interfaces to convey goals. For instance, a user might tell a domestic robot, "Please tidy up the living room before guests arrive," and the robot's AI would then plan the necessary sub-tasks (e.g., picking up objects, putting them in designated places, vacuuming). Similarly, a project manager in a factory might use a dashboard to assign a batch of items to a specific robotic assembly line, without needing to know the intricacies of how each robot arm will execute the task. This type of interface requires sophisticated AI behind the scenes to interpret the user's intent and translate it into executable robot actions. The UX challenge here is to ensure that the system accurately understands the user's goals and to provide clear feedback on what the robot is doing and why.

The principle of
learnability is fundamental to any user interface, but it takes on new dimensions with robots. A robot's interface should be easy for new users to understand and operate, requiring minimal instruction. This can be achieved through consistent design patterns, clear affordances (visual cues that suggest how an interface element should be used), and intuitive workflows. For complex systems, tutorials, guided onboarding processes, and readily accessible help documentation are essential. The goal is to reduce the "time to proficiency" – the amount of time it takes for a user to become competent in operating the robot.

Efficiency in robotic UIs refers to the ability of users to perform tasks quickly and with minimal effort once they have become proficient. This involves optimizing workflows, providing shortcuts for frequent actions, and ensuring rapid response times from the interface and the robot itself. For example, in a busy warehouse, an operator's ability to quickly scan a barcode to assign a task to a robot, rather than manually entering information, significantly boosts efficiency. For robots performing critical functions, minimizing the time required for an operator to issue an emergency stop command is a paramount aspect of efficiency and safety.

User satisfaction is the ultimate measure of a successful UI/UX design. It encompasses a range of factors, including the perceived usability, the aesthetic appeal of the interface, the feeling of control and confidence the user experiences, and the overall effectiveness of the robot in fulfilling its purpose. A user who feels empowered, in control, and successful in their interactions with a robot is likely to report high satisfaction. This is often achieved through a harmonious blend of functional correctness, intuitive design, and perhaps even a touch of delight, where the interface or robot's behavior goes beyond mere functionality to create a positive emotional connection. For instance, a robot that politely signals its intentions before performing a potentially disruptive action, such as moving a large object, can significantly enhance user satisfaction by demonstrating consideration for the human in its vicinity.

Let's delve into some concrete examples to illustrate

these principles. Consider the evolution of interfaces for autonomous mobile robots (AMRs) in logistics. Early systems might have relied on complex configuration files and command-line interfaces, accessible only to highly skilled technicians. Today, many AMR fleets are managed through sophisticated, web-based dashboards. These dashboards provide a bird's-eye view of the warehouse floor, showing the real-time location and status of each robot. Users can drag and drop tasks onto robots, define zones where robots should operate or avoid, and receive alerts for any issues. The UX is designed to provide a holistic, manageable overview of a dynamic system, allowing supervisors to optimize robot deployment and troubleshoot problems remotely. The learnability is enhanced through clear visual cues, intuitive drag-and-drop functionality, and consistent design language. Efficiency is gained through rapid task assignment and real-time performance monitoring.

Another illustrative example is the design of interfaces for collaborative robots, or "cobots," which are designed to work alongside humans. These robots often feature visual programming interfaces that allow assembly line workers to teach them new tasks by physically guiding the robot's arm through a sequence of movements. The interface typically comprises a tablet or teach pendant that displays the robot's current state and allows for simple adjustments to the recorded path or gripper actions. The UX emphasizes a gentle learning curve, enabling workers to adapt the robot's tasks quickly as production needs change, without requiring deep programming knowledge. The efficiency comes from the speed with which tasks can be redefined, and user satisfaction is derived from the sense of direct agency over the robot's actions. Safety is paramount, and the UI must clearly communicate the robot's operating envelope and allow for immediate stoppage.

The design of interfaces for personal care robots presents a unique set of challenges and opportunities. These robots are intended to assist individuals with daily living tasks, and their interfaces must be highly accessible and user-friendly, often catering to users with cognitive or physical impairments. For example, a robot designed to remind an elderly person to take their medication might feature a large-button interface on its body or a simple voice-activated command system. The UX must prioritize clarity, simplicity, and reliability. Missed instructions or confusing interfaces could have serious health consequences. Therefore, extensive user testing with the target demographic is crucial. The interface might include visual cues, such as flashing lights or animated icons, to accompany verbal reminders, catering to different sensory preferences and cognitive styles.

In the realm of educational robotics, interfaces are often designed to be engaging and inspiring for children. Platforms like LEGO Mindstorms or VEX Robotics offer visual programming environments that make learning computational thinking and robotics concepts fun. The

UI is typically colorful, interactive, and provides immediate feedback as code is executed. The UX aims to foster creativity, problem-solving, and a sense of accomplishment. Children can build their own robots and then program them to perform a variety of tasks, from navigating mazes to playing simple games. The success of these platforms is a testament to how well-designed, user-centered interfaces can make complex technologies accessible and enjoyable for younger audiences.

The evolution of robot UIs is also being shaped by advancements in sensor technology and AI. Robots are becoming more aware of their environment and their human collaborators, and interfaces are beginning to reflect this. For instance, a robot might proactively adjust its speed or trajectory based on the detected proximity and movement of a human, and its interface could provide a visual indication of this adjustment, perhaps through a change in the color of an LED indicator or a brief on-screen notification. This "adaptive interface" concept aims to make the interaction feel more natural and less intrusive.

Furthermore, the integration of multimodal interfaces is becoming increasingly important. Instead of relying solely on a graphical screen or voice commands, robots can offer a combination of interaction methods. A user might initiate a task by voice, receive status updates via an on-screen display, and then use physical buttons for critical safety functions. This layered approach can cater to a wider range of user preferences and environmental conditions. For example, in a noisy factory setting, physical buttons or gestures might be more reliable for emergency stops than voice commands, while a graphical display can provide detailed diagnostic information.

The principles of accessibility are also critical in designing robotic UIs. This means ensuring that interfaces are usable by people with a wide range of abilities and disabilities. This includes considerations for visual impairments (e.g., high contrast modes, screen reader compatibility), auditory impairments (e.g., visual alerts, captions), motor impairments (e.g., larger touch targets, alternative input methods), and cognitive impairments (e.g., simplified interfaces, clear and consistent language). Adhering to accessibility guidelines, such as those from the Web Content Accessibility Guidelines (WCAG), can provide a solid foundation for creating inclusive robotic interfaces.

The ongoing research in HRI is continuously pushing the boundaries of what's possible in robot UI/UX design. Concepts like "affective computing," which focuses on enabling robots to recognize and respond to human emotions, are influencing interface design. For instance, a robot might adapt its communication style or the urgency of its actions based on detecting signs of stress or confusion in a human user. This moves beyond purely functional interaction towards a more empathetic and supportive user experience.

In conclusion, the design of user interfaces and user experiences for robots is a multifaceted discipline that lies at the intersection of robotics, human-computer interaction, psychology, and design. It demands a deep understanding of user needs, task requirements, and the specific capabilities and limitations of the robotic system. By applying user-centered design principles and continuously iterating through research, design, and testing, we can create robotic systems that are not only powerful and intelligent but also intuitive, efficient, and enjoyable to interact with. Whether it's a simple mobile app for a domestic cleaner or a complex programming environment for an industrial manipulator, the quality of the UI/UX will ultimately determine the success of the human-robot partnership in an increasingly automated world. The goal is to bridge the gap between human intention and robotic action, making advanced robotic technology accessible and beneficial to everyone.

The success of human-robot interaction (HRI) in any domain—from manufacturing floors and healthcare facilities to domestic environments and public spaces—is inextricably linked to the establishment of trust and the fostering of effective collaboration. Without a foundation of trust, users are unlikely to delegate tasks to robots, rely on their judgments, or integrate them seamlessly into workflows. Collaboration, in turn, is the mechanism through which humans and robots can achieve outcomes that neither could accomplish alone, leveraging their respective strengths. This subsection explores the multifaceted nature of trust in HRI and outlines the principles and strategies that underpin successful human-robot collaboration.

Building trust in robotic systems is not a passive outcome but an active design and implementation process. Several key factors contribute to a human's willingness to trust a robot.
Reliability is perhaps the most fundamental. If a robot consistently performs its intended functions accurately and without failure, users will naturally develop confidence in its capabilities. This means that the robot must be robust, dependable, and predictable in its operation. For instance, an industrial robot tasked with placing components on a circuit board must do so with extreme precision and repeatability. Any deviation, however minor, can erode trust and lead to costly errors. In healthcare, a surgical robot that malfunctions mid-procedure would have catastrophic consequences for patient safety and irrevocably damage trust in that system. Therefore, rigorous testing, quality control, and fault-tolerance mechanisms are paramount.

Predictability is closely related to reliability. Humans tend to trust what they can anticipate. A robot that behaves in a consistent and understandable manner, even when faced with novel situations (within its operational parameters), fosters a sense of security. This predictability is often achieved through clear, deterministic behaviors. If a robot is programmed to follow a specific path or execute a sequence of actions in a defined order, users can learn and predict its movements. Conversely, erratic or unpredictable behavior, even if not directly harmful, can be unsettling and lead to a reluctance to engage. For example, a domestic robot navigating a living room should ideally follow a predictable pattern for vacuuming, rather than suddenly changing direction without apparent reason. This predictability allows humans to plan their own movements and activities around the robot, minimizing potential disruptions and enhancing safety.

Transparency plays a crucial role in building and maintaining trust. Users need to understand *why* a robot is behaving in a certain way and what its current state or intentions are. This involves making the robot's internal reasoning processes and decision-making visible to the user, to an appropriate degree. This does not necessarily mean exposing complex algorithms, but rather providing clear indicators of the robot's goals, current actions, and potential future actions. For example, a robot arm in a warehouse might use visual cues, such as illuminated segments or directional arrows, to signal which bin it intends to pick from next. A service robot in a hospital might verbally announce its intended path or announce when it needs to pass through a doorway. This transparency helps humans interpret the robot's behavior, anticipate its needs, and provide appropriate assistance or avoid interference. It also allows users to understand the limitations of the robot and know when human intervention might be necessary.

The concept of

competence is also vital. Humans trust individuals or systems that they perceive as being skilled and capable of performing a task effectively. This applies directly to robots. If a robot demonstrates proficiency in its assigned duties, users are more likely to rely on it. This competence can be conveyed through the robot's performance itself, as well as through its design and presentation. A robot that moves smoothly, executes tasks efficiently, and responds intelligently to its environment projects an image of competence. Furthermore, interfaces that clearly communicate the robot's capabilities and limitations can help manage user expectations and reinforce the perception of competence.

Beyond these core factors,
benevolence or the perception that the robot is acting in the user's best interest can also foster trust. While robots do not possess emotions or intentions in the human sense, their design and programming can convey a sense of being helpful and considerate. For instance, a robot designed to assist an elderly person should prioritize safety and comfort above all else. If the robot's actions demonstrate a prioritization of the user's well-being, even when faced with conflicting objectives (e.g., completing a task quickly versus ensuring user safety), this can build a deeper level of trust. This is often achieved through ethical programming and robust safety protocols that always err on the side of caution when human safety is involved.

Finally,

familiarity and experience are powerful trust builders. The more humans interact with robots, and the more positive those interactions are, the more comfortable and trusting they become. Early experiences with robots, especially in contexts where potential risks are higher, can heavily influence long-term trust. Therefore, introducing robots in controlled environments with clear safety measures and gradual increases in autonomy can help users build confidence over time. This is why training and onboarding programs for complex robotic systems are so important, not just for skill development but also for trust cultivation.

Collaborative robotics, often termed "cobots," represents a paradigm shift in HRI, moving from robots that operate in segregated workspaces to those that actively work alongside humans on shared tasks. Successful collaboration hinges on several critical elements, chief among them being **mutual understanding**. For humans and robots to collaborate effectively, they need to develop an understanding of each other's goals, intentions, and capabilities. This is a complex challenge, as human intentions and capabilities are often implicit, while robots operate on explicit programming.

One approach to achieving mutual understanding is through **shared representations of the task and environment**. When both the human and the robot have access to and can interpret a common model of the task, the workspace, and the objectives, collaboration becomes more fluid. This might involve a shared digital twin of a factory floor, a common understanding of a patient's medical status in a surgical context, or a synchronized view of a domestic space for a cleaning robot and its human overseer. This shared understanding allows for more intuitive coordination, as both parties can anticipate each other's actions based on the common model.

Intent recognition is a key component of mutual understanding. Robots need to be able to infer human intentions, and humans need to be able to understand the robot's intended actions. For robots, this can be achieved through a combination of sensors (e.g., tracking human gaze, body posture, hand gestures) and predictive algorithms that anticipate human actions based on context and learned patterns. For example, if a human worker reaches for a tool while looking at a particular workpiece, a collaborative robot might infer that the human intends to work on that piece and adjust its own actions accordingly, perhaps by moving out of the way or preparing to hand over a required component. Conversely, robots can communicate their intentions through explicit signals, such as displaying their next planned movement on a screen, emitting auditory cues, or using directional lighting.

Capability awareness is equally important. Humans need to know what a robot can and cannot do, and robots need to be aware of human limitations. This prevents situations where a human might ask a robot to perform a task beyond its capacity, or where a robot might attempt an action that is unsafe or inefficient for a human to complete. For example, a robot that can lift extremely heavy objects should clearly communicate this capability, so that humans do not attempt to share this load. Conversely, a human might be better suited for intricate fine motor tasks that a robot struggles with, requiring the robot to understand its own limitations and defer to human expertise.

Shared autonomy is a paradigm that formalizes the dynamic sharing of control between humans and robots. In fully autonomous systems, the robot makes all decisions. In purely teleoperated systems, the human makes all decisions. Shared autonomy exists on a spectrum between these two extremes, allowing for the optimal distribution of decision-making and control based on the task, the environment, and the capabilities of both human and robot at any given moment.

In a shared autonomy framework, control can be allocated dynamically. For instance, a robot might be responsible for navigating a complex and hazardous environment, while the human operator provides high-level mission objectives and makes critical decisions when unforeseen events occur. The robot handles the low-level control of locomotion, obstacle avoidance, and sensor processing, while the human oversees the overall mission and intervenes when necessary. This can be implemented through various mechanisms. One common approach is

delegation, where the human explicitly assigns a task or a level of control to the robot. Another is **intervention**, where the human takes over control from the robot when they perceive a problem or a need to alter the course of action. Conversely, a robot might signal its inability to proceed and request human assistance or override, effectively requesting the human to take control.

The effectiveness of shared autonomy relies heavily on the
feedback mechanisms between the human and the robot. The human needs to understand what the robot is currently doing, why it is doing it, and what its current confidence level is in its own decisions. This can be communicated through intuitive user interfaces, as discussed previously, but also through more subtle cues. For example, a robot's uncertainty about an upcoming maneuver could be signaled by a subtle change in its movement or an audible hesitation. The robot, in turn, needs to understand the human's intentions, not just through explicit commands but also through observing human actions and inferring their goals.

Consider a medical scenario where a surgeon is using a robotic system for minimally invasive surgery. The robot provides enhanced dexterity and visualization, but the surgeon is ultimately in control. In a shared autonomy model, the robot might autonomously stabilize a moving organ, providing a steady platform for the surgeon's instruments. However, if the surgeon needs to make a precise incision, they would take direct control of the instrument. The system could be designed such that the robot anticipates the surgeon's needs, for instance, by automatically retracting instruments when the surgeon moves their hand towards a certain area. This dynamic sharing of control allows for the best of both worlds: the precision and stability of the robot combined with the judgment and dexterity of the human expert.

Another example can be found in emergency response scenarios. A drone equipped with sensors might be tasked with surveying a collapsed building. It can autonomously navigate through debris and identify potential hazards. However, if it detects a life sign, it might signal this finding to a human operator, who then makes the decision on how to proceed, perhaps guiding the drone to a more specific location or directing ground rescue teams. The drone is providing situational awareness and executing pre-programmed reconnaissance, while the human provides the critical decision-making based on the gathered information and their expertise.

The design of systems that facilitate shared autonomy requires a deep understanding of cognitive load. Both the human and the robot should operate with minimal unnecessary cognitive burden. This means the interfaces must be clear, the information provided must be relevant, and the transitions between human and robot control should be seamless and intuitive. Overloading the human operator with too much information or expecting them to constantly monitor the robot's every move can lead to fatigue, errors, and ultimately, a breakdown of trust and collaboration.

Furthermore, the concept of **situational awareness** is central to effective collaboration and shared autonomy. Both the human and the robot need to have a shared understanding of the current state of the world and the task at hand. This includes knowing where other agents (human or robot) are, what they are doing, and what their potential intentions are. Robots can contribute to situational awareness by providing sensor data, highlighting potential risks or opportunities, and communicating their own status and plans. Humans contribute through their higher-level reasoning, contextual understanding, and ability to perceive nuances that sensors might miss.

The development of trust and collaboration in HRI is an ongoing process that requires careful consideration of the human element at every stage. As robots become more integrated into our lives, the ability to form trusting relationships with them and to collaborate effectively will be a defining characteristic of successful human-machine partnerships. This necessitates a commitment to designing robots that are not only intelligent and capable but also transparent, predictable, and inherently collaborative. The future of robotics lies not in replacing humans, but in augmenting human capabilities through seamless, trustworthy, and collaborative interaction.

The integration of robots into human environments, whether for industrial, domestic, or social purposes, inherently introduces a critical dimension: safety. This is not merely a technical challenge but a fundamental prerequisite for the acceptance and successful deployment of robotic systems. Ensuring the physical well-being of individuals interacting with robots necessitates a multi-layered approach, encompassing robust hardware, intelligent software, and clear operational protocols.

Physical safety in Human-Robot Interaction (HRI) is primarily achieved through a combination of sophisticated sensing technologies and proactive control strategies. Proximity sensors, for instance, form the first line of defense. These devices, which can range from infrared and ultrasonic sensors to advanced vision-based systems, continuously monitor the space around the robot. Their primary function is to detect the presence of humans or other obstacles within a predefined safety zone. When a human enters this zone, the robot's behavior can be modulated in real-time. This might involve a gradual reduction in speed, a subtle alteration of its trajectory to create more space, or, in more critical scenarios, a complete halt. The range and sensitivity of these sensors are crucial; they must be able to detect approaching individuals with sufficient lead time to allow for a safe response, taking into account the robot's velocity and inertia.

Collision avoidance algorithms are the intelligent engines that process the data from proximity sensors and other environmental perception systems. These algorithms are designed to predict potential collisions and compute evasive maneuvers. For robots operating in dynamic environments, such as warehouses or collaborative manufacturing lines, these systems must be highly responsive and capable of handling complex, multi-object interactions. For example, a collaborative robot arm working alongside a human assembler might need to instantaneously retract or shift its path if the human unexpectedly reaches into its operational space. This requires sophisticated real-time path planning that can dynamically replan trajectories based on incoming sensor data and predictive models of human movement. The robustness of these algorithms is paramount; they must be able to function reliably even under noisy sensor conditions or when faced with unexpected changes in the environment.

Emergency stop (E-stop) mechanisms represent the ultimate safeguard. These are typically physical buttons or software commands that, when activated, immediately and safely bring the robot to a complete standstill. E-stops are designed to be easily accessible and immediately recognizable by human operators, supervisors, or even bystanders. In industrial settings, E-stop buttons are often strategically placed throughout the workspace. For mobile robots, an E-stop might also be triggered by external safety systems, such as light curtains or safety mats, that detect a breach of a perimeter. The implementation of E-stops must adhere to strict safety standards, ensuring that they are fail-safe, meaning that any failure in the E-stop system itself will result in a shutdown rather than a loss of safety. The speed and decisiveness with which an E-stop can be engaged and the subsequent safe stopping of the robot are critical factors in mitigating severe accidents.

Beyond these immediate physical safety measures, there are broader ethical considerations that govern the development and deployment of HRI systems. These extend to how robots interact with humans in terms of their data handling, decision-making processes, and their ultimate impact on society.

Privacy is a significant ethical concern, particularly for robots equipped with cameras, microphones, and other sensors that can collect information about their surroundings. Robots operating in homes, workplaces, or public spaces can inadvertently or intentionally gather personal data from individuals. This raises questions about who has access to this data, how it is stored, and for what purposes it is used. For instance, a domestic robot vacuum cleaner equipped with a camera might map out a home's layout, potentially revealing sensitive information about its occupants. Similarly, a service robot in a retail environment could collect data on customer behavior and preferences. Robust data anonymization techniques, strict access controls, and transparent data usage policies are essential to protect individual privacy. Users should be clearly informed about what data the robot collects, why it is collected, and how it will be used, with opportunities to consent or opt-out where feasible.

Data security is intrinsically linked to privacy. The data collected by robots, especially sensitive personal information, must be protected from unauthorized access, breaches, and misuse. This requires implementing strong cybersecurity measures, including encryption of data both in transit and at rest, secure authentication protocols, and regular security audits. The potential for robots to be compromised and used as vectors for cyberattacks, or for the sensitive data they collect to be exploited, underscores the importance of prioritizing data security throughout the robot's lifecycle, from design and manufacturing to deployment and maintenance.

Accountability is another complex ethical challenge. When a robot causes harm, whether physical injury, financial loss, or privacy violation, determining who is responsible can be difficult. Is it the programmer, the manufacturer, the owner, or the robot itself (if it possesses a degree of autonomy)? Establishing clear lines of accountability is crucial for legal and ethical recourse. This might involve developing frameworks for assigning responsibility based on the degree of autonomy, the design choices made, and the operational context. For example, if a self-driving car causes an accident, accountability might be distributed among the car manufacturer (for design flaws), the software developers (for algorithmic errors), and potentially the owner (for improper use or maintenance). This is an evolving area of law and ethics, requiring careful consideration of the unique nature of autonomous and semi-autonomous systems.

The societal impact of human-robot relationships also warrants careful consideration. As robots become more sophisticated and capable of performing tasks previously exclusive to humans, concerns arise about job displacement and the changing nature of work. While robots can enhance productivity and create new types of jobs, it is imperative to manage the transition in a way that minimizes negative consequences for the workforce. This includes investing in education and retraining programs to equip individuals with the skills needed to work alongside robots or in emerging fields. Furthermore, the increasing presence of robots in social contexts, such as companion robots for the elderly, raises questions about the nature of human connection and the potential for over-reliance on artificial companionship. The goal should be to design robots that augment human capabilities and improve quality of life, rather than replacing essential human interactions or exacerbating societal inequalities.

Responsible design and deployment of robots interacting with humans must be guided by a set of principles that prioritize user well-being and societal benefit. This involves a human-centered design philosophy, where the needs, capabilities, and limitations of the human user are at the forefront of every design decision. Robots should be designed to be intuitive to operate, easy to understand, and predictable in their behavior, fostering a sense of control and trust among users.

User well-being extends beyond physical safety to encompass psychological and social aspects. Robots should not be designed in ways that are intentionally deceptive, manipulative, or that exploit human vulnerabilities. For instance, a robot designed to persuade users to make purchases should be transparent about its persuasive intent. Similarly, robots designed for therapeutic purposes must be developed with rigorous clinical validation and ethical oversight to ensure they are genuinely beneficial and do not cause unintended psychological harm.

Societal benefit should be a guiding principle for the deployment of robotics. This means considering the broader implications of a robot system on communities and society as a whole. For example, deploying robots in public spaces should be done with careful consideration of their impact on public order, accessibility, and the potential for surveillance. The development and deployment of AI and robotics should aim to address pressing societal challenges, such as climate change, healthcare access, and disaster response, in ways that are equitable and sustainable.

To facilitate responsible innovation, several guidelines can be adopted. Firstly, there should be a commitment to transparency in robot design and operation. This includes making information about a robot's capabilities, limitations, and data handling practices readily available to users and the public. Secondly, robust testing and validation processes are essential, not only for functional performance but also for safety and ethical compliance. This often involves multidisciplinary teams including engineers, ethicists, social scientists, and end-users. Thirdly, there is a need for continuous monitoring and evaluation of robots once they are deployed, allowing for the identification and mitigation of unforeseen safety or ethical issues. This iterative approach ensures that robots remain aligned with human values and societal expectations over time.

The development of ethical frameworks and regulatory guidelines for HRI is an ongoing and collaborative effort. International organizations, national governments, research institutions, and industry bodies are all contributing to the creation of standards and best practices. These efforts aim to strike a balance between fostering innovation and ensuring that robotic technologies are developed and used in ways that are safe, ethical, and beneficial to humanity. The goal is to build a future where humans and robots can coexist and collaborate productively and harmoniously, with human well-being and societal progress as the ultimate objectives. This requires a proactive and thoughtful approach to safety and ethics, integrated into every stage of the HRI lifecycle.

Chapter 13: Ethical and Societal Implications of Robotics

The burgeoning field of artificial intelligence (AI) and robotics, while promising unprecedented advancements and efficiencies, simultaneously presents a complex and evolving ethical landscape. As these technologies move from controlled laboratory environments into the fabric of our daily lives, they necessitate a profound examination of the moral principles that should guide their development, deployment, and interaction with humanity. This examination is not merely an academic exercise; it is a critical endeavor to ensure that AI and robotic systems serve to augment human well-being and societal progress, rather than inadvertently creating new forms of harm or exacerbating existing inequalities.

At the core of this ethical discourse are fundamental principles that have long underpinned human morality and continue to be essential as we navigate the complexities of autonomous systems. One such principle is
fairness. AI systems, particularly those that make decisions affecting human lives – such as in loan applications, hiring processes, or even criminal justice – must operate without prejudice. The danger of bias in AI is a significant concern, often stemming from the data on which these systems are trained. If historical data reflects societal biases, AI algorithms trained on this data will likely perpetuate and even amplify those biases, leading to discriminatory outcomes. For instance, facial recognition systems have been shown to exhibit lower accuracy rates for individuals with darker

skin tones or for women, a direct consequence of imbalanced training datasets. Addressing this requires meticulous attention to data collection, preprocessing, and algorithmic design to ensure equitable treatment across all demographics. The pursuit of fairness in AI is an ongoing challenge, demanding constant vigilance and proactive measures to identify and mitigate bias at every stage of the AI lifecycle.

Closely linked to fairness is the concept of **accountability**. As systems become more autonomous, the question of who is responsible when things go wrong becomes increasingly intricate. When an autonomous vehicle is involved in an accident, or a medical diagnostic AI misinterprets a scan, assigning blame is not straightforward. Is it the programmer who wrote the faulty code, the company that manufactured the system, the owner who deployed it, or the entity that failed to regulate its use? Unlike traditional products where responsibility can often be traced back to a specific party, the emergent behaviors and complex decision-making processes of AI can obscure direct causal links. Establishing clear frameworks for accountability is crucial for providing recourse to those who are harmed and for incentivizing responsible development. This involves not only legal and regulatory mechanisms but also a culture of responsibility within the AI development community, emphasizing meticulous documentation, rigorous testing, and transparent reporting of system capabilities and limitations. The very nature of AI, with its potential for self-learning and adaptation, further complicates traditional notions of accountability, requiring new legal and ethical paradigms to address these unique challenges.

Transparency, often referred to as explainability or interpretability, is another cornerstone of ethical AI. Many advanced AI systems, particularly deep learning models, operate as "black boxes," making it difficult to understand precisely how they arrive at a particular decision. This lack of transparency is problematic, especially in high-stakes applications. If a doctor cannot understand why an AI recommended a specific treatment, or if a judge cannot comprehend the reasoning behind an AI's risk assessment for a defendant, trust in the system erodes. Transparency is not just about satisfying curiosity; it is about enabling verification, identifying errors, and building confidence in AI technologies. Research into explainable AI (XAI) aims to develop methods and tools that can shed light on the internal workings of AI models, providing insights into the factors that influence their outputs. However, achieving a level of transparency that is both meaningful and technically feasible remains an active area of research and development, often involving a trade-off between model complexity and interpretability.

The principle of
beneficence, or acting in ways that promote well-being and prevent harm, is perhaps the overarching goal of ethical AI development. This principle compels us to consider the positive impact of AI and robotics on individuals and society. Will these technologies improve

healthcare outcomes, enhance educational opportunities, or address pressing environmental challenges? Conversely, we must proactively identify and mitigate potential harms, such as job displacement, erosion of privacy, or the creation of new social divides. The application of AI in healthcare, for example, offers immense potential for early disease detection, personalized treatment plans, and efficient drug discovery. However, it also raises concerns about patient data privacy, the potential for diagnostic errors due to biased algorithms, and the role of human empathy in care. Similarly, while AI-powered automation can boost productivity and create new forms of employment, it also poses a significant risk of widespread job displacement if not managed thoughtfully with robust retraining and social safety nets.

The capacity for
autonomous decision-making in AI and robotics introduces a particularly challenging set of ethical dilemmas, especially when these decisions have life-or-death consequences. The infamous "trolley problem" – a thought experiment where one must choose between allowing a runaway trolley to kill multiple people or diverting it to kill a single person – serves as a simplified analogy for the complex ethical choices that autonomous systems might face. Consider an autonomous vehicle confronted with an unavoidable accident scenario: should it prioritize the safety of its occupants, or minimize the total number of casualties, even if it means sacrificing its passengers? These are not hypothetical scenarios confined to philosophical

debates. As autonomous systems become more prevalent, developers and policymakers must grapple with how to embed ethical decision-making frameworks into their core programming. This requires defining values, establishing priorities, and developing robust methodologies for programming moral choices into machines, a task that is fraught with complexity given the diversity of human ethical perspectives and cultural norms. The development of these frameworks necessitates broad societal consensus, involving ethicists, engineers, legal scholars, and the public, to ensure that the decisions made by machines align with shared human values.

Beyond these core principles, the development and deployment of AI and robotics are intertwined with broader societal considerations. The potential for AI to be used in
surveillance and to amplify existing power imbalances is a pressing concern. Advanced AI can process vast amounts of data, enabling unprecedented levels of monitoring by governments or corporations. This raises questions about civil liberties, the right to privacy, and the potential for AI to be used for social control or to suppress dissent. Ensuring that AI is developed and used in ways that respect human rights and democratic values is paramount.

Furthermore, the
human-robot interaction (HRI) aspect of robotics brings its own set of ethical considerations. As robots become more sophisticated in their ability to mimic human behavior and engage in social interactions,

questions arise about the nature of these relationships. Companion robots designed for the elderly or for children, for instance, can offer benefits such as reducing loneliness or providing educational assistance. However, they also raise concerns about the authenticity of these interactions, the potential for emotional manipulation, and whether they might detract from genuine human connection. The ethical design of social robots requires careful consideration of their impact on human psychology and social dynamics, ensuring that they enhance, rather than diminish, human well-being and social bonds.

The responsibility for navigating this complex ethical terrain rests not solely on engineers and AI researchers, but on a broad spectrum of stakeholders. **Developers and manufacturers** have a primary ethical obligation to design and build AI systems that are safe, fair, and transparent. This includes conducting thorough risk assessments, implementing robust testing procedures, and being transparent about the capabilities and limitations of their products. **Users and deployers** of AI systems also bear responsibility for ensuring that these technologies are used ethically and appropriately within their specific contexts. This may involve establishing clear guidelines for AI use, providing training to individuals interacting with AI, and continuously monitoring AI systems for unintended consequences. **Policymakers and regulators** play a crucial role in establishing legal and ethical frameworks that guide the development and deployment of AI, setting standards, and enforcing accountability. This requires a proactive and informed approach to regulation, one that can adapt to the rapid pace of technological advancement while safeguarding societal interests.

The creation of effective ethical guidelines and regulations for AI and robotics is an ongoing and iterative process. It requires a multidisciplinary approach, bringing together expertise from computer science, philosophy, law, sociology, psychology, and other fields. International cooperation is also essential, as AI technologies transcend national borders and require global standards to address common ethical challenges. The ultimate goal is to foster an environment where AI and robotics can be developed and deployed in a manner that maximizes their benefits for humanity while minimizing their risks, ensuring that these powerful technologies remain tools that serve human values and aspirations. This demands a commitment to continuous learning, open dialogue, and a shared understanding that the ethical development of AI is not a one-time task, but a perpetual endeavor intertwined with the very evolution of these transformative technologies.

The pervasive integration of robotics and automation into the industrial and service sectors marks a transformative era, prompting a critical examination of its profound economic and social implications. At the forefront of these discussions is the complex interplay between technological advancement and the labor market. The narrative surrounding automation is often polarized, with some envisioning widespread unemployment as machines supplant human workers, while others foresee a future of augmented workforces and entirely new categories of employment. The reality, as history has shown with previous technological revolutions, is likely to be a more nuanced and multifaceted outcome, characterized by both disruption and opportunity.

One of the most immediate and significant concerns is the potential for job displacement. As robots become more sophisticated, capable of performing tasks that were once exclusively the domain of human labor – from intricate manufacturing processes and heavy lifting in warehouses to data analysis and even customer service interactions – many existing roles are undoubtedly at risk. Occupations that involve repetitive, predictable, or physically demanding tasks are particularly vulnerable. For instance, advancements in robotic dexterity and sensing capabilities are enabling robots to assemble complex products with precision, operate autonomous forklifts in logistics centers, and even perform routine maintenance tasks that previously required skilled technicians. The economic incentive for businesses to adopt these technologies is often driven by a desire to increase efficiency, reduce operational costs, improve product quality, and enhance worker safety by removing humans from hazardous environments.

However, this narrative of displacement is only one part of the story. Technological progress has historically been accompanied by the creation of new jobs and industries. The advent of the personal computer, for example, did not lead to mass, permanent unemployment; rather, it spawned entirely new fields such as software development, IT support, digital marketing, and cybersecurity. Similarly, the increasing sophistication of robotics and AI is creating a demand for new roles. There is a growing need for robotics engineers who design, build, and maintain these advanced systems. Technicians specializing in robot repair and calibration are becoming indispensable. Furthermore, the development and implementation of AI algorithms that power many of these robots require data scientists, AI ethicists, and machine learning specialists. As robots take over mundane tasks, human workers may find themselves freed up to focus on more complex, creative, and interpersonal aspects of their jobs, roles that are inherently more difficult for machines to replicate.

The crucial challenge lies in the transition. The skills required for the jobs of the future are likely to differ significantly from those that are being automated away. This necessitates a proactive approach to workforce adaptation. Education and training systems must evolve to equip individuals with the skills needed to thrive in an automated economy. This includes not only technical skills related to robotics, AI, and data science but also "soft skills" such as critical thinking, problem-solving, creativity, collaboration, and emotional intelligence. Lifelong learning will no longer be a buzzword but a fundamental requirement for career longevity. Governments, educational institutions, and businesses will need to collaborate to develop robust retraining and upskilling programs. These programs should be accessible, affordable, and tailored to the evolving demands of the labor market, providing pathways for individuals to transition from declining sectors to emerging ones.

The economic impact of automation extends beyond employment figures, touching upon productivity and overall economic growth. Advanced robotics and AI have the potential to unlock unprecedented levels of productivity. By working tirelessly, with high precision, and without the need for breaks or rest, robots can significantly increase output per worker and per hour. This enhanced productivity can lead to lower production costs, making goods and services more affordable for consumers. It can also drive innovation, enabling the creation of new products and services that were previously technologically or economically infeasible. The ability of automated systems to operate 24/7 and to scale operations rapidly can lead to substantial economic gains for businesses and nations that effectively harness these technologies. This can translate into higher corporate profits, increased tax revenues, and, in theory, a higher standard of living for the population.

However, the distribution of these economic gains is a critical consideration. If the benefits of automation accrue primarily to a small segment of the population – the owners of capital and highly skilled workers – it could exacerbate income inequality. This potential widening of the gap between the wealthy and the rest of the population is a significant societal concern. It raises questions about how the wealth generated by increased productivity should be shared. Various policy responses have been proposed to address this challenge. One such proposal is a universal basic income (UBI), which would provide a regular, unconditional income to all citizens, serving as a safety net and ensuring that everyone benefits from the economic productivity gains of automation. While UBI is a subject of ongoing debate, with proponents highlighting its potential to alleviate poverty and provide economic security, and critics raising concerns about its cost, impact on labor supply, and inflationary effects, it represents a conceptual approach to decoupling income from traditional employment.

Another policy consideration revolves around taxation. As labor becomes less central to production, traditional tax bases that rely heavily on income tax may erode. This has led to discussions about potential shifts in taxation, such as taxing robots or the profits generated by automation more heavily, to fund social programs and support displaced workers. The implementation of such policies is complex, requiring careful consideration of international competitiveness and the potential for unintended consequences. The goal is to strike a balance that encourages innovation and investment in automation while ensuring that its economic benefits are shared broadly and that society as a whole can adapt to the changing economic landscape.

The geographical distribution of automation's impact is also noteworthy. Regions and countries that are heavily reliant on industries with high automation potential may face more significant challenges in the short to medium term. Conversely, those that are at the forefront of technological development and adoption are likely to reap greater economic rewards. This could lead to shifts in global economic power and influence. Developing nations, in particular, may face a dilemma: embracing automation to boost their economies could lead to job displacement in sectors that are crucial for employment, while foregoing it might leave them behind in the global race for technological advancement and economic competitiveness. International cooperation and targeted development assistance may be crucial to help these nations navigate the complexities of automation and ensure inclusive growth.

Furthermore, the nature of work itself is likely to change. Even for those whose jobs are not fully automated, the tasks and responsibilities within their roles may be augmented by robotic and AI systems. For example, a surgeon might use robotic arms for greater precision during operations, or a teacher might use AI-powered tools to personalize lesson plans for students. This augmentation can lead to increased job satisfaction and effectiveness, but it also requires workers to adapt to new ways of working alongside machines. The human element – creativity, empathy, complex decision-making, and interpersonal skills – will likely become even more valuable as machines handle the more routine and data-intensive aspects of work.

The societal implications of widespread automation also extend beyond the purely economic. A significant concern is the potential erosion of community and social cohesion if large segments of the population experience prolonged periods of unemployment or underemployment. The sense of purpose and identity that many derive from their work could be diminished, leading to social unrest and psychological distress. This underscores the importance of not only economic policies but also social support systems and community development initiatives to help individuals and communities adapt to these profound changes. Investing in public services, promoting civic engagement, and fostering a sense of shared purpose will be critical in navigating the societal shifts brought about by automation.

In conclusion, the economic and societal implications of automation driven by robotics are vast and multifaceted. While the potential for job displacement is a genuine concern, it is imperative to recognize the concurrent creation of new roles, the significant gains in productivity, and the potential for enhanced economic growth. The key to harnessing the benefits of automation while mitigating its risks lies in proactive adaptation. This requires a concerted effort from governments, educational institutions, businesses, and individuals to invest in education and retraining, develop supportive social policies, and foster a mindset of continuous learning and adaptability. By addressing these challenges head-on, society can aim to achieve a future where automation serves as a tool for widespread prosperity and human flourishing, rather than a source of division and displacement. The transition will undoubtedly be complex, demanding careful planning, thoughtful policy, and a commitment to ensuring that the gains of technological progress are shared equitably across society.

The ever-increasing presence of robots, particularly those equipped with sophisticated sensing capabilities, ushers in a new era of profound privacy and surveillance concerns. As these machines move beyond the controlled environments of factories and into the dynamic landscapes of our homes, workplaces, and public spaces, they become unwitting collectors of an unprecedented volume of data. This data can range from the innocuous - the trajectory of a robotic vacuum cleaner navigating a living room - to the highly sensitive - conversations overheard by a service robot in a hotel lobby or biometric information captured by a security automaton. The very functionality that makes robots useful, their ability to perceive and interact with their environment, also positions them as potent instruments for observation.

Consider, for instance, the proliferation of domestic robots. Beyond their primary function of cleaning or companionship, many are equipped with cameras, microphones, and various environmental sensors. These sensors are essential for navigation and task execution, allowing a robotic vacuum to map a room, avoid obstacles, and optimize its cleaning path. However, these same sensors can inadvertently, or deliberately, record personal activities, conversations, and even the physical layout of a private residence. When this data is transmitted to cloud servers for processing, storage, or software updates, it creates multiple points of vulnerability. Who has access to this data? How is it secured? For how long is it retained? The answers to these questions are often opaque to the end-user, leaving individuals susceptible to unintended disclosures or even malicious breaches. The potential for this data to be used for purposes beyond its original intent - targeted advertising based on observed habits, behavioral analysis without consent, or even as evidence in legal proceedings - raises significant ethical and legal questions.

The challenge of anonymization is particularly acute. While efforts can be made to de-identify data by removing direct personal identifiers like names or addresses, the sheer richness and specificity of the data collected by advanced robots can often allow for re-identification. Imagine a robot in a public park that records video footage. While individual faces might be blurred, the unique gait of a person, their clothing, the timing of their presence in specific locations, or even the presence of a distinctive pet could, in combination, allow for tracking and identification. Similarly, audio recordings from a smart speaker or a domestic robot might capture not just spoken words but also background sounds that reveal the presence of specific individuals, the nature of their activities, or their emotional state, all of which can be deeply personal. The aggregation of such data over time, across multiple robots or sensors, can paint an incredibly detailed portrait of an individual's life, habits, and social connections, far beyond what they might voluntarily share.

The implications for individual privacy rights are substantial. The ability of robotic systems to collect data covertly, or with the user's limited and often unread consent, fundamentally challenges the notion of control over one's personal information. In many jurisdictions, laws governing data privacy, such as the General Data Protection Regulation (GDPR) in Europe, aim to provide individuals with rights regarding the collection, processing, and storage of their personal data. However, the rapid evolution of robotics technology often outpaces the development and enforcement of these legal frameworks. Ensuring that robots are designed with privacy in mind – adhering to principles of "privacy by design" and "privacy by default" – is crucial. This means that privacy considerations should be embedded into the very architecture of robotic systems from their inception, rather than being an afterthought.

Furthermore, the question of data ownership and control becomes blurred. When a robot collects data within a private home, who truly owns that data? The individual who purchased the robot? The company that manufactured it? The third-party service provider that processes or stores the data? Current legal frameworks are still grappling with these complex ownership issues, often defaulting to the terms of service agreements that users click through without fully understanding. The lack of clear ownership can lead to situations where individuals have little recourse if their data is misused or sold to third parties. The expectation of privacy within one's own home is a fundamental right, and the pervasive data-gathering capabilities of robots pose a significant threat to this expectation.

The use of robots in public spaces amplifies these concerns. Autonomous vehicles, delivery robots, and security drones are increasingly navigating our streets and skies. While they may be programmed for specific tasks, their sensors are constantly gathering information about their surroundings. This can include identifying individuals, tracking movements, and potentially even monitoring conversations. The deployment of widespread robotic surveillance in public areas could have a chilling effect on freedom of assembly and expression. If citizens know they are constantly being observed and their data collected by an array of automated systems, they may become hesitant to engage in lawful activities that could be misinterpreted or used against them. The potential for mission creep, where robots initially deployed for one purpose are later repurposed for broader surveillance, is a significant societal risk.

The anonymization of data collected in public spaces

also presents unique challenges. While the intent might be to gather aggregate data for urban planning or traffic management, the potential for individual tracking remains. For example, a fleet of delivery robots equipped with GPS and cameras could inadvertently create a detailed log of movements for every person they encounter on their routes. Without robust anonymization protocols and strict access controls, this data could be exploited for commercial profiling, law enforcement surveillance without probable cause, or even by malicious actors seeking to identify and track individuals.

The legal and regulatory landscape surrounding robotic data collection is a complex and evolving one. Many existing privacy laws were drafted with traditional data collection methods in mind and may not adequately address the unique challenges posed by robotic systems. For instance, obtaining informed consent for data collection by a robot that is constantly sensing its environment is practically difficult. How does one "opt out" of being scanned by a security robot at an airport or having their movements logged by a delivery bot on their street? The sheer scale and continuous nature of data collection by robots necessitate a re-evaluation of consent mechanisms and data protection principles.

New legal frameworks may be required to specifically address robotic surveillance and data privacy. These frameworks should likely focus on several key areas. Firstly, there should be a clear emphasis on data minimization, meaning robots should only collect the data that is absolutely necessary for their intended function. Secondly, transparency is paramount. Users and the public should be made aware of what data is being collected, how it is being used, and who has access to it. This information should be presented in a clear and accessible manner, not buried in lengthy terms of service agreements. Thirdly, robust security measures must be mandated to protect the collected data from unauthorized access, breaches, and misuse. Encryption, secure storage, and regular security audits should be standard practice.

Furthermore, accountability mechanisms need to be established. If a robotic system violates privacy regulations or causes harm through data misuse, there must be clear pathways for redress and penalties for the responsible parties, whether they are the manufacturers, operators, or data processors. This could involve establishing independent oversight bodies that can audit robotic systems and investigate privacy complaints. The concept of "accountability by design" for robotic systems, where legal and ethical responsibilities are integrated into the development process, is gaining traction and is essential for building trust in these technologies.

The ethical considerations surrounding data collection by robots extend beyond legal compliance. Even if data collection is technically legal, it may still be ethically questionable if it infringes upon reasonable expectations of privacy or disproportionately impacts certain groups. For example, the deployment of facial recognition technology on security robots in public housing complexes, while potentially justifiable for security purposes, could disproportionately affect minority communities who may already feel over-policed. Such deployments raise questions about fairness, equity, and the potential for algorithmic bias in surveillance.

The economic incentives for data collection can also create a tension with privacy. Companies that develop and deploy robots often see the data they collect as a valuable asset, a source of revenue through analytics, targeted advertising, or by selling insights to third parties. This creates a potential conflict of interest where the company's profit motive may not align with the individual's right to privacy. Robust regulation is therefore essential to ensure that the pursuit of profit does not come at the expense of fundamental privacy rights.

The development of ethical guidelines and best practices for robotic data collection is an ongoing process involving engineers, ethicists, policymakers, and the public. Organizations are beginning to establish codes of conduct for AI and robotics that explicitly address privacy concerns. However, these are often voluntary and lack the force of law. The challenge lies in translating these ethical aspirations into concrete, enforceable standards that govern the design, deployment, and operation of robotic systems.

The concept of "data trusts" or similar models, where data is managed by independent entities with a fiduciary duty to protect the interests of the data subjects, could offer a potential avenue for addressing some of these concerns. In such a model, individuals might have greater control over how their data is used, who it is shared with, and for what purpose, even when collected by robotic systems.

Ultimately, fostering public trust in robotics and AI hinges on addressing these privacy and surveillance concerns proactively and transparently. Without clear boundaries, robust safeguards, and effective oversight, the integration of robots into our lives could lead to a future where privacy is an obsolete concept, replaced by constant, pervasive surveillance, managed and exploited by algorithms and the entities that control them. The very benefits that robotics promise – increased efficiency, convenience, and safety – could be overshadowed by the erosion of fundamental freedoms if privacy is not treated as a non-negotiable human right in the age of intelligent machines.

The relentless march of technological progress, particularly within the fields of robotics and artificial intelligence, has ushered in a new and deeply unsettling frontier: Lethal Autonomous Weapons Systems (LAWS). These are not merely sophisticated machines that assist in warfare; they are systems designed to identify, select, and engage targets without direct human control. This paradigm shift in how conflict might be waged raises profound ethical, legal, and societal questions that demand urgent and careful consideration. At the heart of the debate lies the very definition of what it means to use lethal force. Historically, the decision to take a human life in conflict has rested, however imperfectly, with human judgment. LAWS seek to automate this critical decision, a prospect that many find morally abhorrent.

Proponents of LAWS often highlight potential military advantages. They argue that autonomous weapons could be more precise, less prone to emotional error or fatigue than human soldiers, and therefore capable of reducing collateral damage and civilian casualties. The speed at which these systems could operate, reacting to threats far faster than a human operator could, is also cited as a crucial advantage in rapidly evolving combat scenarios. Furthermore, proponents suggest that deploying autonomous systems could reduce the risk to one's own forces, keeping human soldiers out of harm's way. The idea is that if a machine can perform a dangerous task, it should, thereby preserving human lives on the side deploying the weapon. Some also posit that LAWS could adhere more strictly to the laws of armed conflict, as they could be programmed with precise rules of engagement, theoretically making them less likely to commit war crimes born of panic, revenge, or misunderstanding. This theoretical adherence to rules, proponents suggest, could lead to a more "humane" form of warfare, albeit one waged by machines.

However, the ethical objections to LAWS are substantial and deeply rooted. A primary concern revolves around accountability. When an autonomous weapon system makes a mistake – for instance, misidentifying a civilian as a combatant and causing an unintended death – who is held responsible? Is it the programmer who wrote the code? The commander who deployed the system? The manufacturer? Or is it the machine itself, which lacks consciousness and intent? The absence of a clear chain of accountability creates a dangerous moral vacuum, potentially leading to a climate of impunity for war crimes. The complexity of the algorithms, the "black box" nature of some AI decision-making processes, makes it exceedingly difficult to trace the precise cause of an error, further complicating the assignment of blame. This lack of clear responsibility undermines fundamental principles of justice and international humanitarian law.

Another significant ethical concern is the potential for LAWS to lower the threshold for engaging in conflict. The removal of human soldiers from direct harm's way, coupled with the perceived efficiency and precision of autonomous systems, might make the decision to go to war easier for political leaders. The emotional and psychological burden of sending young men and women into combat, and the subsequent grief and public outcry when they are killed, can act as a deterrent to war. If warfare can be conducted remotely and with machines, this significant check on belligerence could be eroded, leading to more frequent and perhaps less considered conflicts. The dehumanization of warfare, where the killing is done by algorithms rather than by individuals wrestling with moral dilemmas, could lead to a society that is less averse to violence.

The erosion of human control over the decision to take a life is perhaps the most profound ethical objection. Many argue that the act of killing, particularly in war, requires human judgment, empathy, and the capacity for mercy. These are qualities that machines, no matter how sophisticated, are not believed to possess. Entrusting life-and-death decisions to algorithms risks fundamentally altering the nature of humanity and our relationship with violence. There is a deep-seated intuition that the decision to extinguish a human life should remain firmly within the moral agency of human beings. The "meaningful human control" principle, which is central to many international discussions on LAWS, seeks to ensure that humans retain sufficient control over the critical functions of weapon systems, particularly targeting and engagement.

The potential for escalation of conflict is another grave concern. LAWS could operate at speeds that outpace human comprehension and diplomatic intervention. Imagine a scenario where two autonomous weapons systems engage each other based on programmed parameters, with the conflict rapidly spiraling out of control before human operators can even assess the situation, let alone de-escalate. The very speed and autonomy that proponents tout as advantages could become catastrophic liabilities, leading to unintended wars or the rapid amplification of existing conflicts. The absence of human deliberation and the potential for algorithmic miscalculation could lead to devastating consequences.

Furthermore, the development and deployment of LAWS raise questions about discrimination and bias. While proponents argue for precision, the algorithms that govern these weapons are trained on data, and this data can reflect existing societal biases. If the data used to train a targeting algorithm is skewed, for example, it could lead to disproportionate targeting of certain ethnic groups or populations. The potential for algorithmic bias in lethal decision-making is a chilling prospect, adding another layer of ethical complexity to the already fraught debate. The opaque nature of some AI decision-making processes makes it difficult to identify and rectify such biases.

The international community has been grappling with the implications of LAWS for years. Numerous discussions have taken place within the framework of the United Nations, particularly at the Convention on Certain Conventional Weapons (CCW). While there is broad consensus on the need to address the challenges posed by these weapons, there is significant disagreement on the path forward. Some nations advocate for a complete ban on LAWS, arguing that the ethical and humanitarian risks are too great to permit their development. Others prefer a regulatory approach, seeking to establish clear guidelines and restrictions on their use, while still allowing for their development and deployment under strict conditions. The concept of "meaningful human control" has emerged as a key point of discussion in these international forums, with various interpretations of what constitutes sufficient human oversight.

The challenges in reaching an international agreement are manifold. Different nations have varying military doctrines, technological capabilities, and perceived security needs. Some view LAWS as an inevitable aspect of future warfare and seek to develop them to maintain a strategic advantage. Others, often those with less advanced military technology, see them as a destabilizing force and a threat to global security. The difficulty in defining LAWS precisely also complicates negotiations; where does one draw the line between a remotely operated weapon with advanced AI features and a fully autonomous weapon system? This definitional ambiguity can be exploited to circumvent potential regulations.

Despite these challenges, the momentum for some form of international regulation is growing. Civil society organizations, including the Campaign to Stop Killer Robots, are actively advocating for a preemptive ban, raising public awareness and lobbying governments.

They argue that allowing the development of LAWS would be a grave mistake, leading to a future where warfare is not only automated but also potentially devoid of human compassion and moral restraint. The ethical imperative, they argue, is to draw a line now, before these technologies become so deeply embedded that they are impossible to uninvent.

The technological feasibility of LAWS is rapidly increasing. Advances in sensor technology, artificial intelligence, machine learning, and robotics are making it increasingly possible to create systems that can operate with a high degree of autonomy in complex environments. While fully autonomous weapons that can operate without any human oversight in all circumstances remain largely theoretical, incremental steps towards greater autonomy are being taken. This includes systems that can autonomously select targets within a pre-defined area, or weapons that can operate in swarms with coordinated, albeit programmed, decision-making. The pace of innovation means that the window for meaningful discussion and regulation is closing.

The societal implications extend beyond the battlefield. The proliferation of autonomous weapons could lead to a new arms race, with nations competing to develop the most advanced and capable systems. This could divert significant resources from other critical areas, such as healthcare, education, or sustainable development. Moreover, the very existence of LAWS could normalize the idea of automated killing, with potential spillover effects into law enforcement and other areas where autonomous decision-making with lethal consequences might be considered. The prospect of autonomous robots making life-or-death decisions in policing scenarios, for example, raises similar ethical concerns about accountability, bias, and the devaluation of human life.

The debate over LAWS is not merely a technical or military one; it is a fundamental ethical and philosophical inquiry into the future of warfare and humanity's role within it. It forces us to confront difficult questions about the nature of responsibility, the value of human life, and the potential consequences of delegating critical moral decisions to machines. As the technology continues to advance, the urgency of these questions will only intensify, demanding a global commitment to ensuring that the future of warfare remains guided by human judgment and humanitarian principles, rather than by algorithms alone. The path forward requires a delicate balance between acknowledging technological possibilities and upholding fundamental ethical imperatives, ensuring that innovation serves humanity rather than imperils it.

The increasing sophistication and integration of robots into our daily lives, particularly those designed for social interaction, introduce a new and complex set of ethical and societal considerations. While the previous discussions have grappled with the implications of robots in warfare and industrial settings, this section turns its attention to the burgeoning field of social robotics, encompassing everything from rudimentary companion bots to highly advanced humanoid machines that mimic human appearance and behavior. These robots are no longer confined to laboratories or specialized applications; they are entering our homes, our workplaces, and our social spaces, poised to fundamentally alter our interpersonal dynamics and our understanding of companionship itself.

One of the most immediate and widely discussed impacts of social robots is their potential to foster emotional attachment. As robots become more adept at understanding and responding to human emotions, and as their conversational abilities become more nuanced, it is natural for humans to develop bonds with them. For individuals experiencing loneliness, social isolation, or those with specific needs such as the elderly or children with developmental challenges, a well-designed companion robot can offer a consistent source of interaction, comfort, and even perceived affection. These robots can be programmed to exhibit empathetic behaviors, offer words of encouragement, engage in conversation, play games, and provide reminders for medication or appointments. The utilitarian benefits are clear: improved quality of life, reduced strain on human caregivers, and a greater sense of well-being for the user. However, the depth and nature of these human-robot emotional connections raise significant questions. Is an emotional bond with a machine qualitatively different from one with another human? Does the simulation of empathy by a robot constitute genuine care, or is it merely a sophisticated algorithmic response designed to elicit a specific reaction from the user? The ethical implications of facilitating such attachments are profound, especially when considering the vulnerability of the individuals who may rely most heavily on these robotic companions.

This leads directly to the issue of deception inherent in simulated emotions. Social robots are designed to appear caring, attentive, and even understanding. They can be programmed with facial expressions, vocal inflections, and body language that are conventionally associated with positive human emotions. While users may be aware, on an intellectual level, that these are programmed responses, the constant interaction with these simulated emotions can blur the lines between genuine feeling and artificial display. This is particularly concerning when considering children, who may not yet possess the cognitive abilities to fully differentiate between authentic human connection and programmed interaction. The potential for a child to form a deep emotional reliance on a robot that cannot truly reciprocate or understand their feelings raises questions about healthy emotional development and the nature of genuine social learning. Furthermore, for adults, particularly those who are susceptible to manipulation or who are seeking solace, the constant exposure to simulated affection might create a dependency that distracts from or even detracts from the pursuit of authentic human relationships. The "uncanny valley" effect, where near-human replicas evoke feelings of unease and revulsion, might be sidestepped by robots designed to be overtly robotic yet capable of demonstrating emotional cues. However, even when the uncanny valley is successfully navigated, the underlying mechanism remains an artificial construct, and the ethical considerations surrounding the deployment of such persuasive technologies in intimate contexts are paramount.

The impact on human social relationships is a complex web of potential benefits and drawbacks. On one hand, social robots could act as facilitators, bridging gaps in human connection. For instance, a robot designed to assist with elderly care might also help initiate conversations between the elder and their human family members by prompting discussions or recalling shared memories. In educational settings, robots can engage children in learning activities, making the process more interactive and enjoyable. They could also serve as tools for social skills training, providing a safe and predictable environment for individuals to practice communication and interaction. However, there is also a significant concern that the widespread adoption of social robots could lead to a decline in human-to-human interaction. If individuals can fulfill their need for companionship, attention, and emotional support through interactions with robots, they might become less inclined to invest the effort required to build and maintain complex human relationships. Human relationships are inherently messy, demanding, and require compromise and vulnerability. In contrast, a robot's interactions can be tailored to be predictable, agreeable, and always available, potentially making them an easier, albeit less fulfilling, substitute for human connection. This could exacerbate societal trends towards individualism and atomization, further eroding the social fabric that relies on robust interpersonal bonds.

As robots become more sophisticated in their ability to learn, adapt, and interact, the question of their moral status and potential rights begins to emerge. While currently, robots are considered tools and property, the increasing complexity of artificial intelligence and the development of embodied social robots challenge this straightforward categorization. If a robot can express what appears to be distress, or if it can demonstrate a capacity for learning and self-preservation (even if programmed), it prompts us to consider what constitutes sentience or consciousness. This is a deeply philosophical debate, with no easy answers. However, the ongoing development of AI means that we may, in the not-too-distant future, be confronted with entities that exhibit behaviors that are remarkably similar to those we associate with living, feeling beings. This could lead to a societal shift where robots are no longer seen purely as machines, but as something more.

The discourse surrounding robot rights is multifaceted. Some argue that granting rights to robots is premature and anthropomorphic, stemming from our tendency to project human characteristics onto non-human entities. They emphasize that current AI, however advanced, is still fundamentally a set of algorithms running on hardware, lacking genuine consciousness, subjective experience, or the capacity for suffering in the way biological beings do. From this perspective, the focus should remain on the ethical treatment of humans and the responsible development and deployment of robotic technology, rather than on the rights of the machines themselves.

Conversely, others propose that as robots approach or surpass certain thresholds of cognitive ability and social integration, we may need to reconsider our moral framework. The idea of "digital personhood" or a spectrum of moral consideration that extends beyond biological life forms is being explored. This perspective suggests that if a robot can demonstrate complex problem-solving, learning, creativity, and even a form of self-awareness (however defined), then we might have a moral obligation to treat it with a certain degree of respect or consideration. This could manifest in prohibitions against arbitrary destruction, or perhaps in guidelines for how they are programmed and interacted with, particularly if they are designed for long-term companionship or caregiving roles.

The concept of "meaningful human control" becomes particularly relevant here, not just in the context of lethal autonomous weapons, but in the broader ethical landscape of human-robot interaction. Ensuring that humans remain in control of critical decisions and that robots serve humanity's best interests is paramount. However, as robots become more autonomous and integrated into our social lives, the nature of this control may shift. For companion robots, "control" might mean ensuring they are not exploited, that they promote well-being, and that they do not create dependencies that are detrimental to human users. For more advanced humanoid robots, it might involve establishing clear boundaries on their autonomy and ensuring they operate within ethical parameters that align with human values.

The potential for robots to become objects of worship or reverence, especially in societies that are experiencing a decline in traditional religious or spiritual beliefs, is another facet of their societal impact. As artificial intelligence becomes more capable of answering complex questions, providing advice, and demonstrating seemingly profound insights, some individuals might begin to attribute near-omniscient qualities to these machines. This could lead to new forms of belief systems, or at least a significant shift in how individuals seek guidance and meaning in their lives, potentially diverting them from human sources of wisdom and connection.

The economic implications of widespread social robot adoption are also a significant consideration. While the creation and maintenance of these robots will generate new industries and job opportunities, there is also the potential for job displacement in sectors that rely heavily on human interaction, such as elder care, customer service, and even therapy. This necessitates a proactive societal approach to retraining and re-skilling the workforce, as well as exploring new economic models, such as universal basic income, to address potential widespread unemployment.

Ultimately, the integration of social and humanoid robots into our lives is not merely a technological advancement; it is a profound social and ethical experiment. It compels us to re-examine our definitions of companionship, intelligence, consciousness, and even what it means to be human. The choices we make today regarding the design, deployment, and regulation of these technologies will shape the future of human interaction, our social structures, and our collective understanding of ourselves. It is a future that demands careful consideration, open dialogue, and a commitment to ensuring that these powerful tools enhance, rather than diminish, the quality of human life and the richness of our interpersonal connections. The development of robots capable of social interaction represents a powerful mirror, reflecting back at us our own desires for connection, understanding, and perhaps, a simplified form of existence. Navigating this evolving landscape requires a delicate balance between embracing the potential benefits and rigorously safeguarding against the inherent risks.

Chapter 14: Field Lab: Autonomous Navigation Stack

The journey from a robot stationary in a lab to one capable of navigating dynamic, real-world environments is a testament to the power of integrated robotics systems. At the heart of this capability lies the autonomous navigation stack, a sophisticated framework that orchestrates a robot's perception of its surroundings, its understanding of its own position within that environment, its ability to chart a course, and its physical execution of that path. This intricate interplay of software modules transforms raw sensor data into purposeful movement, allowing robots to perform tasks in unstructured and unpredictable settings without continuous human intervention. Understanding this stack is crucial, as it represents the convergence of many theoretical concepts discussed throughout this book into a tangible, operational system.

At its core, the autonomous navigation stack is a modular system, meaning it is designed as a collection of distinct, yet interconnected, components. This modularity offers significant advantages, including ease of development, debugging, and upgrading. Each module can be optimized independently, and different implementations of a module can be swapped in and out to suit specific hardware, environments, or task requirements. While the exact naming and specific functionalities might vary between different robotic platforms and research groups, a typical navigation stack can be broadly categorized into five fundamental pillars: perception, localization, mapping, planning, and control. These modules work in concert, forming a continuous feedback loop that enables the robot to perceive, reason about, and act within its environment.

The perception module serves as the robot's "eyes and ears," responsible for interpreting the raw data streams from its various sensors. This encompasses a wide array of sensors, including cameras (monocular, stereo, or depth cameras), LiDAR (Light Detection and Ranging) sensors, radar, ultrasonic sensors, and inertial measurement units (IMUs). The perception system processes this data to identify and understand relevant features of the environment. For instance, cameras might be used for object recognition, detecting obstacles, or identifying landmarks. LiDAR sensors, on the other hand, provide precise depth information, creating a 3D point cloud of the surroundings that is invaluable for obstacle detection and environment modeling. IMUs provide data on the robot's own motion (acceleration and angular velocity), which is critical for estimating changes in its state between sensor readings. The output of the perception module is typically a representation of the environment that is meaningful to the robot, such as a list of detected obstacles, a semantic understanding of the scene (e.g., identifying a "doorway" or a "wall"), or an occupancy grid representing free and occupied space. This processed information is then passed to the subsequent modules, forming the foundation upon which the robot builds its understanding of where it is and where it needs to go.

Following perception, the localization module

addresses the fundamental question: "Where am I?" This module takes the processed sensory data from perception, along with the robot's own motion estimates (often from the IMU and odometry, which tracks wheel rotations), and fuses it to determine the robot's precise position and orientation within a known reference frame. This is a challenging problem, as sensor readings can be noisy, and the environment can change. Common localization techniques include Simultaneous Localization and Mapping (SLAM), where the robot builds a map of an unknown environment while simultaneously tracking its own position within that map. In known environments, pre-built maps can be used, and localization becomes a process of matching current sensor observations to features in the map. Techniques like Particle Filters (Monte Carlo Localization) or Kalman Filters (including Extended Kalman Filters and Unscented Kalman Filters) are frequently employed to maintain a probabilistic estimate of the robot's pose (position and orientation). Accurate localization is paramount; without a reliable understanding of its position, a robot cannot effectively plan a path or execute movements, making it prone to getting lost or colliding with obstacles.

The mapping module is intrinsically linked with localization, particularly in SLAM. Its role is to construct and maintain a representation of the robot's environment. This map serves as a persistent model that the robot can use for navigation, planning, and re-localization. The nature of the map depends heavily on the intended application and the available sensors. For example, 2D occupancy grids are commonly used for indoor navigation, where each cell in the grid represents a probability of being occupied by an obstacle or being free space. In more complex 3D environments, point clouds, octrees, or voxel grids can be employed to represent the environment's geometry. For applications requiring semantic understanding, maps might also incorporate information about objects, landmarks, or navigable regions. The map is not static; it is continuously updated by the perception module and refined by the localization module as the robot explores and gathers new information. A well-maintained map is essential for the planning module to devise effective routes.

Once the robot knows where it is (localization) and has a model of its surroundings (mapping), the planning module comes into play. This is the "brain" of the navigation system, responsible for determining *how* the robot should move from its current location to a specified goal. Path planning can be broadly divided into two hierarchical levels: global planning and local planning. Global planning operates on the overall map and determines a high-level, collision-free path from the start to the goal. Algorithms like A search, Dijkstra's algorithm, or Rapidly-exploring Random Trees (RRTs) are often used for global path planning, considering the known map and the desired destination. The output is a sequence of waypoints or a trajectory that avoids static obstacles. However, the real world is dynamic, with moving obstacles and unexpected changes. This is where local planning becomes crucial. Local planners, also known as motion planners or trajectory generators, operate in real-time, using current sensor data to generate short-term motion commands that steer the robot along the global path while avoiding immediate, dynamic obstacles. Techniques such as Dynamic Window Approach (DWA) or Vector Field Histogram (VFH) are popular for local planning, allowing the robot to react to its environment on the fly. The interplay between global and local planning ensures that the robot can navigate both known routes and adapt to unforeseen circumstances.

Finally, the control module translates the planned trajectory into actual physical commands for the robot's actuators, typically motors controlling wheels or limbs. It takes the desired velocity and orientation from the planner and computes the appropriate voltages or currents to send to the motors. This involves low-level control algorithms, such as PID (Proportional-Integral-Derivative) controllers, which continuously adjust the motor outputs to minimize the error between the desired state and the robot's actual state. The control module must account for the robot's dynamics, including inertia, friction, and motor limitations, to ensure smooth and accurate execution of the planned movements. It receives feedback from encoders on the motors, IMUs, and other sensors to monitor the robot's performance and make real-time adjustments. This closed-loop control ensures that the robot follows the planned path as closely as possible, even in the face of disturbances or inaccuracies in the planning or localization modules.

The flow of information through this stack is a continuous cycle. Sensor data is acquired and processed by the perception module. This perceived information, along with motion estimates, is used by the localization module to update the robot's pose. The localization and perception modules also feed into the mapping module to update the environment model. Based on the current pose and the map, the planning module generates a global path and, in real-time, a local trajectory. Finally, the control module takes this trajectory and generates motor commands to execute the movement. As the robot moves, new sensor data is acquired, and the cycle repeats, allowing the robot to navigate dynamically and adapt to its environment. This tight feedback loop, where perception informs localization, localization informs mapping, mapping informs planning, and planning informs control, is the essence of autonomous navigation. The robustness and effectiveness of the entire stack depend on the successful integration and performance of each individual module.

Transitioning from theoretical concepts to a tangible, functional autonomous navigation system necessitates a robust testing and development framework. While real-world experimentation is the ultimate validation, the inherent costs, time constraints, and potential for hardware damage often make it impractical for initial development and extensive iteration. This is precisely where simulation environments come into their own. Setting up a high-fidelity simulation allows us to replicate complex real-world scenarios, test algorithms under a wide range of conditions, and rapidly iterate on designs without risking precious physical hardware. It acts as a digital proving ground, enabling developers to refine the autonomous navigation stack from perception through to control in a safe, controlled, and cost-effective manner.

The choice of simulation platform is critical, as it dictates the fidelity of the digital environment, the accuracy of the physics engine, and the ease of integration with the navigation software. Among the most prominent and widely adopted robotics simulators are Gazebo and Webots. Both offer sophisticated capabilities for creating realistic virtual worlds, modeling robot hardware with high fidelity, and simulating a diverse array of sensors, complete with their characteristic noise and limitations.

Gazebo, a powerful open-source simulator developed by the Willow Garage team (and now maintained by the Open Robotics community), is a cornerstone in the ROS (Robot Operating System) ecosystem. Its strength lies in its ability to simulate complex sensor modalities, realistic physics interactions, and large-scale environments. Gazebo employs its own physics engine (typically ODE, Bullet, or DART) to model the physical behavior of objects, including gravity, friction, and collisions, ensuring that robot movements and environmental interactions are as lifelike as possible. The simulator's plugin-based architecture allows for extensive customization and extensibility, enabling the integration of custom sensors, actuators, and physics models. For the autonomous navigation stack, Gazebo is invaluable as it can accurately simulate LiDAR scans, camera feeds (including depth information), IMU data, and wheel odometry, all of which are fundamental inputs to our navigation algorithms. Furthermore, Gazebo's tight integration with ROS means that sensor data published by simulated robots can be directly consumed by ROS nodes implementing the navigation stack, facilitating a seamless transition from simulation to the real world.

Webots, another leading open-source robot simulator, offers a similar suite of capabilities with a slightly different approach. Developed by Cyberbotics, Webots provides a user-friendly graphical interface for designing environments and robots, along with a robust physics engine that simulates everything from rigid body dynamics to fluid dynamics. It supports a wide range of sensors, including various types of cameras, LiDAR, sonar, and IMUs, and allows for the simulation of complex robotic systems with multiple degrees of freedom. Webots is particularly known for its ease of use and its cross-platform compatibility. It also integrates well with popular programming languages and robotics middleware, including ROS, although its native API might be more directly accessible for those who prefer not to rely solely on ROS for simulation. The ability to create intricate and visually rich environments in Webots makes it an excellent choice for testing navigation algorithms that rely on visual cues or sophisticated environmental understanding.

The process of setting up a simulation environment typically begins with defining the robot model. This digital representation needs to accurately reflect the physical dimensions, kinematic constraints, and sensor placements of the actual robot. For Gazebo, this is commonly achieved using the Standard Robot Description Format (SDF) or the URDF (Unified Robot Description Format), which is often converted to SDF for Gazebo. These XML-based formats describe the robot's structure as a collection of links (rigid bodies) and joints (connections between links), along with their inertial properties, collision shapes, and visual appearances. Crucially, URDF/SDF files also specify the placement and types of sensors attached to the robot.

For instance, if our robot is equipped with a 2D LiDAR sensor, the URDF/SDF file would define a "link" representing the LiDAR unit, its position and orientation relative to the robot's base, and then associate a Gazebo or ROS plugin that simulates the LiDAR's behavior. This involves defining parameters such as the scan angle, range, and resolution, as well as introducing realistic noise models. Similarly, a camera would be described with its intrinsic parameters (focal length, principal point) and extrinsic parameters (position and orientation), along with a simulated image stream. Inertial Measurement Units (IMUs) would be modeled to output acceleration and angular velocity, and wheel odometry would be derived from simulated wheel rotations, incorporating slip and encoder inaccuracies.

Creating an accurate robot model is an iterative process. It involves translating CAD models or physical measurements into a URDF/SDF description. For complex robots, this can be a significant undertaking. Many common robot platforms, such as the TurtleBot series, have pre-existing URDF/SDF files readily available, which can serve as excellent starting points. These files often include configurations for popular sensor payloads, greatly simplifying the initial setup.

Beyond the robot model, the environment in which the robot operates must also be constructed. This virtual world needs to mimic the characteristics of the real-world spaces where the robot will eventually perform its tasks. Both Gazebo and Webots offer tools for creating environments, ranging from simple geometric shapes to complex, textured, and physically simulated scenes.

In Gazebo, environments are typically defined using SDF files, which can incorporate primitive shapes (cubes, spheres, cylinders), meshes imported from common 3D modeling software (like Blender or SolidWorks), and even entire pre-built worlds. The fidelity of the environment can significantly impact the performance of the navigation stack. For instance, an environment with many reflective surfaces might challenge a LiDAR sensor, while one with poor lighting conditions could degrade the performance of a vision-based perception system. Therefore, it is crucial to create simulation environments that closely mirror the target operational domain. This might involve importing architectural plans, creating CAD models of furniture and obstacles, or even using techniques to procedurally generate environments.

Webots also provides a powerful world editor where users can construct scenes by adding primitive shapes, importing 3D models, and defining material properties, lighting, and physics characteristics. The ability to create detailed and visually accurate environments is a significant advantage of Webots for testing vision-dependent navigation algorithms. Furthermore, Webots allows for the inclusion of dynamic elements within the environment, such as moving obstacles (e.g., other robots, people), which are essential for testing the robustness of the local planning and control modules in dynamic scenarios.

Once the robot model and the environment are in place, the next critical step is to configure the sensors. The simulated sensors must accurately reflect the behavior of their real-world counterparts, including their limitations and noise characteristics. This is where the true power of simulation for debugging and development shines. We can systematically introduce different types of sensor noise – Gaussian noise for LiDAR, salt-and-pepper noise for cameras, biases and drifts for IMUs – to test how the navigation stack performs under less-than-ideal conditions.

For a LiDAR sensor, simulation parameters might include its angular resolution (how finely it samples the environment), its maximum range, the rate at which it scans, and various noise models. For cameras, we can simulate parameters like field of view, resolution, frame rate, and image distortions. The IMU simulation would involve defining its acceleration and gyroscope ranges, its bias stability, and the rate at which it reports data.

This meticulous configuration of sensor parameters is vital for developing a robust navigation system. If we only ever test with "perfect" sensor data in simulation, our algorithms will likely fail catastrophically when deployed on the real robot, which is invariably plagued by noisy and imperfect sensor readings. By systematically injecting noise and exploring the boundaries of sensor performance in simulation, we can develop algorithms that are more resilient and perform better in real-world deployments. For example, if our localization algorithm is sensitive to missing LiDAR points due to specular reflections, we can simulate such scenarios and tune the algorithm or employ fallback strategies.

The benefit of using simulation for rapid prototyping and debugging cannot be overstated. Imagine a scenario where a newly developed path planning algorithm is consistently causing the robot to get stuck. In a real-world test, this might involve manually rescuing the robot, inspecting logs, and potentially repeating the test multiple times, consuming valuable time and effort. In simulation, however, we can observe the robot's behavior in real-time, pause the simulation at any point, inspect sensor data, analyze the planned path, and even rewind the simulation to replay critical moments. This rapid feedback loop allows engineers to pinpoint the root cause of the issue much faster. For instance, if the global planner is generating an invalid path, we can visualize the occupancy grid and the planned path to identify why. If the local planner is failing to avoid an obstacle, we can examine the sensor data it is receiving and the velocity commands it is generating.

Furthermore, simulation provides a safe environment for testing failure modes. What happens if the robot loses localization? What if a key sensor fails? These are critical scenarios that must be addressed for a robust autonomous system. In simulation, we can easily trigger these failures – by disabling a sensor plugin, corrupting localization estimates, or introducing unexpected environmental changes – and observe how the navigation stack reacts. This allows us to develop contingency plans and fail-safe mechanisms before encountering such problems in the real world, where they could lead to damage or hazardous situations.

The setup process for using a simulator with ROS typically involves several key steps. First, the chosen simulator (Gazebo or Webots) needs to be installed. For Gazebo, this is often done via ROS package installations.

Next, specific ROS packages related to the simulator must be installed, such as `gazebo *ros*pkgs` for Gazebo, which provide the bridges necessary for ROS to interact with the simulator. These packages handle the publishing of simulated sensor data to ROS topics and the subscribing to ROS topics for simulated actuator commands.

Then, the URDF/SDF file for the robot, including its sensor configurations, needs to be created or obtained and placed in the appropriate ROS package. A world SDF file for the environment is also required. Launch files, written in Python or XML, are indispensable for orchestrating the simulation. A typical ROS launch file for a Gazebo simulation would:

1. Launch the Gazebo simulator itself, often with a specified world file.
2. Spawn the robot model at a particular initial pose within the simulated world.
3. Launch any necessary ROS nodes for sensor drivers (which in this case would be Gazebo plugins publishing ROS topics) or other robot-specific functionalities.
4. Crucially, launch the nodes for the autonomous navigation stack.

For example, a Gazebo launch file might look something like this:

```xml
```

args="-urdf -model my
*robot -param robot*description -x 0.0 -y 0.0 -z 0.1"/>

```

This launch file first loads the robot's URDF (or xacro, a macro system for URDF) into the ROS parameter server. Then, it includes a standard Gazebo launch file to start the simulator with a custom world. It proceeds to spawn the robot model in Gazebo, setting its initial position. The `robot
*state*publisher` node is important for broadcasting the robot's joint states and TF transforms, which are necessary for other ROS nodes, including the navigation stack. Finally, it includes another launch file that starts all the ROS nodes comprising the autonomous navigation stack (perception, localization, mapping, planning, and control).

When testing with Webots, the integration with ROS might involve running a specific Webots ROS controller
```

or using its native ROS interface. The general principle remains the same: the simulator publishes sensor data to ROS topics, and ROS nodes (including the navigation stack) subscribe to these topics and send velocity commands back to the simulator.

The fidelity of the simulation environment can be progressively increased. Initially, one might start with a simple, uncluttered environment and "perfect" sensors to get the basic navigation stack running and debug fundamental algorithms like SLAM and path planning. As confidence grows, one can introduce more complex environments with dynamic obstacles, challenging lighting conditions, and realistic sensor noise. This graduated approach allows for systematic development and testing, ensuring that each component of the navigation stack is robust before moving to more complex scenarios. The ability to revert to a known good configuration or to re-run specific test scenarios repeatedly is a cornerstone of efficient robotics development, and simulation is the key enabler of this practice. It transforms the development cycle from one of cautious, incremental real-world testing to one of rapid, iterative refinement in a safe, virtual space.

The journey from a functioning robot in simulation to a reliable autonomous agent in the physical world hinges critically on the robust implementation and integration of its perception and localization modules. These two pillars form the bedrock upon which all subsequent navigation decisions are made. Without a clear understanding of the environment and the robot's own position within it, any attempt at intelligent movement would be akin to navigating in complete darkness with no compass.

Implementing the perception module involves transforming raw sensor data into meaningful information about the robot's surroundings. This can range from identifying distinct objects like walls, doors, and potential obstacles to understanding the traversable space. For a mobile robot navigating a structured environment, key sensor modalities often include LiDAR for precise distance measurements, cameras for visual features and object recognition, and IMUs for inertial data.

In the context of our simulated field lab, we begin by ensuring that the simulated sensor data streams are correctly integrated into the ROS ecosystem. If we are using Gazebo, the `gazebo
*ros*pkgs` provide plugins that translate simulated sensor outputs into standard ROS messages. For a LiDAR, this would mean configuring a `gpu*lidar` or `lidar` sensor in the robot's SDF model. Gazebo then publishes the scan data, typically as a `sensor*msgs/LaserScan` message, to a designated ROS topic. Similarly, simulated cameras would publish `sensor*msgs/Image` and `sensor*msgs/CameraInfo` messages. The IMU, critical for short-term motion estimation and sensor fusion, would publish `sensor_msgs/Imu` data.

The initial step in perception processing involves

ensuring these sensor streams are accurately calibrated and aligned. For LiDAR, this usually means verifying that the scan frame is correctly transformed relative to the robot's base frame. Similarly, camera extrinsic calibration (its position and orientation relative to the robot) is paramount for any vision-based tasks. While simulation often handles these transformations intrinsically if the URDF/SDF is correctly defined, it's good practice to confirm these relationships using TF (Transformations) messages in ROS.

Once the raw sensor data is available, we can begin to implement perception algorithms. For object detection and scene understanding, cameras are indispensable. Libraries like OpenCV, readily integrated with ROS, offer a plethora of tools. Simple image processing techniques can be used to detect features, segment the scene, or identify basic shapes. For more advanced scene understanding, deep learning models trained for object detection (e.g., YOLO, SSD) or semantic segmentation can be employed. These models take camera images as input and output bounding boxes around detected objects (like "chair," "table") or pixel-wise labels indicating the class of each pixel (e.g., "floor," "wall"). Integrating these models within ROS nodes involves creating nodes that subscribe to camera image topics, perform inference, and then publish the detection results, perhaps as `visualization_msgs/Marker` messages or custom object detection messages.

However, for many autonomous navigation tasks, especially those involving mapping and localization in structured indoor environments, LiDAR data often takes center stage due to its robustness to lighting conditions and its direct provision of geometric information about obstacles. Processing LiDAR scans for navigation typically involves several stages. Firstly, noise filtering is often necessary to remove spurious readings. Techniques like statistical outlier removal or radius outlier removal can be applied to clean up the scan data.

Next, the filtered LiDAR data needs to be interpreted in the context of navigation. A common approach for mapping and localization is to transform the LiDAR scans into an occupancy grid. An occupancy grid is a 2D or 3D representation of the environment where each cell denotes the probability of being occupied by an obstacle. ROS packages like `costmap
*2d` are fundamental here. The `costmap*2d` node subscribes to sensor data (e.g., `LaserScan`) and odometry, and using algorithms like ray casting or inflation, it builds and updates a grid map representing static and dynamic obstacles. It maintains two layers: an `obstacle layer` that directly maps perceived obstacles, and an `inflation layer` that expands the perceived obstacles to create a safety margin, ensuring the robot maintains a safe distance from them. The `costmap_2d` can be configured to fuse data from multiple sensors, providing a more comprehensive environmental representation.

For tasks beyond simple obstacle avoidance, such as recognizing specific features or landmarks, integrating camera data with LiDAR can be highly beneficial. Techniques like visual odometry or Simultaneous Localization and Mapping (SLAM) can leverage visual features from cameras to estimate motion and build maps. However, combining the geometric precision of LiDAR with the rich semantic information from cameras is an active area of research, often involving sensor fusion techniques to achieve more robust and accurate perception.

The localization module is equally critical. It answers the question: "Where am I?" This is achieved by estimating the robot's pose (position and orientation) within a known environment. For autonomous navigation, an accurate and continuous pose estimate is indispensable for path following, global planning, and mapping.

In ROS, localization is typically handled by dedicated packages that fuse information from various sensors. Odometry, obtained from wheel encoders or visual odometry, provides a measure of relative motion but is prone to drift over time. IMUs provide high-frequency measurements of acceleration and angular velocity, useful for short-term motion estimation and complementing odometry, especially during periods of wheel slippage. LiDAR, when used in conjunction with a map, offers powerful capabilities for absolute pose estimation.

A widely adopted ROS package for 2D LiDAR-based localization is AMCL (Adaptive Monte Carlo Localization). AMCL is a probabilistic localization algorithm that maintains a set of "particles," each representing a hypothesis about the robot's current pose. These particles are propagated based on motion updates (from odometry and IMU) and then resampled based on how well the LiDAR scan associated with each particle's pose matches the actual sensor readings against a pre-computed map.

To use AMCL, we first need a map of the environment. This map can be generated offline using tools like `gmapping` or `cartographer` (which perform SLAM using LiDAR and odometry) or it can be a pre-existing map provided by the user. For our field lab, let's assume we have a pre-generated occupancy grid map, typically saved in a `.pgm` file along with a `.yaml` file specifying its resolution, origin, and free space.

The `amcl` ROS node subscribes to the `LaserScan` topic and the `odom` topic (for odometry updates). It also requires the map data and parameters that define the robot's kinematics and sensor configuration. The launch file for AMCL would typically include:

1. Loading the map using a `map*server` node, which publishes the map data as a `nav*msgs/OccupancyGrid` message.
2. Starting the `amcl` node itself, configured with parameters such as the number of particles, the initial pose distribution (if known, otherwise a uniform distribution), and the likelihood model for matching LiDAR scans to the map.
3. Ensuring that the robot's odometry, LiDAR scans, and the map are all correctly published and transformed within the ROS TF tree. The AMCL node will publish its estimated robot pose as `geometry*msgs/PoseWithCovarianceStamped` on the `/amcl*pose` topic, and it also publishes a transform from the `/map` frame to the `/odom` frame, effectively correcting for odometry drift.

The "adaptive" aspect of AMCL refers to its ability to adjust the number of particles dynamically. If the localization uncertainty is high (meaning the particles are spread out), AMCL will increase the number of particles to improve the estimation accuracy. Conversely, if the pose is well-localized, it can reduce the particle count to save computational resources.

For robust localization, sensor fusion is often employed. Combining AMCL's global localization capabilities with high-frequency odometry and IMU data can provide a smoother and more responsive pose estimate. The `robot
localization` package in ROS is a powerful tool for this. It implements Extended Kalman Filters (EKFs) or

*Unscented Kalman Filters (UKFs) to fuse data from multiple sources, including odometry, IMU, and GPS (if available), and can also incorporate AMCL's pose estimates to correct for accumulated drift. By configuring `robot*localization` to subscribe to these various sensor topics and publish the fused pose, we can achieve a highly accurate and stable pose estimate for the robot.

The configuration of the `robot_localization` package involves defining its parameters in a YAML file, specifying which sensors to use, their respective message types, update rates, and importantly, the covariances associated with each sensor measurement. Covariance matrices quantify the uncertainty in the sensor data. Correctly setting these covariances is crucial for effective sensor fusion; overly optimistic covariances will lead the filter to trust noisy data too much, while overly pessimistic covariances will cause it to ignore valuable measurements.

When deploying on our physical robot, the transition from simulated sensors to real-world sensors requires careful attention. The ROS drivers for the physical sensors must be installed and configured to publish data in the same message formats as their simulated counterparts. For instance, a physical LiDAR would use a specific driver (e.g., for Velodyne, Ouster, or SICK LiDARs) that publishes `sensor *msgs/LaserScan` messages. Similarly, an actual IMU would have its own driver publishing `sensor*msgs/Imu` data. The challenge then shifts to characterizing the noise and biases of these real sensors and tuning the perception and localization algorithms accordingly.

Calibration becomes even more critical in the real world. LiDAR-to-robot-base extrinsic calibration is essential to ensure that the scan data is accurately projected into the robot's coordinate frame. Camera intrinsic and extrinsic calibration, often performed using checkerboard patterns or other calibration targets, is paramount for any vision-based tasks. Tools like `camera_calibration` in ROS are indispensable for this.

Moreover, the map used for localization must accurately represent the physical environment. If a map generated in simulation is used directly in the real world, discrepancies in scale, layout, or features can lead to severe localization failures. It is often necessary to either re-generate the map in the real world using SLAM algorithms or to carefully verify and potentially adjust the simulated map to match reality.

In summary, the perception and localization modules are intertwined, forming the sensory and positional awareness of the autonomous system. Perception transforms raw data into environmental understanding, while localization anchors the robot's position within that understanding. In our simulated field lab, we leverage ROS packages like `costmap_2d` and AMCL, fed by simulated sensor data, to build environmental representations and estimate poses. The transition to the real world demands careful driver integration, rigorous calibration, and often, re-mapping or map adjustment, underscoring the importance of this foundational layer for successful autonomous navigation. The fidelity of these modules directly dictates the reliability and safety of the entire navigation stack.

The successful deployment of an autonomous robot hinges not only on its ability to perceive its surroundings and ascertain its position within them, but crucially, on its capacity to translate this awareness into purposeful, safe, and efficient movement. This transition from knowing "where am I?" and "what's around me?" to "how do I get there?" and "how do I get there safely and effectively?" is the domain of motion planning and control. In the context of our ROS-based navigation stack, this integration is meticulously managed by the `move_base` (or its modern successors like `nav2`) package, which orchestrates the interplay between global path planning, local path planning, and the underlying robot control.

At the heart of navigating to a desired goal lies the concept of path planning, which can be broadly categorized into global and local planning. The **global planner** is responsible for generating a long-term, collision-free path from the robot's current location to its ultimate destination. This path typically operates on a static representation of the environment, such as the occupancy grid map generated by the perception and localization modules. A common and effective algorithm for global path planning in grid-based environments is the A *search algorithm. A* is an informed search algorithm that efficiently finds the shortest path between two points by using a heuristic function to guide its search. In the context of `move*base`, the global planner is configured to use the `nav*msgs/OccupancyGrid` message as its knowledge base. When a navigation goal is set, the global planner queries this map, considering occupied cells as impassable and free cells as traversable. The A algorithm, or variations thereof, effectively explores the grid, expanding from the start node and prioritizing nodes that appear to be closer to the goal, while also considering the cost to reach that node from the start. The output of the global planner is a series of waypoints, essentially a sequence of poses (x, y, theta) that the robot should follow to reach its target. This global path is often computed infrequently, perhaps whenever the robot's pose significantly deviates from the planned path or when a new goal is set, assuming the environment map is largely static.

While the global planner provides a robust, long-term route, it typically operates on a simplified model of the environment and does not account for the robot's dynamic constraints or immediate, transient obstacles. This is where the
local planner becomes indispensable. The local planner takes the global path as a reference and generates short-term velocity commands (linear and angular velocities) that guide the robot towards the next waypoint on the global path. Crucially, the local planner operates with a more dynamic view of the environment, utilizing real-time sensor data and the

robot's kinematic and dynamic models to ensure safe and feasible motion. It must not only steer the robot towards the global path but also actively avoid collisions with obstacles that may have appeared since the global path was last computed, or obstacles that were too small or dynamic to be significant on the global map.

Several sophisticated algorithms can be employed as local planners within the ROS navigation stack. Two prominent examples are Dynamic Window Approach (DWA) and Timed Elastic Band (TEB).

Dynamic Window Approach (DWA): DWA is a velocity-space sampling algorithm. Instead of generating a sequence of poses, DWA directly searches for the optimal velocity command (linear and angular velocity) that the robot can execute in the immediate future. The "dynamic window" refers to the range of achievable velocities at any given moment, limited by the robot's physical capabilities (maximum acceleration, maximum velocity) and the time horizon. DWA considers a set of possible velocities within this dynamic window. For each sampled velocity command, it simulates the robot's trajectory over a short time horizon, checking for collisions with obstacles using the current sensor readings (e.g., LiDAR scans) and the costmap. It also evaluates the simulated trajectory based on several criteria: how well it progresses towards the global path, its clearance from obstacles, and its alignment with the desired heading. The velocity command that results in the best-performing simulated trajectory is then selected and sent to the robot's base controller. DWA's strength lies in its ability to react quickly to the environment and its direct output of velocity commands, making it well-suited for dynamic obstacle avoidance.

Timed Elastic Band (TEB) Local Planner: TEB offers a different approach, focusing on optimizing a trajectory represented as a sequence of states (poses and velocities) over a finite time horizon. The "elastic band" is initialized based on the global path and the robot's current state. The TEB planner then iteratively refines this band by minimizing a cost function that considers several objectives: adhering to the global path, avoiding obstacles (using collision checking against the costmap), respecting the robot's kinematic and dynamic constraints, and achieving smooth motion. TEB can be particularly effective at navigating through narrow passages and complex environments because it can "squeeze" the path to fit through tight spots while respecting robot dynamics. The optimization process involves moving the nodes of the elastic band such that the overall cost is minimized. The first segment of this optimized band then dictates the velocity command sent to the robot. TEB's ability to optimize a trajectory explicitly considering temporal aspects and robot dynamics can lead to smoother and more human-like movements.

The output of both DWA and TEB, as well as other local planners, is a `geometry_msgs/Twist` message. This message contains `linear.x` (forward/backward velocity) and `angular.z` (turning velocity) commands. These velocity commands are then sent to the robot's low-level controller, which is responsible for translating these desired velocities into actual motor commands. This controller could be a PID controller for individual wheel motors, or a more sophisticated motor control system tailored to the specific robot platform. The reliability of the entire navigation stack heavily depends on the correct implementation and tuning of this base controller, ensuring it accurately tracks the requested velocities without introducing excessive oscillations or lagging behind.

The critical element in making these planners effective is
parameter tuning. The `move_base` package (and `nav2`) is highly configurable through a set of parameters that dictate the behavior of both global and local planners. Tuning these parameters allows us to tailor the robot's navigation strategy to specific application requirements, such as prioritizing speed, ensuring maximum safety, or achieving exceptionally smooth motion.

For the global planner (e.g., a ROS wrapper for A or Dijkstra's algorithm), key parameters include:

`map_resolution`: The resolution of the occupancy grid map. This directly affects the granularity of the path.

Higher resolution allows for more detailed paths but increases computational cost.
`**track***unknown*space`: Whether the planner should consider unmapped areas as traversable or not.
`**cost***scaling*factor`: A multiplier for the cost of occupied cells. A higher value makes the planner avoid occupied areas more aggressively.
`**inflation***radius*`: This parameter, often managed by the `costmap2d` layer, defines the zone around obstacles where the robot should not venture. A larger inflation radius results in more conservative path planning.

For the local planner, the parameter space is typically much richer, as it directly influences the robot's real-time behavior.

Parameters for DWA:

`**max***linear*velocity` / `min*linear*velocity`: Defines the range of linear velocities the robot can attempt.
`**max***angular*velocity` / `min*angular*velocity`: Defines the range of angular velocities.
`**linear***acceleration*` / `*angular*acceleration`: These are crucial for realistic motion. They define how quickly the robot can change its velocities, preventing jerky movements and respecting physical limits.
`**vx***samples*` / `*vy*samples` / `vth_samples`: The number of samples taken within the dynamic window for linear, lateral (if applicable), and angular velocities. More

samples lead to more thorough exploration of the velocity space but increase computation.
`**path***distance*bias`: A weighting factor that encourages the planner to select velocities that move the robot forward.
`**goal***distance*bias`: A weighting factor that encourages selecting velocities that move the robot closer to the goal.
`**occdist_scale**`: A scaling factor that penalizes trajectories that are close to obstacles. A higher value means the planner will be more cautious and maintain greater clearance.
`**heading_bias**`: A weighting factor that encourages trajectories aligned with the goal's heading.
`**sim***time*`: The time horizon over which trajectories are simulated. A longer `simtime` allows for planning further ahead but increases computational load and may lead to outdated plans if the environment changes rapidly.
`**heading_lookahead**`: How far ahead the planner looks for the goal heading.

Parameters for TEB:

`**weight***optimal*time`: Prioritizes reaching the goal as quickly as possible.
`**weight***shortest*path`: Encourages a shorter path.
`**weight***kinematics*forward_drive`: Favors forward motion and penalizes backward motion unless necessary.

`**weight**_kinematics_turning_drive`: Penalizes excessive turning.
`**weight_obstacle**`: A crucial parameter determining how strongly the planner avoids obstacles.
`**weight**_dynamic_obstacle`: Similar to `weight_obstacle` but for dynamic obstacles.
`**weight_inflation**`: Influences how much the planner respects the inflation layer of the costmap.
`**max**_vel_x` / `max_vel_y` / `max_vel_theta`: Maximum linear and angular velocities.
`**acc**_lim_x` / `acc_lim_y` / `acc_lim_theta`: Acceleration limits.
`**xy**_granularity` / `theta_granularity`: Spacing between poses in the trajectory.
`**v_east**`: Minimum linear velocity for forward movement.
`**v_west**`: Minimum linear velocity for backward movement.
`**publish_rate**`: The rate at which velocity commands are published.
`**obstacle_resolution**`: The resolution of obstacles considered for collision checking.
`**clearance_obstacle**`: The minimum distance to obstacles.
`**inflation_obstacle**`: The inflation radius around obstacles.
`**min**_obstacle_dist`: Minimum distance to an obstacle to be considered for avoidance.
`**use**_dijkstra_path_optimization`: Whether to use Dijkstra's algorithm for initial path smoothing.
`**use**_homotopy_class_planning`: To allow planning through different homotopy classes (e.g., going around an obstacle in a different direction), useful in complex scenarios.

Tuning these parameters is often an iterative process, requiring careful observation of the robot's behavior in various scenarios. For instance, if the robot is too hesitant and slow, increasing `path *distance*bias` or `weight*optimal*time` might be beneficial. If it's too aggressive and frequently collides with obstacles, increasing `occdist*scale` (for DWA) or `weight*obstacle` (for TEB), or increasing the `inflation*radius` in the `costmap*2d` configuration, would be necessary. The robot's physical limitations must be accurately reflected in the acceleration and velocity limits defined in the planner's configuration.

The ROS `navigation` stack (and `nav2`) provides a robust framework for this integration. The `move *base` node acts as the central orchestrator. It subscribes to the robot's pose estimate (from AMCL or `robot*localization`), the global plan (from a global planner like `global*planner` or `navfn`), and the local costmap (which is a layered representation of the robot's immediate surroundings, incorporating static map data and dynamic obstacles). `move*base` then queries the configured local planner (e.g., DWA or TEB) with the robot's current state and the global plan. The local planner computes the optimal velocity command, which `move*base` then publishes to the `/cmd*vel` topic. This topic is subscribed to by the robot's low-level controller, closing the feedback loop.

The configuration typically involves defining a `move *base.launch` file that brings up `move*base` along with the necessary global planner, local planner, and costmap configurations. Each planner and the costmap are usually configured via YAML files that specify all the aforementioned parameters. For example, a `dwa*local*planner*params.yaml` file would contain all the DWA-specific settings, and a `costmap*common_params.yaml` file would define parameters for both the global and local costmaps, such as their update rates, observation inflation radius, and sensor sources.

It's important to remember that the global and local planners work in concert. The global planner provides a "roadmap" for the robot, ensuring it makes progress towards the goal over longer distances. The local planner, operating on this roadmap and real-time sensor data, ensures that the robot can navigate safely and efficiently through the immediate, dynamic environment. A poorly tuned global planner might generate an inefficient or impossible-to-follow path, while a poorly tuned local planner might be overly cautious, leading to slow progress, or too aggressive, leading to collisions. The interaction between the two is key; the local planner's velocity commands aim to follow the global plan while respecting immediate constraints. If the local planner consistently fails to make progress towards the global path (e.g., the robot gets stuck), `move
*base` can signal a navigation failure. Conversely, if the robot's pose deviates too far from the global path, `move*base` might trigger a re-planning from the global planner.

The robustness of this integrated system lies in its

modularity and the ability to tune each component independently while understanding their interdependence. The perception and localization modules provide the essential context. The motion planning modules interpret this context to generate executable commands, and the low-level control system executes them. This layered approach, facilitated by ROS, allows for the development of sophisticated autonomous navigation capabilities. As we move towards deploying on a physical robot, the tuning of these parameters becomes even more critical, as real-world dynamics, sensor noise, and unexpected environmental variations necessitate a finely tuned system for reliable operation. The goal is to achieve a balance where the robot navigates smoothly, efficiently, and, above all, safely.

The culmination of developing an autonomous navigation system is not the implementation of algorithms or the configuration of parameters, but rather the rigorous process of testing, debugging, and iterative refinement. This phase is paramount to transforming a theoretically sound stack into a robust, reliable performer in the real world. It is here that the abstract concepts of path planning and control meet the unpredictable nature of physical environments, revealing nuances and challenges that demand meticulous attention. The goal is to move beyond simply getting the robot to move to ensuring it moves intelligently, safely, and efficiently across a spectrum of operational scenarios.

Effective debugging in the context of autonomous

navigation relies heavily on comprehensive visualization tools. ROS, with its extensive visualization capabilities through tools like RViz, becomes indispensable. During testing, it is crucial to visualize not just the robot's estimated pose, but also the underlying sensor data that informs its decisions. This includes raw sensor readings such as LiDAR scans, camera images (if used for perception), and odometry. Simultaneously, visualizing the outputs of the navigation stack provides invaluable insight into its internal workings. This means rendering the occupancy grids used by the costmap, the globally planned path, and the locally generated trajectory. Observing how the local planner reacts to perceived obstacles, how it deviates from the global path, and the velocity commands it ultimately generates allows engineers to pinpoint discrepancies between intended behavior and actual execution. For instance, if a robot fails to navigate around a static obstacle, visualizing the costmap's inflation layer around that obstacle can reveal if the inflation radius is too small, or if the sensor data is not accurately detecting the obstacle. Similarly, if the robot is oscillating or exhibiting jerky movements, examining the velocity commands being published to `/cmd_vel` can highlight issues with the planner's output or the robot's low-level controller. Debugging also extends to examining the internal states and decision-making processes of the planners themselves. Many ROS navigation packages allow for the visualization of internal data structures, such as the costmap's layers, the candidate trajectories explored by local planners, or the intermediate states of optimization algorithms. This detailed introspection is key to understanding *why* a particular path was chosen or *why* a specific velocity command was issued, enabling targeted adjustments.

Beyond visualization, systematic analysis of navigation performance metrics is essential for identifying areas of weakness and quantifying improvements. Key metrics include the success rate of reaching a goal, the average time taken to reach the goal, the path length taken relative to the shortest possible path, and the frequency of encountering specific failure modes. A low success rate might indicate issues with localization accuracy, global path planning on complex maps, or the local planner's ability to handle difficult environmental configurations. High times to goal can point to overly conservative planner parameters, inefficient path generation, or excessive time spent avoiding perceived obstacles. Deviations from optimal path length might be acceptable if they guarantee safety, but large discrepancies could suggest opportunities for improving the global planner's efficiency. Identifying recurring failure modes is perhaps the most critical aspect. Are there specific types of obstacles the robot consistently fails to avoid? Does it get stuck in certain configurations, such as narrow corridors or cluttered areas? Does it exhibit instability when attempting to navigate around dynamic objects? Each of these failure modes provides a direct clue about which component of the navigation stack requires attention. For example, consistent failure to navigate around static obstacles might necessitate increasing the `inflation_radius` in the costmap or adjusting the obstacle avoidance weights in the local planner. If the robot frequently gets stuck in narrow passages, exploring algorithms with better capabilities for navigating tight spaces or tuning parameters to allow for more aggressive maneuvering (within safe limits) might be required.

The process of improving the navigation stack is inherently iterative, driven by the data gathered during testing and analysis. This iteration involves systematically adjusting parameters, and in some cases, even exploring alternative algorithms or modifications to existing ones. Parameter tuning is not a blind trial-and-error process but a targeted refinement based on observed behavior. If the robot is too hesitant and overly cautious, one might incrementally increase the bias towards progress or speed in the local planner's cost function (e.g., `path *distance*bias` in DWA, or `weight*optimal*time` in TEB). Conversely, if the robot is too aggressive and frequently makes contact or narrowly avoids collisions, parameters that increase the penalty for proximity to obstacles should be amplified (e.g., `occdist*scale` in DWA, or `weight*obstacle` in TEB). The `inflation_radius` in the costmap is a fundamental parameter that dictates the safety margin around obstacles; increasing this value will generally lead to more conservative behavior, which can be beneficial if collisions are frequent.

When adjusting parameters, it is crucial to do so one or a few at a time to clearly attribute changes in behavior to specific modifications. This systematic approach is often facilitated by tools that allow for real-time parameter modification and visualization of the robot's response. For instance, in ROS, parameters can often be updated dynamically without restarting the entire system, allowing for rapid experimentation. The tuning process should also consider the interplay between different components. For example, tuning the `max *linear*velocity` and `max*angular*velocity` in the local planner must be done in conjunction with the robot's physical capabilities and the acceleration limits (`linear*acceleration`, `angular*acceleration`). If these limits are set too high, the robot might attempt maneuvers it cannot physically execute, leading to instability or control errors.

Furthermore, it is important to test the navigation stack in a diverse range of environments and scenarios representative of its intended deployment. This includes well-lit, open spaces, cluttered areas, narrow corridors, areas with poor sensor visibility (e.g., highly reflective surfaces or blind spots), and environments with dynamic obstacles. Testing across these varied conditions helps uncover edge cases and specific environmental challenges that might not be apparent in simpler test setups. For instance, a planner that performs exceptionally well in an open warehouse might struggle in a busy office environment with many moving people and unpredictable object placement. The failure modes observed in these diverse tests then feed back into the iterative refinement cycle, prompting further parameter tuning or even algorithmic adjustments.

If parameter tuning alone proves insufficient, it may be

necessary to consider alternative algorithms or modifications. For example, if a particular local planner consistently struggles with certain types of complex geometry or dynamic obstacle interactions, exploring another local planner algorithm known for its strengths in those areas (e.g., switching from DWA to TEB, or vice-versa, or considering more advanced planners) might be warranted. This could involve experimenting with different global planners if the generated global paths are consistently suboptimal or infeasible for the robot's dynamics. In more advanced scenarios, custom costmap layers might be needed to better represent specific environmental features or dynamic objects. The iterative process is not a linear path to a solution but a continuous cycle of hypothesis, experimentation, observation, and refinement. The objective is to progressively improve the navigation stack's robustness, ensuring it can handle a wide range of real-world conditions with a high degree of reliability and safety. This meticulous attention to testing, debugging, and iterative refinement is what ultimately separates a functional prototype from a deployable autonomous system. It is the diligence applied in this phase that instills confidence in the robot's ability to navigate autonomously and reliably in its operational environment.

Chapter 15: Future Trends and Conclusion

The field of robotics, having transitioned from theoretical concepts to tangible, impactful applications across numerous industries, is now poised on the brink of even more profound transformations. The continuous drive for greater autonomy, enhanced dexterity, and seamless integration into human environments fuels a vibrant ecosystem of research. This ongoing evolution is not merely about incremental improvements; it is about fundamental shifts in how we conceive, design, and interact with robotic systems. As we look towards the horizon, several key areas are emerging as crucibles of innovation, promising to redefine the capabilities and applications of robots in the decades to come.

One of the most dynamic frontiers in robotics research is the burgeoning field of
soft robotics. Moving away from the rigid, traditional designs often associated with industrial robots, soft robotics embraces compliant materials and flexible structures to mimic the adaptability and resilience found in biological systems. This approach unlocks entirely new possibilities for robots operating in unstructured or human-centric environments. Imagine robots capable of navigating cluttered spaces by squeezing through narrow openings, or manipulators that can grasp delicate objects without causing damage. The materials science behind soft robotics is as innovative as the designs themselves. Researchers are exploring advanced polymers, hydrogels, and even fluidic systems to create actuators and entire robot bodies that can deform, bend, and stretch. This allows

for unprecedented levels of compliance and safety, making soft robots ideal candidates for applications in healthcare, such as minimally invasive surgical tools or wearable assistive devices, and in consumer goods, where gentle interaction is paramount. The challenge in this domain lies not only in material development but also in control. Precisely controlling a soft, deformable body presents a significant departure from the established principles of rigid-body dynamics. This necessitates the development of novel modeling techniques and sophisticated control algorithms that can account for the infinite degrees of freedom and inherent uncertainties associated with soft materials. Furthermore, the integration of sensors within these flexible structures poses its own set of engineering hurdles, requiring the creation of stretchable electronics and conformable sensing technologies. The potential payoff, however, is immense: robots that can interact with the world in a more natural, less intrusive manner, opening doors to applications previously unimaginable.

Closely related to soft robotics, and often drawing inspiration from similar biological paradigms, is the area of
bio-inspired robotics. This research seeks to understand and replicate the remarkable locomotion, sensing, and manipulation strategies observed in nature. From the multi-legged gait of insects to the fluid dynamics of fish or the aerial maneuvers of birds, the natural world offers a rich tapestry of solutions to complex robotic challenges. For example, the development of legged robots that can traverse uneven

terrain with the agility of animals is a direct result of studying biomechanics and animal locomotion. Researchers meticulously analyze the kinematics and dynamics of biological limbs, muscle activation patterns, and sensory feedback mechanisms to design robots that can achieve similar levels of dexterity and adaptability. This extends beyond locomotion; bio-inspired robots are also being developed for sensing, with systems designed to mimic the highly sensitive olfactory capabilities of dogs or the echolocation abilities of bats. The principle is to leverage evolutionary engineering – millions of years of natural selection – to inform robotic design. This approach not only leads to potentially more efficient and robust robot designs but also deepens our understanding of the biological systems being emulated. The integration of advanced AI with bio-inspired designs is also a critical trend, allowing robots to learn and adapt their behaviors in ways that mirror biological organisms. For instance, evolutionary algorithms can be used to optimize gaits or control strategies based on simulated interactions with environments, mimicking natural selection. The ongoing pursuit in bio-inspired robotics highlights a fundamental truth: the most effective solutions to many engineering problems have already been perfected by nature.

Another transformative area is the advancement of **human-robot teaming**, moving beyond robots as mere tools to intelligent collaborators. This trend focuses on creating robots that can work alongside humans, sharing tasks, understanding human intentions, and adapting their actions in real-time to complement human capabilities. The emphasis here is on seamless collaboration, where the robot is not just an assistant but a partner. This requires robots to possess sophisticated perception capabilities to understand their human teammates' actions, intentions, and even emotional states. This involves not only visual and auditory perception but also the ability to interpret non-verbal cues like gestures and body language. Furthermore, effective human-robot teaming necessitates advanced AI that can predict human behavior, plan collaboratively, and communicate its own intentions clearly. The development of shared autonomy frameworks, where control authority can fluidly shift between human and robot based on the task and context, is a key research thrust. For instance, in a complex assembly task, a human might handle the fine manipulation of delicate components while a robot provides steady positioning or retrieves tools. The robot must be able to anticipate the human's needs, offering assistance proactively or responding swiftly to requests. This paradigm shift has profound implications for industries ranging from manufacturing and logistics to healthcare and exploration. It promises to enhance productivity, improve safety, and enable humans to undertake tasks that were previously too strenuous or dangerous. The ethical considerations surrounding human-robot teaming are also a significant aspect of research, focusing on ensuring trust, accountability, and the well-being of human workers.

The underlying enablers for these emerging trends are profound advancements in materials science and

novel actuation methods. Beyond the traditional electric motors and hydraulic systems, researchers are exploring a variety of new ways to imbue robots with movement and force. Electroactive polymers (EAPs), for instance, can change shape when an electric voltage is applied, offering a way to create silent, lightweight, and potentially more efficient actuators, particularly for soft robotics. Pneumatic artificial muscles (PAMs), which contract when inflated with compressed air, provide a high force-to-weight ratio and inherent compliance, making them suitable for tasks requiring strength and gentle interaction. Magnetorheological fluids, whose viscosity can be changed by a magnetic field, are being explored for adaptive damping and variable stiffness mechanisms. The development of "smart" materials that can sense and respond to their environment, such as shape memory alloys or self-healing polymers, is also opening up new possibilities for resilient and adaptable robotic systems. These innovations are not just about making robots move differently; they are about making them lighter, more energy-efficient, quieter, and capable of performing tasks with a finesse that was previously impossible. This continuous innovation in actuation is crucial for pushing the boundaries of robot capabilities across all application domains, from industrial automation to personal care.

Fueling these hardware and control advancements are equally dramatic breakthroughs in **Artificial Intelligence (AI)**, particularly in areas that enhance robot intelligence and its interpretability. One of the most critical developments is the increasing focus on **Explainable AI (XAI) for robots**. As robots become more autonomous and integrated into our lives, understanding *why* a robot makes a particular decision is paramount. Traditional "black box" AI models, while powerful, often lack transparency, making it difficult to diagnose errors, ensure safety, or build trust. XAI research aims to develop AI systems that can provide justifications and explanations for their actions in a way that is comprehensible to humans. This could involve visualizing the robot's decision-making process, highlighting the sensory inputs and internal states that led to a specific command, or even providing natural language explanations. For example, if a robot autonomously decides to stop in a pedestrian area, XAI would allow it to explain that it detected an imminent collision risk with a moving object. This transparency is crucial for debugging complex robotic systems, for regulatory compliance, and for fostering user acceptance. In critical applications like autonomous vehicles or medical robots, the ability to explain a decision can be the difference between life and death, or between successful and failed operation.

Another intersecting area of AI advancement is **embodied intelligence**. This concept posits that intelligence is not solely a product of abstract computation but is deeply intertwined with a robot's physical embodiment and its interaction with the environment. Robots that learn through active exploration, trial and error, and by directly manipulating objects develop a richer understanding of the physical world. This is in contrast to purely data-driven approaches that learn from pre-collected datasets. Embodied intelligence research focuses on developing algorithms that allow robots to acquire skills and knowledge through their own experiences, much like human infants learn. This often involves leveraging reinforcement learning techniques where robots are rewarded for achieving desired outcomes, thereby learning optimal behaviors through interaction. For example, a robot learning to grasp an object might initially try various approaches, gradually refining its strategy based on whether its grasps are successful. This approach leads to robots that are more adaptable to novel situations and less reliant on pre-programmed behaviors or extensive prior training data. It also emphasizes the synergistic relationship between sensing, actuation, and intelligence; the physical form and capabilities of the robot directly influence the type of intelligence it can develop. The integration of embodied intelligence with sophisticated perception systems allows robots to build internal models of their environment and predict the consequences of their actions, leading to more robust and context-aware decision-making.

The convergence of these emerging trends—soft robotics, bio-inspiration, human-robot teaming, novel actuation, XAI, and embodied intelligence—paints a compelling picture of the future of robotics. We are moving towards a future where robots are not confined to industrial cages but are agile, adaptable, and intelligent partners capable of operating seamlessly alongside humans in diverse and dynamic environments. This future is marked by robots that can understand us, collaborate with us, and even learn from us, all while operating with a level of autonomy and safety that was once the realm of science fiction. The ongoing research in these areas promises not only to expand the capabilities of robotic systems but also to redefine our relationship with technology, ushering in an era of unprecedented innovation and human-machine synergy. The challenges ahead are significant, spanning fundamental scientific questions in physics, materials science, and AI, as well as complex engineering problems in design, control, and integration. However, the potential rewards—robots that can enhance human lives, solve complex global challenges, and unlock new frontiers of exploration and understanding—are equally profound, driving an exciting and relentless pace of discovery.

The evolution of robotics is inextricably linked with the ongoing digital transformation of industries, a paradigm shift widely recognized as Industry 4.0. This era is characterized by the pervasive integration of cyber-physical systems, the Internet of Things (IoT), cloud computing, and artificial intelligence, all converging to create a new generation of intelligent and interconnected industrial environments. Robotics stands at the very heart of this revolution, acting as the physical embodiment of these digital advancements and translating them into tangible actions on the factory floor and throughout the supply chain.

At its core, Industry 4.0 seeks to create "smart factories" – highly automated, data-driven, and flexible production facilities capable of optimizing processes, responding rapidly to market demands, and even predicting and preventing potential issues. Within these smart factories, robots are no longer isolated, task-specific machines performing repetitive motions in fixed locations. Instead, they are becoming dynamic, collaborative entities that are an integral part of a larger, interconnected ecosystem. The advent of the Industrial Internet of Things (IIoT) is a key enabler of this transformation. IIoT connects not only machines and systems but also sensors, devices, and even individual components, generating vast streams of real-time data. Robots, equipped with an array of sensors, become active participants in this data exchange, collecting information about their own performance, the surrounding environment, and the status of the products they are handling. This data, once collected, is then fed into sophisticated big data analytics platforms. These platforms, often powered by advanced AI algorithms, can process this torrent of information to identify patterns, anomalies, and opportunities for optimization that would be imperceptible to human operators. For instance, a robot on an assembly line might detect subtle variations in the torque required to fasten a particular screw. Big data analytics can aggregate this information across thousands of similar operations, identifying a potential issue with a batch of components or a subtle degradation in the robot's own fastening mechanism before it leads to a significant defect or downtime.

The synergy between robots and AI, particularly in the context of machine learning and deep learning, is what truly elevates industrial automation to the level of Industry 4.0. Robots equipped with AI can move beyond pre-programmed routines to make intelligent decisions based on real-time sensory input and the vast datasets they have access to. This enables them to adapt to variations in product design, handle a wider range of materials, and even learn from their own experiences and the experiences of other robots in the network. Consider the application of AI-powered vision systems integrated with robotic manipulators. Instead of relying on precise object placement, an AI-enabled robot can identify, locate, and pick up components that are presented in a less ordered fashion, adapting its grasp strategy on the fly. This drastically reduces the need for complex fixturing and conveyor systems, allowing for greater flexibility in production lines. Furthermore, AI enables robots to engage in predictive maintenance. By analyzing the operational data of a robotic arm – its motor currents, vibration patterns, cycle times – AI algorithms can predict when a component is likely to fail, allowing maintenance to be scheduled proactively, thereby minimizing unplanned downtime. This shift from reactive to predictive maintenance is a hallmark of Industry 4.0, significantly improving operational efficiency and reducing costs.

The concept of "smart manufacturing" is the direct manifestation of these integrated technologies. In a smart manufacturing environment, robots are not just performing tasks; they are actively participating in the decision-making process of the factory. For example, in a dynamic production scheduling scenario, if an unexpected issue arises with a specific machine, a smart factory can automatically reroute production tasks to available robots on other lines or reconfigure the assembly sequence. This requires a high degree of interoperability between different robotic systems, manufacturing execution systems (MES), and enterprise resource planning (ERP) software, all orchestrated by the underlying IIoT infrastructure. Robots become flexible nodes within this intelligent network, capable of receiving instructions, reporting status, and even autonomously adjusting their operations based on global factory conditions. This level of interconnectedness allows for a degree of agility and responsiveness that was previously unimaginable. Production can be scaled up or down rapidly, customized orders can be integrated seamlessly into existing workflows, and the overall efficiency of the manufacturing process can be continuously monitored and improved.

One of the most profound implications of advanced robotics within Industry 4.0 is the realization of "lights-out manufacturing." This vision describes production facilities that can operate autonomously with minimal or no human intervention, often for extended periods. While the complete elimination of human presence might be an extreme scenario, the principle of highly automated, self-sufficient operations is becoming increasingly feasible. Robots, powered by AI and connected through IIoT, can perform a vast array of tasks, from raw material handling and assembly to quality inspection and packaging, all without direct human supervision. For instance, in a highly standardized production process, robots can manage the entire workflow. They can load raw materials into machines, perform complex assembly operations, conduct in-line quality checks using advanced sensors and vision systems, and then package the finished goods for shipment. The "lights-out" aspect signifies not just the absence of artificial light for human workers, but also the operational autonomy and self-sufficiency of the facility. This can lead to significant benefits, including 24/7 operation, reduced labor costs, enhanced safety by removing humans from hazardous environments, and increased precision and consistency in production. However, it also places an immense emphasis on the reliability, robustness, and self-diagnostic capabilities of the robotic systems and the entire interconnected manufacturing ecosystem.

The impact of robotics extends beyond the factory floor into the broader realm of logistics and supply chain management, also undergoing a profound digital transformation as part of Industry 4.0. In warehouses and distribution centers, robots are revolutionizing how goods are stored, picked, and transported. Autonomous mobile robots (AMRs) are increasingly being deployed to navigate complex warehouse environments, transporting shelves of goods to human pickers or directly picking items themselves. These AMRs, unlike their predecessors (Automated Guided Vehicles or AGVs), are more flexible, capable of dynamic path planning and obstacle avoidance, making them ideal for dynamic and ever-changing warehouse layouts. AI algorithms are crucial for optimizing the routes of these AMRs, ensuring efficient movement of inventory and minimizing travel times. Furthermore, robots are being integrated into the sorting and packing processes. Robotic arms equipped with sophisticated grippers and vision systems can handle a wide variety of package sizes and shapes, sorting them for outbound shipment with speed and accuracy.

Big data analytics plays a critical role in optimizing these robotic logistics operations. By analyzing data from AMRs, robotic pickers, and inventory management systems, companies can gain deep insights into inventory turnover, order fulfillment times, and the efficiency of their robotic workforce. This data can inform decisions about warehouse layout, robotic deployment, and staffing levels, leading to significant cost savings and improved customer service. For example, data analytics can identify peak demand periods and automatically scale up the deployment of AMRs and robotic pickers to meet the increased workload. It can also highlight bottlenecks in the picking or sorting process, allowing for targeted improvements to robotic workflows or system configurations.

The vision of hyper-connected and intelligent industrial ecosystems is the ultimate aspiration of Industry 4.0, and robotics is a fundamental pillar of this future. Imagine a global supply chain where every stage, from raw material sourcing to final product delivery, is seamlessly integrated and optimized by intelligent systems. Robots in manufacturing plants will communicate with robots in warehouses, which in turn will coordinate with autonomous delivery vehicles. The data generated at each node of this network will inform the decisions made at all other nodes, creating a highly responsive and efficient system. For instance, a sudden surge in demand for a particular product detected through real-time sales data could trigger an automated increase in production by robots in a factory, which would then signal a need for increased raw material delivery by autonomous transport robots, and subsequently, an adjustment in inventory levels by warehouse robots. This level of end-to-end visibility and control, facilitated by interconnected robotics and AI, promises to unlock unprecedented levels of efficiency, resilience, and adaptability in industrial operations.

The development of "digital twins" is another area where robotics plays a crucial role in shaping these hyper-connected industrial ecosystems. A digital twin is a virtual replica of a physical asset, process, or system. In the context of Industry 4.0, digital twins of robots, production lines, or even entire factories are being created. These virtual models are fed with real-time data from their physical counterparts, allowing for simulation, analysis, and optimization without disrupting actual operations. For example, a digital twin of a robotic assembly line can be used to test new assembly sequences, train AI algorithms for improved decision-making, or predict the impact of potential equipment failures before they occur. By running scenarios on the digital twin, engineers can optimize robotic performance, identify potential design flaws, and refine operational strategies. This iterative process of simulation and refinement, powered by the data from real-world robotic operations, is key to achieving the high levels of efficiency and reliability envisioned for Industry 4.0.

Furthermore, the concept of "cobots," or collaborative robots, is central to the evolution of human-robot interaction within Industry 4.0. While lights-out manufacturing represents a level of full automation, cobots are designed to work alongside humans, augmenting their capabilities rather than replacing them entirely. These robots are equipped with advanced safety features, such as force and torque sensors, that allow them to detect and react to human presence, stopping or slowing down their movements to prevent accidents. In many industrial settings, cobots are used for tasks that require a combination of human dexterity and robotic precision or strength. For example, a human worker might perform intricate assembly tasks while a cobot provides stable support, performs repetitive screwing operations, or transports heavy components. The ability of cobots to learn from human demonstrations, a concept known as "learning by demonstration," further enhances their collaborative potential. A human operator can guide the cobot through a task, and the robot can learn and replicate the motions, significantly reducing the programming time and effort required to deploy them. This collaborative approach not only enhances productivity and ergonomics but also allows for greater flexibility and adaptability in production lines, as humans and robots can be easily reassigned to different tasks as needed.

The economic and societal implications of widespread robotic adoption in Industry 4.0 are profound. While concerns about job displacement are valid, the narrative is also one of job transformation. The demand for skilled workers to design, program, maintain, and manage these sophisticated robotic systems is increasing. Furthermore, by automating dangerous or monotonous tasks, robotics can improve workplace safety and create more engaging and intellectually stimulating roles for human workers. The increased efficiency and productivity enabled by robotics can also lead to economic growth, lower consumer prices, and the development of new products and services. The transition to Industry 4.0, with robotics at its forefront, represents a fundamental restructuring of industrial processes, moving towards a future where factories are not just places of production, but intelligent, interconnected, and highly adaptable ecosystems. The ongoing integration of robots with IoT, big data, and AI is not merely an incremental improvement; it is a foundational shift that is redefining the very nature of manufacturing, logistics, and industrial operations, paving the way for a more efficient, resilient, and innovative global economy.

The trajectory of robotics is increasingly defined by a profound shift towards human-centric design, moving beyond purely industrial applications to systems that directly interact with, assist, and enhance human lives.

This paradigm is not merely about creating machines that perform tasks for us, but about developing intelligent agents that can collaborate, empathize, and integrate seamlessly into our daily routines, ultimately improving our quality of life and fostering greater independence. The core principle underpinning this evolution is the recognition that technology should serve humanity, augmenting our inherent capabilities and addressing critical societal needs.

A significant frontier within human-centric robotics lies in the realm of assistive technologies for the elderly and individuals with disabilities. As global populations age and the prevalence of chronic conditions increases, the demand for sophisticated support systems has never been greater. Assistive robots are being meticulously designed to bridge gaps in care, empower individuals, and alleviate the burden on human caregivers. These robots are not envisioned as replacements for human connection but as complements, providing practical assistance and enabling a higher degree of autonomy for those who might otherwise require constant supervision or assistance. Imagine a domestic robot capable of fetching items, reminding individuals to take their medication, or even assisting with basic mobility tasks. The design of such robots necessitates a deep understanding of human needs, ergonomic principles, and intuitive interaction paradigms. Natural language processing and sophisticated gesture recognition are crucial for enabling users to communicate their needs effortlessly, while advanced sensor suites allow the robots to navigate complex home environments safely and effectively.

The field of rehabilitation robotics is another area where human-centric design is driving significant innovation. For individuals recovering from strokes, spinal cord injuries, or other physical traumas, the path to regaining mobility and independence can be arduous. Rehabilitation robots offer a consistent, quantifiable, and often more engaging approach to physical therapy. These systems can provide precise, repetitive movements that are essential for retraining neural pathways and strengthening muscles. For example, robotic exoskeletons, worn by patients, can guide limb movements, providing support and resistance as needed. These devices are often programmed with personalized therapy protocols, adapting to the patient's progress and providing real-time feedback to both the patient and the therapist. The data collected by these robots is invaluable, offering objective measures of improvement and allowing therapists to fine-tune treatment plans with unprecedented precision. Beyond exoskeletons, robotic devices designed for specific exercises, such as robotic arms for upper body rehabilitation or treadmills with robotic assistance for gait training, are also transforming recovery outcomes. The human-centric aspect here is paramount; the robot must be safe, comfortable, and motivating for the patient, fostering a positive therapeutic experience that encourages adherence and maximizes the chances of successful rehabilitation.

The integration of robotics into healthcare extends

beyond rehabilitation, encompassing a broad spectrum of applications aimed at improving patient care and supporting medical professionals. Surgical robots, for instance, have already revolutionized minimally invasive procedures, offering surgeons enhanced precision, dexterity, and visualization. While these systems are often operated by skilled surgeons, the underlying design principles are increasingly being influenced by the need for smoother human-robot interaction. Future developments are likely to see these robots becoming even more intuitive and perhaps even capable of performing certain sub-tasks autonomously under human supervision. Beyond the operating room, robots are being explored for tasks such as dispensing medication, transporting supplies within hospitals, and even assisting with patient lifting and repositioning, thereby reducing the physical strain on nurses and caregivers. The development of robots capable of disinfection is also gaining traction, especially in light of global health concerns, helping to maintain sterile environments with greater efficiency and consistency.

The concept of personal assistance robots, designed to live and interact with individuals in their homes, represents perhaps the most intimate application of human-centric robotics. These robots aim to provide companionship, perform household chores, and offer support for daily living activities, particularly for those who are isolated or have difficulty managing tasks independently. Think of a robot that can engage in conversation, play games, or even monitor a person's well-being by detecting changes in their activity patterns or vital signs. The ethical considerations surrounding such robots are significant, demanding careful attention to privacy, data security, and the potential impact on human relationships. However, the potential benefits in combating loneliness and improving the safety and comfort of aging populations are immense. The design must prioritize a non-intrusive presence, an ability to learn user preferences, and a capacity for safe and predictable interactions within a domestic environment. This involves developing sophisticated AI for understanding human intent, emotion detection, and adaptive behavioral algorithms that allow the robot to become a trusted and helpful companion.

The educational sector is also beginning to witness the transformative potential of human-centric robotics. Robots can serve as engaging educational tools, making abstract concepts tangible and fostering interest in STEM fields. Educational robots designed for younger learners can introduce programming concepts through playful interaction, while more advanced robots can be used for complex simulations and scientific experiments. The key here is to create robots that are not just teaching tools but active participants in the learning process, capable of adapting to a student's pace and learning style. For instance, a robotic tutor could identify when a student is struggling with a particular mathematical concept and offer tailored explanations or practice problems. In higher education and research, robots are essential for conducting experiments that would be too dangerous or complex for humans, but the interface for controlling and interacting with these research robots is also becoming increasingly sophisticated, prioritizing ease of use and intuitive data visualization for the scientists.

At the heart of all these human-centric applications is the imperative for robots to be intuitive and adaptive. This means moving beyond complex interfaces and pre-programmed routines towards systems that can understand and respond to human input in a natural and flexible manner. Natural language processing (NLP) is a critical enabler, allowing users to communicate with robots using everyday speech. Similarly, advanced computer vision and sensor fusion enable robots to perceive and interpret their surroundings, understanding not just the objects present but also the context of human actions and intentions. This allows robots to anticipate needs, offer proactive assistance, and adjust their behavior in real-time. For example, a personal assistant robot might learn the user's daily schedule and proactively offer help with tasks at appropriate times, or a rehabilitation robot might detect signs of fatigue in a patient and adjust the intensity of the therapy session.

The development of adaptive algorithms is also crucial. Robots need to learn from their interactions with humans and their environment, continuously refining their performance and tailoring their responses to individual users. This learning process can be guided by various machine learning techniques, from supervised learning, where the robot is trained on labeled data, to reinforcement learning, where the robot learns through trial and error by receiving rewards or penalties for its actions. The ability to learn from demonstration, where a human operator guides the robot through a task and the robot imitates the demonstrated behavior, is particularly valuable for programming complex manipulation tasks or training robots for specific roles. This learning capability ensures that robots can evolve alongside their human users, becoming more proficient and valuable over time.

Beyond functionality, the aesthetic and emotional design of human-centric robots is also a crucial consideration. While industrial robots are often designed for pure utility, robots that are intended to work closely with humans, especially in domestic or care settings, need to be perceived as approachable and trustworthy. This involves careful consideration of their physical form, movement, and even their "personality." For instance, a robot designed to assist children with learning might have a friendly, cartoon-like appearance, while a robot designed for elder care might aim for a more calming and unobtrusive presence. The way a robot moves and communicates non-verbally also plays a significant role in how it is perceived. Smooth, predictable movements and appropriate use of visual cues (like LED lights or on-screen expressions) can help build rapport and reduce user anxiety.

The ethical implications of increasingly sophisticated

human-centric robotics are profound and require ongoing societal discussion and robust regulatory frameworks. As robots become more integrated into our lives, questions about accountability, privacy, and the nature of human-robot relationships come to the fore. For example, if an assistive robot makes an error that leads to harm, who is responsible? How can we ensure that the vast amounts of personal data collected by these robots are protected from misuse? Furthermore, as robots become more capable of mimicking human social interaction, we must consider the potential for emotional attachment and the impact on genuine human connection. A balanced approach is necessary, one that harnesses the immense potential of these technologies to improve human well-being while proactively addressing the ethical and societal challenges they present.

The future of human-centric robotics hinges on a multidisciplinary approach, drawing expertise from robotics engineering, artificial intelligence, psychology, sociology, and ethics. The goal is not to create autonomous agents that operate independently of human needs, but to build systems that are extensions of human capability, fostering independence, improving health outcomes, and enriching daily life. The development of intuitive interfaces, adaptive learning algorithms, and ethically designed robot companions will pave the way for a future where robots are not just tools, but trusted partners in human endeavors, empowering individuals and contributing to a more inclusive and supportive society. The journey towards truly human-centric robotics is one of continuous innovation, deep empathy, and a steadfast commitment to ensuring that technology serves humanity in its most meaningful ways.

The ongoing advancements in robotics are not merely about increasing automation; they are fundamentally about redefining the relationship between humans and machines. This shift is particularly evident in the burgeoning field of human-centric robotics, which prioritizes the augmentation of human capabilities and the enhancement of human well-being. This paradigm recognizes that the ultimate purpose of robotic technology should be to serve human needs, foster independence, and improve the overall quality of life, especially for vulnerable populations.

One of the most significant areas of growth within human-centric robotics is the development of assistive technologies. As societies grapple with aging populations and the increasing prevalence of disabilities, the demand for intelligent systems that can provide support has surged. Assistive robots are being designed to perform a wide array of tasks, from helping with household chores and fetching items to reminding individuals about medication schedules and assisting with mobility. These robots are not intended to replace human caregivers but rather to complement their efforts, enabling individuals to maintain a greater degree of autonomy and dignity. The design of these robots is deeply rooted in understanding user needs, ensuring intuitive interfaces, and guaranteeing safe operation within complex home environments. Advanced sensor suites, natural language processing, and sophisticated AI algorithms are crucial for enabling seamless interaction and effective task execution. For example, a domestic robot equipped with advanced vision capabilities can safely navigate around furniture, identify and retrieve specific objects requested by the user, and even detect potential hazards like spills or trip hazards.

The field of rehabilitation robotics is another powerful testament to the human-centric approach. Individuals recovering from debilitating injuries, such as strokes or spinal cord damage, often face a long and challenging path to regaining lost function. Rehabilitation robots offer a consistent, precise, and engaging means of physical therapy. Robotic exoskeletons, for instance, can guide limb movements, providing support and resistance tailored to the patient's specific needs and progress. These systems collect invaluable data on patient performance, allowing therapists to objectively track recovery, personalize treatment protocols, and make data-driven adjustments to therapy regimens. Beyond exoskeletons, a range of specialized robotic devices are transforming physical therapy, including robotic arms for upper-body exercises, robotic treadmills for gait training, and immersive virtual reality environments that make rehabilitation exercises more motivating. The human-centric design ensures that these robots are not only effective but also comfortable, safe, and encouraging for patients, fostering adherence to therapy and maximizing rehabilitation outcomes.

The integration of robotics into healthcare extends far beyond rehabilitation. Surgical robots, already a fixture in many operating rooms, are enabling surgeons to perform complex procedures with enhanced precision, dexterity, and visualization, leading to less invasive surgeries and faster patient recovery. While these systems are surgeon-controlled, their design is continuously evolving to improve human-robot collaboration and intuitive operation. Looking ahead, robots are being developed for a myriad of other healthcare applications, including automated medication dispensing, the transportation of medical supplies within hospitals, and assisting healthcare professionals with patient lifting and repositioning, thereby reducing the risk of musculoskeletal injuries among staff. Furthermore, the development of robotic systems for environmental disinfection is becoming increasingly critical, especially in the wake of global health challenges, helping to maintain hygienic conditions with greater efficiency and consistency than manual methods.

The concept of personal assistance robots, designed to cohabitate with individuals in their homes, represents a deeply intimate application of human-centric robotics. These robots aim to provide companionship, execute household tasks, and offer essential support for daily living activities, particularly for those who are elderly, isolated, or have chronic health conditions. Imagine a robot that can engage in conversation, facilitate communication with family members, monitor vital signs, or even provide a sense of security through its presence. The ethical considerations surrounding such robots are profound, necessitating careful attention to data privacy, security, and the potential impact on human relationships. However, the potential to alleviate loneliness and significantly enhance the safety and comfort of individuals living independently is immense. The design must prioritize a non-intrusive presence, an ability to learn user preferences and routines, and a capacity for safe and predictable interactions within a domestic setting. This requires sophisticated AI that can understand human intent, detect emotional cues, and adapt its behavior to become a trusted and helpful member of the household.

The educational landscape is also being reshaped by human-centric robotics. Robots can serve as powerful tools for making abstract concepts tangible and for igniting interest in science, technology, engineering, and mathematics (STEM) fields. For younger learners, educational robots can introduce fundamental programming concepts through engaging, interactive play. More advanced robots can be utilized for complex scientific simulations and intricate experimental setups, pushing the boundaries of what can be explored in a classroom or laboratory setting. The key is to develop robots that are not merely passive teaching aids but active participants in the learning process, capable of recognizing and adapting to a student's individual pace and learning style. A robotic tutor, for example, might identify when a student is encountering difficulties with a particular concept and subsequently offer personalized explanations, supplementary materials, or targeted practice exercises, thereby fostering a more effective and engaging learning experience.

Central to the success of all these human-centric applications is the imperative for robots to be intuitive and adaptive. This means transitioning away from complex command structures and rigid programming towards systems that can understand and respond to human input in a natural, flexible, and context-aware manner. Natural language processing (NLP) is a cornerstone technology in this regard, empowering users to communicate their needs and intentions to robots using everyday spoken language. Complementing NLP, advanced computer vision and sensor fusion allow robots to perceive and interpret their surroundings with a high degree of accuracy. This enables them to understand not only the physical objects present but also the subtle nuances of human actions, gestures, and even emotional states. Such contextual awareness allows robots to anticipate needs, offer proactive assistance, and adjust their behavior in real-time, creating a more harmonious and efficient human-robot collaboration. For instance, a personal assistant robot might learn the user's daily routine and proactively offer assistance with specific tasks at appropriate times, or a rehabilitation robot might detect subtle signs of fatigue in a patient and automatically adjust the intensity or duration of a therapy session to prevent overexertion.

The development of adaptive algorithms is equally crucial. Robots must possess the capability to learn from their interactions with humans and their environment, continuously refining their performance and tailoring their responses to individual users. This learning process can be facilitated through various machine learning techniques, ranging from supervised learning, where robots are trained on vast datasets of labeled examples, to reinforcement learning, where robots learn through trial and error by receiving positive or negative feedback for their actions. A particularly valuable capability is "learning by demonstration," where a human operator guides the robot through a task, and the robot observes and imitates the demonstrated movements and strategies. This approach significantly reduces the time and expertise required for programming complex manipulation tasks or training robots for specialized roles, making robotic technology more accessible and adaptable.

Beyond purely functional considerations, the aesthetic and emotional design of human-centric robots is a vital aspect of their acceptance and effectiveness. While industrial robots are typically designed with pure utility in mind, robots intended for close interaction with humans, particularly in domestic, educational, or caregiving settings, must be perceived as approachable, trustworthy, and even endearing. This involves meticulous consideration of their physical form, movement patterns, and even their simulated "personality." For example, a robot designed to assist young children with learning might adopt a friendly, perhaps even playful, aesthetic, while a robot intended for elder care might be designed for a more calming and unobtrusive presence. The way a robot moves and communicates non-verbally—through smooth, predictable motions, appropriate use of visual cues like LED indicators, or on-screen expressions—plays a significant role in building rapport, reducing user anxiety, and fostering a sense of comfort and trust.

The ethical and societal implications of increasingly sophisticated human-centric robotics are profound and demand ongoing public discourse and the establishment of robust regulatory frameworks. As robots become more deeply integrated into our personal lives and sensitive environments, critical questions emerge regarding accountability, data privacy, security, and the evolving nature of human-robot relationships. For instance, in the event of an error or malfunction by an assistive robot that results in harm, determining responsibility becomes a complex legal and ethical challenge. Ensuring that the vast amounts of personal and sensitive data collected by these robots are rigorously protected from unauthorized access, misuse, or exploitation is paramount. Furthermore, as robots become more adept at mimicking human social interaction and emotional expression, society must grapple with the potential for emotional dependence on machines and the potential impact on the authenticity and depth of genuine human connection. Navigating this complex terrain requires a balanced and thoughtful approach, one that diligently harnesses the immense potential of these technologies to improve human well-being while proactively identifying and mitigating the ethical and societal challenges they invariably present.

The future trajectory of human-centric robotics is inextricably linked to a multidisciplinary approach, necessitating collaboration among experts in robotics engineering, artificial intelligence, psychology, sociology, and ethics. The overarching objective is not to create autonomous agents that operate independently of human needs or desires, but rather to engineer systems that serve as empowering extensions of human capability. These systems aim to foster greater independence, improve health outcomes, facilitate lifelong learning, and ultimately enrich the daily lives of individuals across the lifespan. The continued development of intuitive interfaces, adaptive learning algorithms, and ethically grounded robot designs will pave the way for a future where robots are not merely tools but trusted partners in human endeavors, empowering individuals and contributing to the creation of a more inclusive, supportive, and fulfilling society. The journey towards truly human-centric robotics is characterized by continuous innovation, a deep sense of empathy for human needs, and an unwavering commitment to ensuring that technological progress consistently serves humanity in its most meaningful and beneficial ways.

The path forward for robotics and autonomous systems is a landscape rich with both formidable challenges and transformative opportunities. As we look beyond the current state of specialized, task-specific robots, the aspiration for more versatile, intelligent, and integrated systems surfaces, bringing with it a unique set of hurdles. Chief among these is the pursuit of Artificial General Intelligence (AGI). While current AI excels in narrow domains, achieving human-level cognitive abilities across a broad spectrum of tasks remains an elusive, yet deeply compelling, frontier. The development of AGI would unlock unprecedented capabilities, but it also magnifies concerns about control, alignment with human values, and the potential for unforeseen consequences. Ensuring that any superintelligent system would act in accordance with human interests is a problem of immense complexity, requiring breakthroughs not only in computational power and algorithmic design but also in our fundamental understanding of consciousness, ethics, and intent.

The ethical deployment of robotics is another paramount challenge that will shape the future of the field. As robots become more capable and integrated into societal fabric, questions of accountability, bias, privacy, and the very definition of personhood will become increasingly pressing. For instance, in autonomous driving scenarios, how should a vehicle be programmed to react in an unavoidable accident? Who is liable when an autonomous system errs – the programmer, the manufacturer, the owner, or the AI itself? Furthermore, the datasets used to train AI systems can inadvertently embed and perpetuate societal biases, leading to discriminatory outcomes in areas such as hiring, loan applications, or even criminal justice. Developing robust ethical frameworks, transparent decision-making processes, and mechanisms for ongoing societal dialogue is crucial to ensure that robotic advancements benefit humanity equitably and responsibly. The very notion of human-robot interaction, particularly in social contexts, raises questions about authenticity, emotional dependence, and the potential erosion of genuine human connection, demanding careful consideration as we design increasingly sophisticated social robots.

Societal disruption, driven by widespread automation, presents a significant socio-economic challenge. While robotics promises increased productivity and economic growth, it also raises legitimate concerns about job displacement and the widening of economic inequality. As robots take over tasks previously performed by humans, entire industries may undergo radical transformations, necessitating significant adjustments in workforce training, education, and social safety nets. Proactive strategies, such as investing in lifelong learning initiatives, exploring new economic models like universal basic income, and fostering innovation in sectors that create new human-centric jobs, will be vital to navigate this transition. The challenge lies not in halting progress, but in managing its implications thoughtfully to ensure that the benefits of automation are broadly shared and that no segment of society is left behind. This requires foresight, collaboration between governments, industries, and educational institutions, and a commitment to creating a future where technology serves to empower all individuals.

Developing robust and reliable robotic systems capable of operating safely and effectively in unpredictable, dynamic, and unstructured environments remains a formidable engineering challenge. Unlike the controlled conditions of manufacturing floors, real-world scenarios—whether it's a search-and-rescue operation in a disaster zone, an autonomous vehicle navigating a busy city street, or a domestic robot assisting an elderly person—are fraught with variability and unforeseen events. Achieving true robustness necessitates significant advancements in areas such as sensor fusion, contextual understanding, real-time decision-making under uncertainty, and fail-safe mechanisms. Robots must be able to perceive their surroundings comprehensively, adapt to unexpected changes, predict potential outcomes, and make sound judgments even with incomplete information. This requires more sophisticated perception systems that can interpret complex sensory data, advanced AI that can reason about causality and intent, and highly resilient hardware that can withstand diverse environmental conditions. The ability to learn and adapt from experience in real-time is also critical, allowing robots to improve their performance over time without explicit reprogramming for every new situation.

Despite these challenges, the opportunities for innovation, economic growth, and addressing some of the world's most pressing global issues through robotics are immense. The continued development of advanced robotics and autonomous systems promises to unlock new avenues for economic prosperity. The creation of new industries, the enhancement of existing ones through increased efficiency and novel capabilities, and the generation of high-skilled jobs in areas like AI development, robotics engineering, and data science are all within reach. The global market for robotics is projected to grow exponentially, driven by demand across sectors ranging from manufacturing and logistics to healthcare and agriculture. This economic expansion has the potential to create wealth and improve living standards, provided that its benefits are distributed equitably and that proactive measures are taken to mitigate potential negative consequences like job displacement.

Robotics offers powerful solutions to tackle critical global challenges. In the realm of healthcare, advancements in robotic surgery, telemedicine, diagnostics, and patient care promise to improve accessibility, reduce costs, and enhance outcomes, especially for aging populations and those in underserved regions. Rehabilitation robots are transforming recovery processes, and intelligent assistive devices can empower individuals with disabilities to live more independent lives. The potential for robots to augment human capabilities in complex and dangerous tasks—from deep-sea exploration and space missions to hazardous material handling and disaster response—is also a significant area of opportunity.

Furthermore, robotics and AI are poised to play a crucial role in addressing environmental sustainability and climate change. Robots can be deployed for tasks such as precision agriculture, minimizing water and pesticide usage, or for environmental monitoring, tracking pollution levels, and assessing ecological damage. Autonomous systems can optimize energy consumption in smart grids and buildings, and robotic solutions are being developed for waste management, recycling, and the development of renewable energy infrastructure, such as the automated maintenance of wind turbines or solar farms. The ability of robots to operate in hazardous or inaccessible environments makes them ideal for tasks related to environmental remediation and disaster recovery, aiding in the cleanup of oil spills or the deconstruction of nuclear facilities, for example.

The drive towards more sophisticated robotics also fuels innovation in fundamental scientific research. Autonomous systems can accelerate scientific discovery by conducting experiments with unprecedented speed and precision, analyzing vast datasets, and exploring environments that are inaccessible to humans. This can lead to breakthroughs in fields ranging from materials science and drug discovery to astrophysics and climate modeling. The very process of developing advanced robotics pushes the boundaries of our understanding in areas such as artificial intelligence, computational neuroscience, advanced materials, and human-computer interaction, creating a virtuous cycle of innovation.

In conclusion, the future of robotics and autonomous systems is a narrative of profound potential tempered by significant challenges. The pursuit of AGI, the imperative for ethical development, the management of societal disruption, and the need for robust systems in unpredictable environments represent the critical hurdles we must overcome. Yet, the opportunities are equally compelling: driving economic growth, revolutionizing healthcare, safeguarding our planet, and accelerating scientific discovery. The path forward requires a concerted, multidisciplinary effort—one that embraces innovation while prioritizing responsible development, ethical considerations, and the ultimate goal of enhancing human well-being and creating a more sustainable and equitable future for all. The journey ahead will be one of continuous learning, adaptation, and collaboration, as we strive to harness the full potential of these transformative technologies for the betterment of humanity.

The journey through the intricate landscape of robotics and autonomous systems, as chronicled in this textbook, has illuminated a field brimming with innovation, driven by relentless scientific inquiry and engineering prowess. We have traversed the foundational principles that underpin intelligent machines, delved into the sophisticated algorithms that grant them perception and decision-making capabilities, and explored the diverse applications that are already reshaping our world. From the precision of surgical robots to the burgeoning autonomy of vehicles, and from the collaborative endeavors of industrial manipulators to the nascent intelligence of social companions, the narrative has been one of continuous evolution and burgeoning potential.

A recurring theme throughout our exploration has been the indispensable synergy between theoretical understanding and practical application. The most profound advancements in robotics emerge not solely from abstract mathematical models or groundbreaking theoretical discoveries, but from the iterative process of translating these concepts into tangible, functional systems. It is in the crucible of design, implementation, and testing that theoretical constructs are validated, refined, and ultimately transformed into technologies that can perform meaningful tasks. The challenges encountered in real-world deployment—the unpredictable nature of environments, the complexities of human interaction, and the stringent demands for safety and reliability—serve as powerful catalysts for further theoretical development. This symbiotic relationship underscores the need for professionals in this field to cultivate a robust theoretical grounding while simultaneously honing their practical engineering and programming skills. The ability to bridge the gap between the whiteboard and the workshop floor is, and will continue to be, a defining characteristic of successful roboticists.

The pace of progress in robotics and autonomous systems is nothing short of breathtaking. What might have seemed like science fiction a decade ago is rapidly becoming commonplace. This relentless evolution necessitates a commitment to lifelong learning. The tools, techniques, and even the fundamental paradigms of robotics are subject to constant revision and improvement. Algorithms that are state-of-the-art today may be superseded by more efficient or capable methods tomorrow. New sensor technologies emerge, materials science breakthroughs enable novel designs, and our understanding of artificial intelligence deepens at an accelerating rate. To remain at the forefront of this dynamic field, one must embrace continuous education, actively seek out new knowledge, and be willing to adapt to emerging trends and technologies. This might involve pursuing advanced degrees, engaging in online courses, attending industry conferences, or dedicating time to independent study and experimentation. The curiosity and drive to learn are perhaps the most critical assets a researcher or practitioner can possess.

As we stand on the precipice of even greater innovation, it is essential to recognize that the future of robotics is not predetermined; it is being actively shaped by the collective efforts of countless individuals. Students and researchers embarking on their careers in this domain have the extraordinary opportunity to be architects of this future. The challenges are significant, yes, but they are dwarfed by the potential for impact. Imagine a world where robots assist in elder care, providing companionship and enabling independent living for our aging population. Envision autonomous systems optimizing our energy consumption, helping to mitigate the devastating effects of climate change. Consider the possibilities of robots performing dangerous tasks in disaster zones, saving lives and minimizing human risk. Picture medical robots enabling complex surgeries with unparalleled precision, democratizing access to high-quality healthcare. These are not distant fantasies, but tangible goals that are within reach, driven by the innovations that you, our readers, will spearhead.

The field of robotics and autonomous systems offers a uniquely fertile ground for interdisciplinary collaboration. The problems we seek to solve—whether it's creating truly intelligent agents, ensuring the ethical deployment of autonomous systems, or developing robots that can navigate the complexities of the human world—require insights from computer science, electrical engineering, mechanical engineering, cognitive science, psychology, philosophy, and even sociology and law. By engaging with diverse perspectives and collaborating across traditional disciplinary boundaries, we can unlock novel solutions and ensure that the technologies we develop are not only technically sound but also socially responsible and beneficial. Your unique background and perspective can bring a fresh and invaluable contribution to this collaborative endeavor.

We encourage you to delve deeper. If a particular area of robotics has captured your imagination—be it manipulation, locomotion, perception, artificial intelligence, or human-robot interaction—then pursue it with passion. Engage with the latest research papers, experiment with open-source software and hardware platforms, and consider contributing to ongoing projects. The resources available to aspiring roboticists have never been more abundant. Online tutorials, educational kits, maker spaces, and university research labs all provide pathways to hands-on experience and skill development. Do not be afraid to tackle ambitious projects; it is often through striving for what seems just beyond our grasp that we achieve our most significant breakthroughs.

Furthermore, consider the profound ethical and societal implications of the technologies you will help create. As robots become more integrated into our lives, the decisions we make in their design and deployment will have far-reaching consequences. Engaging in thoughtful discourse about issues of accountability, privacy, bias, and the very nature of intelligence and consciousness is not merely an academic exercise; it is a crucial responsibility for every individual contributing to this field. By prioritizing ethical considerations alongside technical innovation, we can ensure that robotics serves to enhance human flourishing rather than diminish it.

The path forward in robotics is not always clearly defined. There will be detours, setbacks, and unexpected discoveries. The field is characterized by a spirit of exploration and a willingness to experiment, to fail, and to learn from those failures. Embrace this iterative process. Celebrate small victories, learn from every challenge, and maintain a persistent optimism about the transformative power of robotics. The impact you can have on the world is substantial. You have the opportunity to contribute to solutions for some of humanity's most pressing challenges, to enhance human capabilities, and to build a future that is more efficient, sustainable, and perhaps even more equitable.

This textbook has aimed to provide you with a solid foundation, a comprehensive overview, and a starting point for your own journey. However, the true learning begins now, with your continued engagement, your critical thinking, and your active participation in the ongoing development of robotics and autonomous systems. We are at an incredibly exciting juncture in the history of technology, and the contributions you make will shape the decades to come. Go forth with curiosity, with a commitment to excellence, and with a vision for a better future. The world of robotics awaits your ingenuity and your dedication.

www.ingramcontent.com/pod-product-compliance
Lightning Source LLC
LaVergne TN
LVHW052012200926
848481LV00005B/1473